CLEARING

THE *Air*

ASTHMA AND
INDOOR AIR EXPOSURES

Committee on the Assessment of
Asthma and Indoor Air

Division of Health Promotion and
Disease Prevention

INSTITUTE OF MEDICINE

NATIONAL ACADEMY PRESS
Washington, D.C.

NATIONAL ACADEMY PRESS • 2101 Constitution Avenue, N.W. • Washington, D.C. 20418

NOTICE: The project that is the subject of this report was approved by the Governing Board of the National Research Council, whose members are drawn from the councils of the National Academy of Sciences, the National Academy of Engineering, and the Institute of Medicine. The members of the committee responsible for the report were chosen for their special competences and with regard for appropriate balance.

Support for this study was provided by the U.S. Environmental Protection Agency (contract no. X825863-01-3). The views presented in the book are those of the Institute of Medicine Committee on the Assessment of Asthma and Indoor Air and are not necessarily those of the funding organization.

Library of Congress Cataloging-in-Publication Data

Institute of Medicine (U.S.). Committee on the Assessment of Asthma and
    Indoor Air. Clearing the air : asthma and indoor air exposures /
    Committee on the Assessment of Asthma and Indoor Air, Division
    of Health Promotion and Disease Prevention, Institute of Medicine.
        p. cm.
    Includes bibliographical references and index.
    ISBN 0-309-06496-1 (case)
        1. Asthma. 2. Indoor air pollution. 3. Asthma—Government policy—
    United States. I. Title.
    RA645.A83 I55 2000
    362.1'96238—dc21   00-025801

Clearing the Air: Asthma and Indoor Air Exposures is available for sale from the National Academy Press, 2101 Constitution Avenue, N.W., Box 285, Washington, DC 20055. Call 800-624-6242 (202-334-3313 in the Washington DC metropolitan area) or visit the NAP's on-line bookstore at **www.nap.edu.**

For more information about the Institute of Medicine, visit the IOM home page at **www.iom.edu.**

Printed in the United States of America.

The serpent has been a symbol of long life, healing, and knowledge among almost all cultures and religions since the beginning of recorded history. The image adopted as a logotype by the Institute of Medicine is based on a relief carving from ancient Greece, now held by the Staatliche Museen in Berlin.

THE NATIONAL ACADEMIES

National Academy of Sciences
National Academy of Engineering
Institute of Medicine
National Research Council

The **National Academy of Sciences** is a private, nonprofit, self-perpetuating society of distinguished scholars engaged in scientific and engineering research, dedicated to the furtherance of science and technology and to their use for the general welfare. Upon the authority of the charter granted to it by the Congress in 1863, the Academy has a mandate that requires it to advise the federal government on scientific and technical matters. Dr. Bruce M. Alberts is president of the National Academy of Sciences.

The **National Academy of Engineering** was established in 1964, under the charter of the National Academy of Sciences, as a parallel organization of outstanding engineers. It is autonomous in its administration and in the selection of its members, sharing with the National Academy of Sciences the responsibility for advising the federal government. The National Academy of Engineering also sponsors engineering programs aimed at meeting national needs, encourages education and research, and recognizes the superior achievements of engineers. Dr. William A. Wulf is president of the National Academy of Engineering.

The **Institute of Medicine** was established in 1970 by the National Academy of Sciences to secure the services of eminent members of appropriate professions in the examination of policy matters pertaining to the health of the public. The Institute acts under the responsibility given to the National Academy of Sciences by its congressional charter to be an adviser to the federal government and, upon its own initiative, to identify issues of medical care, research, and education. Dr. Kenneth I. Shine is president of the Institute of Medicine.

The **National Research Council** was organized by the National Academy of Sciences in 1916 to associate the broad community of science and technology with the Academy's purposes of furthering knowledge and advising the federal government. Functioning in accordance with general policies determined by the Academy, the Council has become the principal operating agency of both the National Academy of Sciences and the National Academy of Engineering in providing services to the government, the public, and the scientific and engineering communities. The Council is administered jointly by both Academies and the Institute of Medicine. Dr. Bruce M. Alberts and Dr. William A. Wulf are chairman and vice chairman, respectively, of the National Research Council.

Staff

DAVID A. BUTLER, Study Director
JAMES A. BOWERS, Research Assistant
JENNIFER A. COHEN, Research Assistant
ROSE MARIE MARTINEZ, Director, Division of Health Promotion and Disease Prevention (as of December 1999)
KATHLEEN R. STRATTON, Director, Division of Health Promotion and Disease Prevention (through November 1999)
DONNA D. DUNCAN, Division Assistant
ANDREA COHEN, Financial Associate

PREFACE

The tremendous burden of disease imparted by asthma, the alarming escalation of asthma prevalence, and the doubling of the asthma mortality rate in the United States since the 1970s have attracted increased attention from those concerned about the health of the American public, including Congress. Many agencies of the U.S. government have been charged to increase their efforts at addressing at least one facet of the problem, from research into pathogenesis by the National Institutes of Health to examination of environmental factors by the Environmental Protection Agency (EPA). It is highly likely that such a rapid rise in asthma prevalence is due to a change in some factor or factors in the environment. Identifying these factors could allow remediation, and—perhaps—prevention. Within this context the EPA sought the guidance of the Institute of Medicine (IOM) in evaluating the quality and nature of the scientific data relating constituents of indoor air and the occurrence of asthma.

The multidisciplinary committee convened by the IOM to respond to this charge, with considerable help from IOM staff, spent many hours in literature review and discussion. Our goal was to reach consensus about how strongly the research data implicated various components of indoor air as causes of asthma. A second goal was to evaluate the state of the scientific evidence concerning specific strategies for exposure mitigation and prevention.

Common problems frustrated our efforts, starting with the imprecise and variable definition of asthma used in research studies, and followed immediately by what it means to "cause" asthma. An unfortunate configuration of genes that influence the immune or inflammatory responses might be said to be the primary "cause" of asthma. But even identical twins may differ as to whether they have asthma or not. Exposure to some environmental factor or factors is required to elicit the clinical expression of asthma, i.e., cause the *development* of asthma. The same or other agents may then cause *exacerbation* of asthma symptoms in these individuals. Thus, the committee divided its analysis into whether an agent might cause asthma development or exacerbation of symptoms.

Considering how the indoor environment might be modified to reduce the risk of asthma development or exacerbation presented a particularly formidable challenge. It has been known for a long time that changing the environment of an asthmatic individual can reduce symptoms, at least temporarily. Mitigation is possible for individual patients. However, little data are available to allow firm conclusions about specific mitigation techniques applied as a public health measure. Although interventions that might reduce the severity of asthma in individual patients emerge from the committee's review, it is hard to escape the overwhelming conclusion that more research is desperately needed to form the basis for public health interventions. Too much ignorance remains regarding the biologic changes that permit the disease to emerge and recur, the environmental "causes" that may underlie the increased prevalence, the socioeconomic differences in rates of morbidity and mortality, and the means of effective exposure mitigation and prevention. Although it will be essential to gain a better understanding of the relationship between particular agents and asthma, no single agent or factor has yet been identified as a necessary or sufficient cause of asthma. Until a more fundamental understanding is available, multifaceted approaches will be needed to address the interrelationships among biologic, environmental, and socioeconomic factors that permit expression of this disease.

The committee exercised final responsibility for all content of the report, but we were not its only contributors. In fact, we could not have completed our task satisfactorily without the substantial help of the individuals cited in the Acknowledgments section. We are especially indebted to Peter J. Gergen, Donald K. Milton, William B. Rose, and Kathleen Kreiss, who furnished text and discussions that were essential to our deliberations on certain subjects. The committee has also recognized that the report could never have been developed without the work of the extraordinary staff assigned to us by the Institute of Medicine—David Butler, James Bowers, Jennifer Cohen, Donna Duncan, Andrea Cohen, and Kathleen Stratton. In particular, David Butler, Study Director, with intelligence, patience, persistence, and hard work, expedited and channeled our deliberations through a dismaying array of subjects, from exposure assessment to pathophysiology, to their final expression as the report.

Richard B. Johnston, Jr., M.D.
Chair

ACKNOWLEDGMENTS

Preparation of this report could not have happened without the guidance and expertise of numerous individuals. Although it is not possible to mention by name all of those who contributed to this committee's work, the committee wants to express its gratitude to a number of individuals for their special contributions.

Sincere thanks go to all of the participants at the workshops convened on January 18 and March 22, 1999. The intent of these workshops was to gather information regarding exposure to specific indoor air allergens and irritants, and asthma pathogenesis, the triggering of asthma attacks, and the exacerbation of symptoms in asthmatics. The speakers, who are listed in Appendix B, gave generously of their time and expertise to help inform and guide the committee's work.

We are particularly appreciative of the efforts of four individuals who contributed text that was used or formed the basis of discussions in this report: Peter J. Gergen, M.D., M.P.H.; Kathleen Kreiss, M.D.; Donald Milton, M.D., Dr.P.H.; and William B. Rose, M.Arch. Their contributions greatly aided the committee and had a significant positive impact. The Committee on the Assessment of Asthma and Indoor Air has, of course, final responsibility for all content in the report.

The committee extends special thanks to the dedicated and hard working staff at the Institute of Medicine (IOM). The expertise and leadership of Kathleen Stratton, Director of the Division of Health Promotion and Disease Prevention, helped to ensure that this report met the highest standards for quality.

This report has been reviewed in draft form by individuals chosen for their diverse perspectives and technical expertise, in accordance with procedures approved by the NRC's Report Review Committee. The purpose of this independent review is to provide candid and critical comments that will assist the institution in making the published report as sound as possible and to ensure that the report meets institutional standards for objectivity, evidence, and responsiveness to the study charge. The review comments and draft manuscript remain confidential to protect the integrity of the deliberative process. We wish to thank the following individuals for their participation in the review of this report: Eula Bingham, University of Cincinnati; Noreen Clark, University of Michigan School of Public Health; Peyton Eggleston, Johns Hopkins University School of Medicine; Leslie Grammer, Northwestern University Medical School; Jonathan Samet, Johns Hopkins University School of Hygiene and Public Health; Olli Seppänen, Helsinki University of Technology; and Scott Weiss, Harvard University School of Medicine. While the individuals listed above have provided constructive comments and suggestions, it must be emphasized that responsibility for the final content of this report rests entirely with the authoring committee and the institution.

Finally, the committee would like to thank the chair, Richard Johnston, Jr., M.D., for his outstanding work, leadership, and dedication to this project.

CONTENTS

CLEARING
THE *Air*

EXECUTIVE SUMMARY

T he statistics are disturbing.

The Centers for Disease Control and Prevention (CDC) estimates that asthma affected about 17.3 million individuals in the United States in 1998. It is the most common chronic illness among children in the United States and one of the most common chronic illnesses overall in the country. Although by many measures the health of Americans is improving, CDC notes the self-reported prevalence rate for asthma increased 75% from 1980 to 1994. Studies show that asthma mortality is disproportionately high among African Americans and in urban areas that are characterized by high levels of poverty and minority populations. Nor is the phenomenon limited to the United States. The prevalence of asthma in some other parts of the world—including Australia, New Zealand, Ireland, and the United Kingdom—exceeds that of the United States.

Researchers have wondered whether the indoor environment may play a role in the increasing asthma problem. There is ample justification for this speculation. We know, for example, that individuals spend nearly all of their time indoors—most of it in their own homes—and that many of the exposures thought to be associated with asthma occur predominately indoors. If the indoor environment plays a role, then interventions to limit or eliminate exposures there have the potential to help asthmatics and perhaps result in primary prevention of the illness.

Against this backdrop, the U.S. Environmental Protection Agency (EPA) is developing an outreach strategy focused on reducing asthma-related morbidity and mortality potentially associated with exposure to indoor environments. To help ensure that such efforts are based on sound science, EPA requested that the National Academies undertake an assessment of asthma and its relationship to indoor air quality. The EPA charged the committee with two primary objectives:

1. To provide the scientific and technical basis for communications to the public on the health impacts of indoor pollutants related to asthma, and mitigation and prevention strategies to reduce these pollutants.

2. To help determine what research is needed in these areas.

This report presents the results of that assessment.

ORGANIZATION AND FRAMEWORK

The content of this report reflects the committee's goal to speak to a wide-ranging audience of science, health, and engineering professionals; government officials; and interested members of the public. The material presented thus covers a broad range of topics in order to establish a common base of knowledge for the reader. The scope of this material is far too vast for any one book to deal with comprehensively. Other publications, cited throughout the report, go into greater detail on specific issues.

The major topics addressed in the report are the following:

• the definition of asthma and the characteristics of its clinical presentation (Chapter 1);
• methodologic issues in evaluating the evidence regarding indoor air exposures and asthma, including the categorizations used to summarize the evidence and the framework for considering exposure to indoor sources (Chapter 2);

TABLE 1 Indoor Exposures Addressed in This Report

Biological

Animals	Fungi or molds
Cats	Houseplants
Dogs	Pollen
Rodents	Infectious agents
Cows and horses	Rhinovirus
Domestic birds	Respiratory syncytial virus
Cockroaches	*Chlamydia trachomatis*
House dust mites	*Chlamydia pneumoniae*
Endotoxins	*Mycoplasma pneumoniae*

Chemical

NO_2, NO_X (nitrogen oxides)	Plasticizers
Pesticides	Volatile organic compounds
Ozone*	Formaldehyde
Particulate matter with sources	Fragrances
other than ETS*	Environmental Tobacco Smoke (ETS)
SO_2, SO_X (sulfur oxides)*	

*An outdoor air pollutant potentially associated with asthma that can penetrate the indoor environment and that may in some cases have indoor sources. Since the committee's mandate was to address indoor air pollutants, the discussion of this agent is less detailed than others in the report and no conclusions are drawn concerning outdoor exposures and asthma outcomes.

- patterns of asthma morbidity and mortality (Chapter 3);
- the pathophysiology of asthma—that is, the molecular mechanisms that underlie the structural and functional changes in the lungs and airways of asthmatics (Chapter 4);
- the committee's review of the state of the scientific literature regarding indoor air exposures and the exacerbation and development of asthma—Table 1 lists the biologic and chemical exposures addressed in this report. (Chapters 5–7);
- the scientific literature on general exposures in indoor environments (Chapters 8–9); and
- how indoor exposures to pollutants associated with the incidence or symptoms of asthma are affected by building ventilation and particle air cleaning (Chapter 10).

The committee faced a significant challenge in conducting its review—research on asthma is burgeoning and significant new papers are constantly being published. Although the committee did its best to paint an accurate picture of the state of the science at the time the report was completed, it is inevitable that research advances will overtake its conclusions.

CONCLUSIONS ABOUT THE RELATIONSHIP BETWEEN INDOOR EXPOSURES AND ASTHMA

The committee used a uniform set of categories to summarize its conclusions regarding the association between exposure to an indoor agent and asthma development and exacerbation, and the effectiveness of exposure mitigation and prevention measures. Box 1 lists the definitions of these categories. The distinctions among categories reflect the committee's judgment of the overall strength, quality, and persuasiveness of the scientific literature evaluated. Chapter 2 details the methodologic considerations underlying the categorizations and their definitions.

The sections below are a synopsis of the committee's findings. Chapters 5 through 10 address the reasoning underlying the conclusions and present the findings in greater detail.

Exposure Settings

The indoor exposures considered in this report are highly dependent on the characteristics of the outdoor and indoor environment and its occupants. For example, house dust mites are a very common exposure in temperate and humid regions. They are found primarily within residences, concentrated in the bedroom. Cockroaches, which also thrive in temperate and humid regions, are an important exposure in some urban environments. They are found primarily near food sources. Fungi are ubiquitous and have been the primary source of allergen for several studied populations. Endotoxins may be found in humidifiers and in bacteria from other indoor, as well as outdoor sources. In some environments, exposure to animal allergens; molds; environmental tobacco smoke (ETS); indoor combustion products; and chemicals used in cleaning, building materials, and furnishings may be im-

BOX 1
Categories of Evidence Used in This Report

Sufficient Evidence of a Causal Relationship
Evidence is sufficient to conclude that a causal relationship exists between the action or agent and the outcome. That is, the evidence fulfills the criteria for "Sufficient Evidence of an Association" below and in addition satisfies criteria regarding the strength of association, biologic gradient (dose–response effect), consistency of association, biologic plausibility and coherence, and temporality used to assess causality.

Sufficient Evidence of an Association
Evidence is sufficient to conclude that there is an association. That is, an association between the action or agent and the outcome has been observed in studies in which chance, bias, and confounding can be ruled out with reasonable confidence. For example, if several small studies that are free from bias and confounding show an association that is consistent in magnitude and direction, there may be sufficient evidence of an association.

Limited or Suggestive Evidence of an Association
Evidence is suggestive of an association between the action or agent and the outcome but is limited because chance, bias, and confounding cannot be ruled out with confidence. For example, at least one high-quality study shows a positive association, but the results of other studies are inconsistent.

Inadequate or Insufficient Evidence to Determine Whether or Not an Association Exists
The available studies are of insufficient quality, consistency, or statistical power to permit a conclusion regarding the presence or absence of an association; or no studies exist that examine the relationship. For example, available studies have failed to adequately control for confounding or have inadequate exposure assessment.

Limited or Suggestive Evidence of No Association
Several adequate studies are mutually consistent in not showing an association between the action or agent and the outcome. A conclusion of "no association" is inevitably limited to the conditions, level of exposure, and length of observation covered by the available studies. *In addition, the possibility of a very small elevation in risk at the levels of exposure studied can never be excluded.*

portant. Many of these pollutants are also present in outdoor air, and indoor exposures can result from the infiltration of outdoor air into buildings.

Indoor Air Exposures and Asthma Exacerbation

Studies of asthma can be divided into those dealing with factors leading to the development of asthma and those dealing with factors that exacerbate the illness in known asthmatics. Most of the research on this topic addresses "asthma exacerbation," the onset or worsening of symptoms—some combination of shortness of breath, cough, wheezing, and chest tightness—in someone who already has developed asthma.

Epidemiologic investigations, challenge studies, and clinical experience have yielded solid information on the potential for many indoor exposures to exacerbate asthma. The committee found **sufficient evidence to conclude that there is a causal relationship** between

- exposure to the allergens produced by cats, cockroaches, and house dust mites, and exacerbations of asthma in sensitized individuals; and
- ETS exposure and exacerbations of asthma in preschool-aged children.

There is **sufficient evidence of an association** between several exposures and exacerbations of asthma. Dog allergen exposure is associated with exacerbation of asthma in individuals specifically sensitized to these allergens. Fungal exposure is associated with exacerbation in sensitized asthmatics and may be associated with nonspecific chest symptoms. Research indicates that rhinovirus infection is associated with wheezing and exacerbations in asthmatics. There is also sufficient evidence to conclude that brief high-level[1] exposures to NO_2 and increased airway responses among asthmatic subjects to both nonspecific chemical irritants and inhaled allergens.

---

[1]At concentrations that may occur only when gas appliances are used in poorly ventilated kitchens.

Damp conditions are associated with the presence of symptoms considered to reflect asthma; symptom prevalence among asthmatics is also related to dampness indicators. The factors related to dampness that may actually lead to asthma exacerbation are not yet confirmed, but probably relate to dust mite and fungal allergens. There is sufficient evidence that some nonresidential buildings provide exposures that exacerbate asthma. However, the specific agents responsible for such exacerbations are as yet unstudied.

Limited or suggestive evidence was found for an association between exposures to domestic birds and exacerbation of asthma, although it is unclear what portion of this association is attributable to an allergic asthmatic response to the mites harbored by these birds. There is also limited or suggestive evidence of a relationship between

- exposure to the infectious agents respiratory syncytial virus (RSV), *Chlamydia pneumoniae,* and *Mycoplasma pneumoniae,* and exacerbation of asthma;
- chronic ETS exposure and exacerbation of asthma in older children and adults;
- acute ETS exposure and exacerbation of asthma in individuals responsive to this exposure;
- nonacute, nonoccupational formaldehyde exposure and wheezing and other respiratory symptoms; and
- exposure to certain fragrances and the manifestation of respiratory symptoms in asthmatics sensitive to such exposures.

Inadequate or insufficient information was identified to determine whether or not exacerbations of asthma result from nonacute, nonoccupational exposures to cow, horse, and rodent allergens; endotoxins; houseplants[2] or cut flowers; the bacterial agent *Chlamydia trachomatis*; pesticides; plasticizers; and volatile organic compounds (VOCs) other than formaldehyde. Some of these same agents do or may play a role in asthma resulting from

---

[2]Mites and fungi associated with houseplants could be involved in asthma outcomes but no studies document this connection.

exposures in occupational settings, a topic outside the purview of this study.

Although there is sufficient evidence to conclude that pollen exposure is associated with exacerbation of existing asthma in sensitized individuals, and pollen allergens have been documented in both dust and indoor air, there is inadequate or insufficient information to determine whether *indoor* exposure to pollen is associated with exacerbations of asthma.

These findings are summarized in Table 2.

Indoor Air Exposures and Asthma Development

The second outcome reviewed by the committee was the development of asthma—the initial onset of the illness. Asthma is defined by the manifestation of a set of symptoms rather than by any one objective test. With asthma symptoms ranging from clearly episodic to nearly continuous, from mild to severe, and from coughing without other respiratory symptoms to a loud wheeze, the initial diagnosis of the illness can be complicated and subject to controversy. It is thus difficult to study the determinants of and influences on asthma development. An additional complication stems from the fact that some of the most provocative evidence regarding development comes from studies of infants. Prior to the age of approximately 3, children may exhibit symptoms that are characteristic of asthma, but they may not exhibit persistent asthmatic symptoms or other related conditions such as bronchial reactivity or allergy later in life. Chapter 1 discusses the definitions of asthma and the characteristics of its clinical presentation.

Saying that a particular agent may be associated with the development of asthma does not mean it is the sole factor determining whether an individual will manifest the illness. Most scientists believe that some individuals have a prior, underlying predisposition that permits the evolution of clinical asthma. The development of this predisposition to asthma is dependent on a complex—and at present poorly understood—combination of factors, which are partially inherited and partially acquired later in life.

After careful consideration of the scientific literature, the com-

TABLE 2 Summary of Findings Regarding the Association Between Indoor Biologic and Chemical Exposures and the *Exacerbation* of Asthma in Sensitive Individuals

Biological Agents	Chemical Agents
Sufficient Evidence of a Causal Relationship	
Cat	ETS (in preschool-aged children)
Cockroach	
House Dust Mite	
Sufficient Evidence of an Association	
Dog	NO_2, NO_x (high-level exposures*)
Fungi or molds	
Rhinovirus	
Limited or Suggestive Evidence of an Association	
Domestic birds	ETS (in school-aged and older children, and
Chlamydia pneumoniae	in adults)
Mycoplasma pneumoniae	Formaldehyde
Respiratory Syncytial Virus (RSV)	Fragrances
Inadequate or Insufficient Evidence to Determine Whether or Not an Association Exists	
Cow and horse	Pesticides
Rodents (as pets or feral animals)	Plasticizers
Chlamydia trachomatis	VOCs
Endotoxins	
Houseplants	
Pollen exposure in indoor environments	
Insects other than cockroaches	
Limited or Suggestive Evidence of No Association	
(no agents met this definition)	

*At concentrations that may occur only when gas appliances are used in poorly ventilated kitchens

mittee concluded there is **sufficient evidence of a causal relationship** between exposure to house dust mite allergen and the development of asthma in susceptible children. This conclusion was based on the preponderance of several lines of evidence, including the results of clinical studies and population-based, case-control, and prospective epidemiologic investigations; the consistency of the association in different racial and ethnic groups; and the presence of a dose–response relationship between exposure to dust mite allergen and sensitization. Chapter 5 delineates the reasoning underlying this conclusion in greater detail.

There is **sufficient evidence to conclude that there is an association** between ETS exposure and the development of asthma in younger children. In the limited number of studies that have been able to separate the effects of maternal active smoking during pregnancy from the effects of ETS exposure after birth, evidence suggests that—although both exposures are detrimental—maternal smoking during pregnancy has the stronger adverse effect.

Limited or suggestive evidence exists for associations between

- cockroach allergen exposure and development of asthma in preschool-aged children; and
- infection with RSV and development of asthma in preschool-aged children.

The impact of exposure to these agents has been the subject of great research interest in the past few years, and efforts presently under way may clarify their role in asthma development.

Published case reports, public health surveillance of physician reporting, and cross-sectional studies of building occupants with indoor air quality complaints also provide limited or suggestive evidence of an association between aspects of the nonindustrial indoor environment and the development of asthma, with a building occupancy-related pattern of symptoms and in some instances objective abnormalities. What is lacking for the most part, however, is knowledge of specific etiologic agents in these nonindustrial indoor environments that might be responsible for new work-related asthma cases.

Inadequate or insufficient evidence exists to determine whether or not the other indoor exposures listed in Table 1 are associated with the development of asthma. This lack of information points to a gap in present-day knowledge concerning asthma—one that will be challenging to resolve.

There is **limited or suggestive evidence of no association** between infection with rhinovirus—the medical term for the large and ubiquitous group of viruses responsible for a variety of respiratory infections including those referred to as "the common cold"—and asthma development.

Table 3 summarizes these findings.

TABLE 3 Summary of Findings Regarding the Association Between Indoor Biologic and Chemical Exposures and the *Development* of Asthma

Biologic Agents	Chemical Agents
Sufficient Evidence of a Causal Relationship	
House dust mite	(no agents met this definition)
Sufficient Evidence of an Association	
(no agents met this definition)	ETS (in preschool-aged children)
Limited or Suggestive Evidence of an Association	
Cockroach (in preschool-aged children)	(no agents met this definition)
Respiratory Syncytial Virus (RSV)	
Inadequate or Insufficient Evidence to Determine Whether or Not an Association Exists	
Cat	NO_2, NO_x
Cow and horse	Pesticides
Dog	Plasticizers
Domestic birds	VOCs
Rodents	Formaldehyde
Cockroaches (except for preschool-aged children)	Fragrances
Endotoxins	ETS (in school-aged and older
Fungi or molds	children, and in adults)
Chlamydia pneumoniae	
Chlamydia trachomatis	
Mycoplasma pneumoniae	
Houseplants	
Pollen	
Limited or Suggestive Evidence of No Association	
Rhinovirus (adults)	(no agents met this definition)

Effectiveness of Indoor Environmental Interventions in Limiting Exposures and Affecting Asthma Outcomes

Patients with asthma and the parents of children with asthma need reliable information on which measures are likely to be most effective for improving indoor air quality. Specific recommendations are found in each chapter but there are general principles that should be kept in mind. Agents that can exacerbate asthma may generally be thought of in two categories: specific allergens and nonspecific respiratory tract irritants. Exposure to nonspecific irritants, such as cigarette smoke, may lead to asthma symptoms in any person with asthma; while allergens are only problems for individuals who are allergic to them. For example, if a person with asthma is allergic to cats, exposure to cats may cause wheezing; but if that person is not allergic to cats, exposure to them will not cause any problems. Therefore, reducing indoor airborne exposure to irritants is likely to help all asthmatic individuals to some degree while reductions in allergen exposure would only be expected to help individuals who are allergic to the allergens being reduced.

While the report identifies a number a mitigation strategies that are or may be effective in reducing exposure to potentially problematic agents, the committee found only a small number for which there is presently evidence that proper implementation of the strategy results in an improvement of symptoms or lung function in asthmatics. It is important to remember, though, that the absence of evidence does not mean an absence of effect. The science regarding indoor environmental interventions, exposure limitation, and effects on asthma outcomes is not nearly as well developed as that regarding the health effects of exposures. Exposure assessment[3] is often the weakest link in environmental health studies because it is difficult to do and is given inadequate attention by many researchers.

---

[3]Classically, "exposure assessment" involves specifying the population that might be exposed to the agent of concern; identifying the routes through which exposure can occur, and estimating the magnitude, duration, and timing of the dose that individuals might receive as a result of their exposure (NAS, 1994).

Nonetheless, the committee was able to identify well-conducted, rigorous studies on which to base conclusions.

Sufficient evidence of an association was found between the use of a combination of physical measures and a reduction in indoor **dust mite** allergen levels in dust samples. As detailed in Chapter 5, strategies for the effective control of mite growth vary by climate. Such measures have been shown to be effective at reducing symptoms in controlled trials and should be part of normal management of asthma in mite-allergic individuals. Several studies now under way are evaluating whether aggressive allergen avoidance regimes have an effect on the subsequent development of asthma. The results of these and other studies will inform the question of whether primary prevention of dust mite-induced asthma is possible. Two related issues that will have to be addressed are (1) the feasibility of implementing such comprehensive interventions and (2) whether these interventions result in lower rates of sensitization to a particular exposure or all exposures.

The committee found limited or suggestive evidence that the combined use of **cockroach** extermination and control of potential reservoirs of allergen in beds, carpets, furnishings, and clothing through cleaning can achieve a short-term decrease in cockroach allergen levels in indoor environments. Extermination alone appears ineffective because significant allergen levels remain in settled dust; cleaning alone in the absence of complete extermination does not eliminate the sources of the allergen. There was inadequate or insufficient evidence to determine whether or not an association exists between any cockroach mitigation or prevention strategy and transient or long-term improvement of symptoms or lung function in cockroach-allergic asthmatics. However, since evidence does suggest that dust mite mitigation strategies result in improvement of symptoms or lung function, mitigation of cockroach exposures would appear to be a sensible course of action in the absence of more definitive information.

Although the strategy may be unpopular, there is limited or suggestive evidence of an association between removal of a **cat** from the home and improvement of symptoms or lung function in cat-allergic asthmatics. Concomitant removal or isolation of known reservoirs of cat allergen (carpets, upholstery, mattresses, pillows) may be required to diminish allergen levels to those commonly measured in homes without cats. Limited or suggestive evidence indicates that some measures short of removal (e.g., washing the animal) may result in transient reduction in allergen levels. However, there is inadequate or insufficient evidence to determine whether or not an association exists between measures short of removal of a cat from the home and improvement in symptoms in cat-allergic asthmatics. Data on the effectiveness of interventions for **other animals** are too sparse to draw informed conclusions.

It is possible to physically remove accessible growing **fungi** from indoor environments. The entry of fungal spores from outdoors can be substantially reduced in mechanically ventilated buildings by pressurizing them and filtering incoming air; closing windows should also reduce indoor concentrations from outdoor sources. Although there is limited or suggestive evidence that such steps may result in a reduction in the levels of fungi in the indoor environment, the health impact of such reduction has not been studied. Fungi are difficult to kill, and dead fungal material probably contains allergens that can become airborne, although this has not been thoroughly tested.

There is relatively little information on the impact of ventilation and air-cleaning measures on indoor **pollen** levels, although it is clear that shutting windows and other measures that generally limit the entry rate of unfiltered outdoor air can be effective.

No general conclusions about means of altering exposure to low levels of **endotoxin** can be made at the present time. However, avoiding the use of cool mist humidifiers would appear to be a simple and effective means of eliminating risk of high-level exposure to endotoxin at home as well as exposure to organisms associated with hypersensitivity pneumonitis.

Source control—that is, stopping smoking—appears to be the only reliably effective means of preventing **environmental tobacco smoke** exposure. There is sufficient evidence to conclude that increased ventilation is *technologically capable* of reducing the indoor concentration of ETS particles and gases, and that particle air-cleaning methods are *technologically capable* of reducing the indoor concentration of ETS particles. However, evidence is lacking on whether interventions designed to encourage the use of the requisite ventilation and air cleaning methods would be associated with a reduction in asthma development or exacerbation.

Control options for **chemical and particulate pollutants** in indoor environments include source modification (removal, substitution, or emission reduction), ventilation (exhaust or dilution), or pollutant removal (filtration). The various forms of pollutant source modification are usually the most effective. For most gaseous pollutants—NO_2 for example—removal via air cleaning is not presently practical.

No intervention studies clearly document that any form of **dampness** control works effectively to reduce symptoms or to reduce the chances of asthma development. However, given its relationship to factors (such as dust mites and fungal growth) associated with asthma, steps to reduce dampness may be appropriate. For homes, these measures include powered mechanical ventilation to remove or dilute occupant-generated moisture, proper installation of vapor barriers, channeling ground water away from foundations, sealing below-ground walls to prevent water intrusion, protecting ground-level concrete slabs from moisture intrusion, and constructing crawl spaces to prevent water intrusion.

There are both theoretical evidence and limited empirical data indicating that feasible modifications in **ventilation** rates can decrease or increase[4] concentrations of some of the indoor pollutants associated with asthma by up to approximately 75%. Limited or suggestive evidence exists to indicate that particle **air cleaning** is associated with a reduction in the exacerbation of asthma symptoms. Theoretical and limited empirical data indicate that particle air cleaners are most likely to be effective in reducing the exacerbation of asthma symptoms associated with particles smaller than approximately 2 μm, such as ETS particles[5] and some airborne cat allergen. There is insufficient evidence to determine whether or not the use of particle air cleaners is associ-

---

[4]The indoor concentrations of some pollutants from outdoors—particulate matter and ozone, for example—may increase with the ventilation rate.

[5]Particle air cleaners are *not* effective in reducing concentrations of the gaseous components of ETS.

ated with decreased asthma development. It should also be noted that microorganisms can grow on some air-cleaning equipment such as filter media; thus, improperly maintained air cleaners are also a potential source of indoor pollutants.

Inadequate or insufficient information was available regarding several other interventions. These are discussed in Chapters 5 through 10.

It is difficult to draw general conclusions regarding effective indoor environmental interventions. However, the committee is able to offer some observations. For many allergens, effective strategies consist of integrated approaches consistently applied over time. The two primary components of an integrated approach are (1) removal or cleaning of allergen reservoirs and (2) control of new sources of exposure. Source removal—where it is possible—is typically the most effective control measure and may be the only effective measure for some agents. Avoidance of exposure through source removal, substitution, or emission reduction is usually the most successful approach for chemical agents.

GENERAL RESEARCH
RECOMMENDATIONS AND CONCLUSIONS

Asthma is a complex illness. The many variables that determine its development and severity defy simple summary. Although great strides have been made over the past few years in elucidating mechanisms and understanding the role of environmental and genetic influences, much work remains to be done. Importantly, we still do not know whether or to what extent the reported increases in asthma can be attributed to indoor exposures.

Subsequent chapters of this report contain specific recommendations for further research on the biologic and chemical agents addressed and on the characteristics of indoor environments that may influence asthma outcomes. A digest of these recommendations is contained in Chapter 11. Some general observations are offered below.

The factors that determine the predisposition to sensitivity to certain agents and lead to the development of asthma are still not well understood. There is a great need for studies that rigorously examine the role of prenatal exposure and whether the age of first exposure influences the development of sensitization. The interaction of different environmental exposures with genetic susceptibilities—a topic of great interest but little research progress—also has to be pursued.

A major problem in choosing and implementing an intervention to mitigate an exposure is the generally limited data available. The limitations exist in regard to both the quantity and the quality of research data. Many of the studies reported are not based on rigorous protocols. Definition of clinical outcome (especially in infants), measurement of exposure, rigorous study design, appropriate population selection, and generalizability of the findings are among the issues that are often not adequately addressed. Indoor environments typically include exposures to multiple potentially problematic agents—dust mites and fungi, for example, are ubiquitous. It has proven difficult to assess the individual roles of the factors implicated in existing studies because complete characterization of exposures has not been done. Therefore, it is often not possible to determine with confidence whether any effects noted are indeed the results of specific exposures studied or of confounders.

The poor and inner city residents are vulnerable populations for asthma development, morbidity, and mortality. As such, there is great interest in identifying effective means to address prevalent exposure problems. Although some research on interventions has been directed at these populations, some of the strategies tried may not be practical to implement unless the subjects are part of an organized protocol providing guidance and funds. Further, individuals living in public or rental housing, or in multifamily units, may not have control over parts of their indoor environment that would be desirable to modify, such as carpeting, excessive moisture, and comprehensive pest management. Future research has to address more effectively the feasibility and generalizability of intervention programs on target populations.

Finally, to date there has been little connection between the scientific literature regarding asthma and the scientific literature regarding the characteristics of healthy indoor environments (for example, building design and operation; and sources, transport, control methods, and exposures to indoor pollutants). Relatively

little of the existing medical and epidemiologic literature on
asthma quantifies indoor environmental conditions such as humidity, ventilation, and pollutant concentrations or exposures in
sufficient detail. The effectiveness of exposure limitation strategies in reducing exposures and asthma development or exacerbation has, in general, been inadequately studied. These are areas of
research that have the potential to impact public health significantly. The committee believes that better communication between medical, public health, behavioral science, engineering,
and building professionals is likely to result in more informed
studies on the causes of asthma and the means to limit problematic exposures. The committee encourages efforts to bring these
groups together to educate one another on their areas of expertise. Although considerable work has been done and is being done
on asthma per se, increased research efforts are needed to address
the characteristics of healthy indoor environments. Asthma research clearly needs interdisciplinary involvement—not only of
clinicians, immunologists, and researchers in related biologic areas—but also of engineers, architects, materials manufacturers
and others who are responsible for the design and function of
indoor environments. Collaborations should be fostered, and consideration should be given to formulating model research protocols that include indoor environmental characteristics.

REFERENCES

Benson V, Marano MA. 1998. Current estimates from the National Health
    Interview Survey, 1995. National Center for Health Statistics. Vital Health
    Statistics Series 10 No. 199. DHHS Publication PHS 98-1527.
Carr W, Zeitel L, Weiss K. 1992. Variations in asthma hospitalizations and deaths
    in New York City. American Journal of Public Health 82:59–65.
Lang DM, Polansky M. 1994. Patterns of asthma mortality in Philadelphia from
    1969 to 1991. New England Journal of Medicine 331:1542–1546.
Mannino DM, Homa DM, Pertowski CA, Ashizawa A, Nixon LL, Johnson CA,
    Ball LB, Jack E, Kang DS. 1998. Centers for Disease Control and Prevention.
    Surveillance for Asthma Prevalence—United States, 1960–1995. Morbidity and
    Mortality Weekly Report. 47(No. SS-1):1–28.
National Academy of Sciences (NAS). 1994. Science and Judgement in Risk
    Assessment. National Academy Press: Washington, DC.
Rappaport S, Boodram B. 1998. Forecasted state-specific estimates of self-reported
    asthma prevalence—United States, 1998. Morbidity and Mortality Weekly
    Report 47(47):1022–1025.

1
MAJOR ISSUES IN
UNDERSTANDING *Asthma*

The purpose of this chapter is to provide background information helpful to the understanding of the material covered in the report. It contains a summary of the 1993 Institute of Medicine report *Indoor Allergens*, which covered some of the same topics examined here. It also addresses some of the major issues in understanding the medical condition called asthma: the controversy over the definition of the illness; the characteristics of its clinical presentation in children, adolescents, and adults; and the concepts of the "development of asthma" and "exacerbations of asthma." Finally, the chapter presents brief discussions of four topics addressed in greater detail later in the report: risk factors, trends in prevalence, pathophysiology, and tools for evaluating the effectiveness of interventions to reduce asthma.

ORIGIN OF THE STUDY

In 1993, as a result of joint funding between the U.S. Environmental Protection Agency (EPA) Indoor Air Division (IAD) and several agencies within the Department of Health and Human Services, the Institute of Medicine (IOM) issued a major report: *Indoor Allergens: Assessing and Controlling Adverse Health Effects* (hereafter called *Indoor Allergens*). *Indoor Allergens* received wide public and press attention and helped to focus public health policy on the dramatic increases in asthma, especially in children. The report also pointed out the role of indoor allergens such as dust mites, cockroaches, fungi, and pet dander in the etiology of asthma. At about the same time, IAD and the EPA Office of Research and Development issued a major assessment of the health impacts of environmental tobacco smoke (ETS), which found a correlation between exposure to ETS and asthma in children.

EPA's asthma and indoor air initiatives are now the responsibility of the Indoor Environments Division (IED), a part of the Office of Radiation and Indoor Air. IED has been actively involved in public outreach efforts on asthma and its relationship to indoor environmental pollutants. These include educational campaigns in high-risk communities, the *Indoor Air Quality (IAQ) Tools for Schools* Action Kit, and cooperative efforts with other government agencies.

EPA is currently developing an outreach strategy focused on reducing asthma-related morbidity and mortality associated with exposure to indoor environments. To help ensure that such efforts are based on sound science, EPA requested that the National Academies undertake an assessment of asthma and its relationship to indoor air quality. This report presents the conclusions of that research effort.

SUMMARY OF THE *INDOOR ALLERGENS* REPORT

The early 1990s saw an increase in the level of concern about the potential adverse health effects of indoor air quality. Motivated by this concern, several agencies of the federal government asked the IOM to undertake an assessment of the public health significance of indoor allergens. The IOM responded by assembling a committee of experts in such fields as allergy and immunology, epidemiology, mycology, engineering, industrial health, pulmonology, education, and public policy. The study undertaken by the committee had three primary objectives:

1. to identify airborne biological and chemical agents found indoors that can be directly linked to allergic diseases;
2. to assess the health impacts of these allergens; and

3. to determine the adequacy of the knowledge base that is currently available on this topic.

That report described what was then known about the adverse human health effects caused by indoor allergens, the magnitude of the problem nationally, the specific causative agents and their sources, the testing methods used for identifying allergens and diagnosing related diseases, and associated educational and research needs. The committee responsible for the study identified and developed a list of research agenda items and priority recommendations. The recommendations focused primarily on the need to improve awareness and education, while the research agenda focused on the longer-term, more expensive, and more technical aspects of fundamental research and data collection.

Indoor Allergens details the 1993 study committee's conclusions. Although some of these conclusions address medical conditions and topics outside the scope of the present report, several address issues related to asthma and the impact of indoor air exposures.

The *Indoor Allergens* committee recommended that steps be taken to improve estimates of allergenic disease incidence and prevalence, and to establish effective mechanisms for medical professionals to acquire assessments of potential exposure to indoor allergens in residential environments. It called for improvements in heating, ventilating, and air-conditioning (HVAC) equipment in order to minimize allergen reservoirs and amplifiers, and for the development of consensus standards for controlling moisture in buildings to help control microbial and arthropod aeroallergens and allergen reservoirs. The committee also recommended several educational initiatives including the development of focused intervention programs for allergic populations with different socioeconomic and educational characteristics. It called for efforts to inform architects, engineers, contractors, building maintenance personnel, and others responsible for the design and maintenance of indoor environments about the magnitude and severity of diseases caused by indoor allergens and the health implications of the design, construction, and operation of buildings.

Among the report's research agenda items were calls to

• better characterize rates of initial asthma sensitization, incidence, prevalence, and morbidity, and clarify the relation between these and socioeconomic status, race, and other factors;

• identify, characterize, and determine the health impact of indoor allergens, specifically suggesting research on allergenic chemicals, arthropods, dust mites, fungi, and indoor animals and plants;

• conduct dose–response studies in humans to determine both the relationship between allergen concentration and immunologic response and the threshold environmental exposure concentration for sensitization; and

• evaluate the effectiveness and cost-effectiveness of a broad variety of environmental control measures on patient symptoms, and determine whether long-term allergen avoidance has a positive effect on quality of life.

DEFINITIONS OF ASTHMA

Although patients of all ages are routinely diagnosed with asthma, finding a widely accepted definition for this disease has proven to be problematic (Samet, 1987; Toelle et al., 1992, 1997). One commonly used definition of asthma (Murphy, 1997) states:

> Asthma is a chronic inflammatory disorder of the airways in which many cells and cellular elements play a role, in particular, mast cells, eosinophils, T lymphocytes, macrophages, neutrophils, and epithelial cells. In susceptible individuals, this inflammation causes recurrent episodes of wheezing, breathlessness, chest tightness, and coughing, particularly at night or in the early morning. These episodes are usually associated with widespread but variable airflow obstruction that is often reversible either spontaneously or with treatment. The inflammation also causes an associated increase in the existing bronchial hyper-responsiveness to a variety of stimuli. (Murphy, 1997)

There are two important concerns with this definition: (1) the definition implies that asthma is a single disease entity, although much of the contemporary evidence suggests that asthma is a syndrome caused by several different mechanisms (Borish, 1999); and (2) many interpret this definition to mean that asthma results from an aberration or variation of the immune system leading to chronic inflammation. There is general agreement that asthma is always associated with inflammation within the lungs, and the intensity of the inflammation is related to the severity of respiratory symptoms and the degree of bronchial hyperresponsiveness (Clough and Dow, 1987; Ingram, 1991; Pattemore and Holgate, 1993). (Hyperresponsiveness refers to the abnormally large response of the lungs to the inhalation of minor irritants such as cold air.) There is also consensus that inflammation is the cause of hyperresponsiveness (Ingram, 1991; Jeffrey et al., 1989; Richmond et al., 1996; Woolley et al., 1996). Discovering the origin or origins of the inflammatory response, however, remains a critical unanswered question for researchers.

The absence of a universally accepted definition of asthma makes it especially difficult to arrive at a consistent operational definition for epidemiologic studies. One of the most commonly used definitions of asthma in epidemiology is a "physician's diagnosis" (Barbee et al., 1985; Dodge et al., 1986; Samet, 1987; Yunginger et al., 1992). This term is imprecise since there is little information about the reasoning and consistency used by physicians when making this diagnosis. A variety of definitions, based on questions about symptoms, have been proposed. The validity of these symptom-based definitions has rarely been rigorously evaluated (Toelle et al., 1997).

For the purposes of this report, asthma is understood to be a chronic disease of the airways characterized by an inflammatory response involving many cell types. Both genetic and environmental factors appear to play important roles in the initiation and continuation of the inflammation. Although the inflammatory response may vary from one patient to another, the symptoms are often episodic and usually include wheezing, breathlessness, chest tightness, and coughing. Symptoms may occur at any time of the day but are more commonly seen at night. These symptoms are associated with widespread airflow obstruction that is at least partially reversible with pharmacologic agents or time. Many persons with asthma also have varying degrees of bronchial hyperresponsiveness (Britton, 1992; Ingram, 1991). Research has shown that after long periods of time this inflammation may cause a gradual alteration or remodeling of the architecture of the lungs that cannot be reversed with therapy (Jeffrey et al., 1989; Kamm and Drazen, 1992; Murphy, 1997; Richmond et al., 1996).

CLINICAL PRESENTATION OF ASTHMA

Asthma may present at any age, but most studies suggest that in the majority of patients, asthma will present before puberty (Barbee et al., 1985; Martinez et al., 1995). Discussing the presentation of asthma is complicated by a lack of consensus on criteria for defining the onset of asthma. Many children who are sick with respiratory infections will experience asthma-like symptoms. In some children the symptoms will diminish, whereas in others the symptoms will persist (Brooke et al., 1995; Dodge et al., 1996; Martinez et al., 1995; Ross et al., 1995; Williams and McNicol, 1969). A diagnosis of asthma is dependent upon the recurring nature of these symptoms over a period of time. It appears that the more frequently these episodes occur, the more likely the child is to have asthma (Dodge et al., 1996; Martin et al., 1982; Martinez et al., 1995; Williams and McNicol, 1969). Unfortunately, no clear criteria mark the transition from recurrent wheezing with infections to asthma, and there are no tests capable of confirming a diagnosis (Brooke et al., 1995; Dodge et al., 1996). Tests of pulmonary function are very helpful in the diagnosis of asthma in adolescents and adults, but testing the lung function of children ages 1–6 is very difficult and possible only in a small number of research settings. This means that defining the onset of asthma depends on the variable skills and criteria applied by different physicians.

In most children, asthma begins as episodes of prolonged coughing, with or without wheezing, within the first few years of life. In young children, these symptomatic episodes are almost always associated with infections of the respiratory tract. The agents most often associated with these respiratory infections are common viral respiratory pathogens (Busse, 1989, 1995; Folkerts and Nijkamp, 1995; Martinez, 1995; Pattemore et al., 1992) . The potential roles of other infectious agents including mycoplasma and chlamydia have been questioned but not defined (Hahn et al., 1991, 1998; von Hertzen et al., 1999). In the majority of children these symptomatic episodes resolve with time. For others, the episodes will continue and will gradually begin to occur without the concomitant presence of infection (Brooke et al., 1995; Martinez et al., 1995). The frequency and severity of these episodes appear to increase with exposure to tobacco smoke and other forms of airborne pollutants (Arlian et al., 1993; Gidding and Schydlower, 1994; Menon et al., 1991). A family history of asthma and allergy and a personal atopic predisposition increase the likelihood that asthma will develop (Brooke et al., 1995; Martinez et al., 1995; Williams and McNicol, 1969).

With asthma symptoms ranging from clearly episodic to nearly continuous, from mild to severe, and from an isolated cough to a loud wheeze, diagnosing patients accurately can prove to be very difficult. In some children, asthma presents as distinct episodes of wheezing and difficulty breathing, whereas in others, cough may be the only complaint. When episodes are distinct events and wheezing is a prominent symptom, a diagnosis of asthma is relatively easy to make. However, when symptoms are less episodic and when wheezing is minimal or absent, asthma can be misdiagnosed or missed altogether. In many children the episodes of symptoms are diagnosed as recurrent bronchitis, bronchiolitis, or pneumonia (Brooke et al., 1995; Dodge et al., 1996; Sherman et al., 1990). When treated with antibiotics, these illnesses seemingly "resolve." If these episodes occur only a few times each year, the true asthma diagnosis may be missed for years (Davis, 1976; Martin et al., 1982; Schwartz et al., 1990). Physician recognition of asthma symptoms is further complicated by the parents' and child's perceptions, expectations, and abilities to describe the symptoms. When a diagnosis of asthma has been made, many parents of asthmatic children often state that the symptoms they now recognize as coming from asthma were present months or years before a diagnosis was made.

These variations in the presentation of asthma lead to confusion between the concepts of the "development of asthma" and the "exacerbations of asthma." The concept of developing asthma is that the lungs of a normal individual go through a process in which they develop characteristic, chronic, eosinophilic inflammation. The eosinophilic inflammation is associated with symptoms, such as cough and wheezing, that are recognized as asthma. Because there is no distinct finding or test that allows precise identification of these changes, it is usually impossible to pinpoint when asthma actually begins or develops. The term asthma exacerbation is used when referring to the sudden onset of symptoms in someone who already has developed asthma. In older children and some adults, the onset of asthma can be defined by the first exacerbation that brought the patient to medical attention. However, even in these individuals there was presumably a gradual process that took days, weeks, or months before the exacerbation appeared. There is essentially no information concerning whether changes in lung function or immune process in the lung can be detected prior to or during the development of asthma.

Another important but difficult concept is referred to as "growing out of asthma," which originated from the experiences of many parents and physicians. As already mentioned, many infants and young children wheeze in association with viral respiratory infections (Brooke et al., 1995; Martinez et al., 1995; Ross et al., 1995). As these infants grow older they cease to wheeze with this type of infection. If the child had been diagnosed as having asthma, he or she has now "outgrown" that asthma. There are also children who have recurrent episodes of wheezing during childhood and are diagnosed as having asthma, but cease to wheeze during adolescence, often during the years of puberty. The probability that the symptoms of asthma will remit appears to be higher if the child has little evidence of allergic disease (Martinez et al., 1995; Ross et al., 1995; Williams and McNicol, 1969). Some of the children whose asthma has remitted will redevelop asthma symptoms in adulthood. Because of the length of time required, there have been few prospective studies of the risk of asthma recurring once remission has occurred; hence there is no information about risk factors for redeveloping asthma.

As children move into adolescence and adulthood, respiratory infections remain a common cause of symptomatic episodes (Busse, 1995; Martinez, 1995). Additionally, symptoms may occur "spontaneously" or with exercise. Spontaneous symptoms are often discovered to be the result of exposure to either an allergen or a potent airborne irritant. When symptoms of coughing and wheezing are associated only with exercise, it is often difficult to distinguish whether these are new symptoms or unrecognized symptoms that have been present for years. Symptoms associated with exercise are more easily recognized in older children because they typically have a higher level of physical activity and have developed the verbal means to describe their symptoms. In some adolescents and adults, symptoms are perceived more as chest tightness or chest pain than as difficulty breathing. People who experience these symptoms may not recognize them as coming from the chest and complain instead of chronic fatigue or of becoming fatigued rapidly during the day (Brooke et al., 1995; Dodge et al., 1993).

An interesting change that takes place between childhood and adulthood is in the ratio of males to females with asthma. In childhood, boys with asthma outnumber girls by 1.5–2 to 1. By 20–30 years of age, women with asthma outnumber men by 1.5–2 to 1, a complete reversal of the childhood ratio (Barbee et al., 1985; Clough, 1993). Although it is tempting to speculate that this change, which occurs over the years of sexual maturation, results from hormonal changes, little is known about the actual cause.

Although asthma can develop anytime, there are times in a person's life when it can be especially troublesome. For some women, asthma first appears or markedly increases during pregnancy. Because poorly managed asthma is associated with an increased risk of maternal and fetal complications, the prompt recognition and appropriate treatment of asthma during pregnancy are important. During pregnancy, the onset of asthma and changes in the severity of preexisting asthma are presumed to be related to the major hormonal changes that occur, but the exact cause is unknown.

Occupational asthma is another complex problem (Cartier, 1994; Park et al., 1986). In some individuals, asthma symptoms first develop as a result of an occupation-related exposure. In most cases the agent responsible for the onset is an allergen to which the worker has become sensitized. In other cases the agent is a strong respiratory irritant. Occupational problems are beyond the scope of this report, however, and are therefore not discussed in detail.

A final important concern about the progression of asthma throughout life is the relationship between asthma and chronic obstructive pulmonary disease. As mentioned earlier, the chronic inflammatory process of asthma appears to ultimately result in irreversible obstructive changes in the lungs. The progression from a largely reversible airway obstruction, asthma, to an essentially irreversible state, chronic obstructive lung disease, appears to be highly variable and dependent upon a variety of factors such as smoking. The contribution of indoor sources of allergens and irritants to respiratory disease is important because chronic obstructive pulmonary disease is an important cause of increased morbidity and mortality among adults.

RISK FACTORS FOR ASTHMA

As early as the 1920s, studies demonstrated that a familial predisposition to asthma existed, suggesting that genetics may play a role in asthma development. This genetic influence has remained constant in subsequent studies; however it explains only 30–80% of the asthma risk. The remaining risk appears to be related to environmental exposure. The recently noted increase in the prevalence of asthma suggests a change in some environmental influence since it is hard to imagine a significant change in human genetics in such a short time (Borish, 1999).

Major studies have been conducted to better define the genes related to the development of asthma. These studies reveal complex relationships between genes and asthma (Borish, 1999). It appears that asthma results from the effects of multiple genes, not a single gene. Further complicating the association between genetics and asthma has been the discovery that genes closely linked to asthma in one population may not be significantly linked in another population. Some of these discrepancies appear to be racial, and some appear to be related to the peculiarities of relatively isolated and therefore somewhat inbred populations. These findings raise the question of whether asthma is best thought of as a single disease entity, a syndrome, or a final common manifestation of several different disease processes (Borish, 1999).

Many different environmental variables have been evaluated in relationship to asthma. Some of these are nonspecific such as an increase in global pollution, a decrease in exercise or outdoor play because of television and computer games, fewer childhood infections because of immunizations, more childhood respiratory infections because of day care, alterations in microbial flora because of the frequent use of antibiotics, or changes in indoor environments. Alternatively, there are relatively specific factors, such as increased exposure to dust mite allergens, that are discussed in more detail in Chapters 5–8.

TRENDS IN THE PREVALENCE OF ASTHMA

Many studies have shown that the prevalence of asthma has been increasing in the United States for the past 30 years. Although this subject is explored further in Chapter 3, it is important to summarize some of the most important aspects of asthma trends to provide a better understanding of the problem of asthma in the United States today.

In 1998, the Centers for Disease Control and Prevention (CDC) published a study concerning the change in the prevalence of asthma across the United States from 1960 to 1995 (Mannino et al., 1998). That report combined information from several sources to produce a broad picture of the changes in asthma prevalence. Population estimates were based on the 1960, 1970, 1980, and 1990 censuses and the 1996 intercensal estimate. Each data set was stratified by region, sex, race, and age group. Self-reported asthma data came from the National Health Interview Survey, which is conducted annually. Data on visits to physicians' offices for asthma have been collected since 1975 by the National Center for Health Statistics (NCHS). The data have been gathered on five different occasions within the study interval, and each time approximately 2,000 physician offices or about 30,000–60,000 patient encounters, were evaluated. Also since 1992, NCHS has gathered annual data on hospital emergency and outpatient department visits. Hospitalizations attributable to asthma were estimated from 1979 to 1994 from the National Hospital Discharge Survey. Only cases with a primary discharge diagnosis of asthma were included. Finally, mortality was estimated from the Underlying Cause of Death data set from NCHS for 1960 through 1995 to identify all deaths in which asthma was selected as the underlying cause.

The results of this study show an increase in the prevalence of asthma and death rates from asthma over 15 years both nationally and regionally. Regional differences were found for some end points such as hospitalization but not for others such as prevalence. The self-reported prevalence of asthma increased by 75% from 1980 to 1994. In 1993–1994, an estimated 13.7 million residents of the United States reported asthma during the preceding 12 months. A significant increase in asthma was found for all races, both sexes, and for all age groups. The increase was most prominent among children 0–4 years (160%) and 5–14 years (74%). During 1993–1994, the self-reported prevalence of asthma was slightly higher among children ≤14 years of age than among persons ≥15 years of age. In this same year, prevalence rates were similar in all four regions of the country.

When office visits for asthma are considered, the estimated annual number of visits increased from 4.6 million to 10.4 million between 1975 and 1993–1994. Again, the increasing rates were found for all race strata, both sexes, and all age groups. During 1993–1994, the rate of office visits for asthma was lowest for the 15–34 year age group.

Data on emergency room (ER) visits for asthma were available only from 1992–1995, and during this interval there was no significant change. In 1995, there were in excess of 1.8 million ER visits for asthma. African Americans had consistently higher rates of ER visits than whites. The rate of ER visits decreased with increasing age.

From 1979–1980 to 1993–1994, the estimated number of asthma-related hospitalizations increased, but the rate of hospitalization did not change over this interval. Hospitalization rates were consistently higher among African Americans than among whites. In 1993–1994, the age-adjusted asthma hospitalization rates were higher in the Northeast than in the West. In each time interval examined, the hospitalization rates were highest among children 0–4 years, lowest among persons aged 15–34 years, and intermediate for those ≥35.

Mortality rates from asthma are confused by the changes in the International Classification of Diseases (ICD) coding criteria. Asthma death rates declined from 1960–1962 to 1975–1978, and then began to rise again by 1993–1994. The 1960–1962 death rate was 28.2 per 100,000 in contrast to a rate of 17.9 per 100,000 in 1993–1994. The lowest death rate was in 1975–1978 at 8.2 per 100,000 or less than half the rate of 1993–1994.

These national statistics are consistent with many local and regional reports showing an increase in the prevalence of asthma predominantly in children (Gergen et al., 1988; Gerstman et al., 1993; Vollmer et al., 1998; Yunginger et al., 1992). This problem has been found in all regions of the United States and affects all races. Even though all races are affected, most studies have consistently shown a greater impact of asthma in African Americans in comparison to whites (Cunningham et al., 1996; Gergen, 1996; Gergen et al., 1988; Gerstman et al., 1993). Similar trends have been observed in most developed countries. This increase in the prevalence of asthma over the relatively short interval of approximately 30 years strongly suggests that some as yet unidentified environmental or behavioral change is responsible.

MECHANISMS OF ASTHMA

Details of the pathophysiology of asthma are presented in Chapter 4; thus, only the major concepts are discussed here. Figure 1-1 attempts to illustrate these major concepts in a schematic form. Asthma appears to present in two different forms, allergic and nonallergic asthma, illustrated on the left and right sides of the figure, respectively. The critical difference is that in persons with allergic asthma, inhalation of allergens initiates an inflammatory response that leads to hyperreactivity of the airways and symptoms of asthma. In persons with nonallergic asthma, the inflammatory process and airway hyperreactivity appear the same as in individuals with allergic asthma, but allergic responses, defined by the presence of immunoglobulin E (IgE) antibodies specific for allergens, cannot be demonstrated. As shown in the figure, allergic asthma is the result of allergen exposure leading to allergic sensitization of a genetically predisposed individual. The development of allergic sensitization during allergen exposure may be influenced by other environmental effects such as the frequency and type of respiratory infections, passive exposure to tobacco smoke, or intensity of allergen exposure. Airway hyperreactivity may be a direct result of allergic airway inflammation or may result from one or more genes. Once asthma is present, a variety of exposures may result in acute or chronic asthma symptoms. In addition to allergen exposure, the exposures capable of

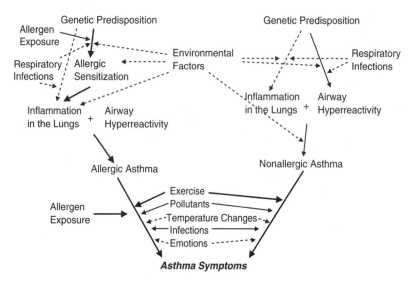

FIGURE 1-1 Development of asthma symptoms. This figure attempts to illustrate schematically the interrelationships of a number of factors thought to be important in the development of both allergic and nonallergic asthma. The weight of the lines attempts to provide some information about the strength of the evidence for the relationships: bold lines where there is ample evidence for a relationship and regular lines where the relationships have been demonstrated or at least strongly suggested. Dashed lines illustrate relationships that are likely to exist but for which there is little direct evidence at present. The phrase "genetic predisposition" is used to convey all of the genes in the human genome that are likely to be directly related to the outcome.

causing symptoms include airborne irritants, infections, and exercise.

The development of nonallergic asthma is somewhat harder to explain, although there are suggestions that chronic infections may be related to asthma in some nonallergic individuals (von Hertzen et al., 1999). As in the case of allergic asthma, inflammation occurs in the lungs and is typically accompanied by airway hyperreactivity. A variety of exposures may lead to wheezing in the nonallergic asthmatic; however, the origin of the inflammation cannot be directly identified. It is probable that various genetic influences on the immune system lead to inflammation in the lungs following some as yet undefined environmental exposure. It is possible that nonallergic asthma is the result of an unidentified allergen, but for many reasons this appears unlikely.

Estimates suggest that approximately 80% of asthma in children is allergic asthma. In adults, the percentage of those with allergic asthma is lower, in the 30–50% range depending on the population studied (Burr, 1993; Eriksson, 1990; von Mutius, 1996; Wever and Wever-Hess, 1993). The lower prevalence of allergic asthma in adults may arise in many different ways. The prevalence of allergic sensitivity is lower in adults than in children. It is possible that the inflammatory response that started as an allergic response in a child may become self-perpetuating by adulthood even as the allergic sensitization is lost. It is also possible that these are two different disease processes that merely have similar clinical appearances.

EVALUATING THE EFFECTIVENESS OF INTERVENTIONS TO REDUCE ASTHMA

Before an asthma intervention can be evaluated, the goals of the intervention must be defined. The goal of asthma therapy is ultimately to prevent a patient's asthma from altering or unduly influencing their life. Specific goals proposed in the *Guidelines for the Diagnosis and Management of Asthma* include (1) prevention of chronic and troublesome symptoms; (2) maintenance of "normal" pulmonary function; (3) prevention of recurrent exacerbations of asthma and minimization of the need for emergency department visits or hospitalizations; (4) provision of optimal pharmacotherapy with minimal or no adverse effects; and (5) meeting patients' and families' expectations and satisfaction with asthma care (Murphy, 1997).

Once goals are agreed upon, tools must be selected to evaluate how close the intervention comes to achieving all of the goals. In a typical clinical practice, the physician simply asks patients how they are doing. This is imprecise, and responses can vary depending upon how well the person can perceive his or her symptoms. Since many patients perceive their symptoms poorly, other tools are necessary to assess the adequacy of asthma control.

Pulmonary function tests, primarily spirometry, are essential tools for evaluating a patient with asthma. Good control of asthma is reflected in normal or near-normal spirometry. Sequential spirometry allows the physician to follow the course of disease and judge the adequacy of therapy. In children old enough to perform spirometry, lung growth can be monitored and assessed more easily.

There has been increasing recognition that control of asthma involves more than the absence of symptoms and the normality of lung function tests. Important aspects of asthma management involve the total impact of the disease on a person's life. How much does it affect what a person tries to accomplish? How does it affect daily activities? Do people avoid certain situations for fear of an attack? These diverse aspects have been grouped under the term "quality of life," and questionnaires have been developed that attempt to capture and estimate these aspects of disease control (Juniper et al., 1993; Rowe and Oxman, 1993).

Other methods for evaluating asthma are constantly being developed and evaluated. One of the most important areas in need of an adequate means of testing is the intensity of lung inflammation. In research settings it is possible to directly sample the linings of the airways and the numbers of inflammatory cells within the lungs, but these techniques are very expensive and involve some risk to the patient. Techniques are needed that could estimate lung inflammation accurately and yet be simple and inexpensive enough for use on a routine clinical basis.

The present state of the art in assessing any form of intervention for the control of asthma typically involves four components: (1) patient symptom scores or checklists, (2) physician assessments, (3) spirometry (often supplemented with home peak flow monitoring), and (4) completion of quality-of-life questionnaires. Given the lack of precision of these instruments, it is usually necessary to apply them with a relatively large number of persons for weeks or months before the effect of the intervention can be estimated adequately. In some cases, other measurements can be used to supplement the components listed above. These measures could include some attempt at estimating lung inflammation, determining allergen-specific IgE levels in serum, or assessing exercise tolerance.

REFERENCES

Arlian LG, Vyszenski-Moher DL, Fernandez-Caldas E. 1993. Allergenicity of the mite, *Blomia tropicalis*. Journal of Allergy and Clinical Immunology 91(5):1042–1050.

Barbee RA, Dodge R, Lebowitz ML, Burrows B. 1985. The epidemiology of asthma. Chest 87(1 Suppl.):21S–25S.

Borish L. 1999. Genetics of allergy and asthma. Annals of Allergy, Asthma, and Immunology 82(5):413–426.

Britton J. 1992. Airway hyperresponsiveness and the clinical diagnosis of asthma: histamine or history? Journal of Allergy and Clinical Immunology 89(1 Pt 1):1922.

Brooke AM, Lambert PC, Burton PR, Clarke C, Luyt DK, Simpson H. 1995. The natural history of respiratory symptoms in preschool children. American Journal of Respiratory and Critical Care Medicine 152(6 Pt 1):1872–1878.

Burr ML. 1993. Epidemiology of allergy. Monographs in Allergy 31:80–102.

Busse WW. 1989. The relationship between viral infections and onset of allergic diseases and asthma. Clinical and Experimental Allergy 19(1):1–9.

Busse WW. 1995. Viral infections in humans. American Journal of Respiratory and Critical Care Medicine 151(5):1675–1677.

Cartier A. 1994. Definition and diagnosis of occupational asthma. European Respiratory Journal 7(1):153–160.

Clough JB. 1993. The effect of gender on the prevalence of atopy and asthma. Clinical and Experimental Allergy 23:883–885.

Clough JB, Dow L. 1987. Epidemiological approach to bronchial responsiveness. Clinical Allergy 17(4):265–269.

Cunningham J, Dockery DW, Speizer FE. 1996. Race, asthma and persistent wheeze in Philadelphia schoolchildren. American Journal of Public Health 86(10):1406–1409. [Comment in Am J Public Health 1996. 86(10):1361–1362.]

Davis JB. 1976. Asthma and wheezy bronchitis in children. Skin test reactivity in cases, their parents and siblings. A controlled population study of sex differences. Clinical Allergy 6(4):329–338.

Dodge R, Cline MG, Burrows B. 1986. Comparisons of asthma, emphysema, and chronic bronchitis diagnoses in a general population sample. American Review of Respiratory Disease 133(6):981–986.

Dodge R, Burrows B, Lebowitz MD, Cline MG. 1993. Antecedent features of children in whom asthma develops during the second decade of life. Journal of Allergy and Clinical Immunology 92(5):744–749.

Dodge R, Martinez FD, Cline MG, Lebowitz MD, Burrows B. 1996. Early childhood respiratory symptoms and the subsequent diagnosis of asthma. Journal of Allergy and Clinical Immunology 98(1):48–54.

Eriksson NE. 1990. Allergy screening with Phadiatop and CAP Phadiatop in combination with a questionnaire in adults with asthma and rhinitis. Allergy 45(4):285–292.

Folkerts G, Nijkamp FP. 1995. Virus-induced airway hyperresponsiveness. Role of inflammatory cells and mediators. American Journal of Respiratory and Critical Care Medicine 151(5):1666–1674.

Gergen P. 1996. Social class and asthma—distinguishing between the disease and the diagnosis. American Journal of Public Health 86(10):1361–1362. [Comment on Am J Public Health 1996. 86(10):1406–1409.]

Gergen PJ, Mullally DI, Evans R III. 1988. National survey of prevalence of asthma among children in the United States, 1976 to 1980. Pediatrics 81(1):1–7.

Gerstman BB, Bosco LA, Tomita DK. 1993. Trends in the prevalence of asthma hospitalization in the 5- to 14-year-old Michigan Medicaid population, 1980 to 1986. Journal of Allergy Clinical Immunology 91(4):838–843.

Gidding SS, Schydlower M. 1994. Active and passive tobacco exposure: A serious pediatric health problem. A statement from the Committee on Atherosclerosis and Hypertension in Children, Council on Cardiovascular Disease in the Young, American Heart Association. Circulation 90(5):2581–2590.

Hahn DL, Dodge RW, Golubjatnikov R. 1991. Association of *Chlamydia pneumoniae* (strain TWAR) infection with wheezing, asthmatic bronchitis and adult-onset asthma. Journal of the American Medical Association 266(2):225–230. [Comment in JAMA 1991. 10;266(2):265.]

Hahn DL, Bukstein D, Luskin A, Zeitz H. 1998. Evidence for *Chlamydia pneumoniae* infection in steroid-dependent asthma. Annals of Allergy, Asthma, and Immunology 80(1):45–49.

Ingram RH Jr. 1991. Asthma and airway hyperresponsiveness. Annual Review of Medicine 42:139–150.

Jeffery PK, Wardlaw AJ, Nelson FC, Collins JV, Kay AB. 1989. Bronchial biopsies in asthma. An ultrastructural, quantitative study and correlation with hyperreactivity. American Review of Respiratory Disease 140(6):1745–1753.

Juniper EF, Guyatt GH, Ferrie PJ, Griffith LE. 1993. Measuring quality of life in asthma. American Review of Respiratory Disease 147(4):832–838.

Kamm RD, Drazen JM. 1992. Airway hyperresponsiveness and airway wall thickening in asthma. A quantitative approach. American Review of Respiratory Disease 145(6):1249–1250. [Comment on Am Rev Respir Dis 1992. 145(6):1251–1258.]

Mannino DM, Homa DM, Pertowski CA, Ashizawa A, Nixon LL, Johnson CA, Ball LB, Jack E, Kang DS. 1998. Centers for Disease Control and Prevention. Surveillance for Asthma Prevalence—United States, 1960–1995. Morbidity and Mortality Weekly Report 47(SS-1):1–28.

Martin AJ, Landau LI, Phelan PD. 1982. Predicting the course of asthma in children. Australian Paediatric Journal 18(2):84–87.

Martinez FD. 1995. Viral infections and the development of asthma. American Journal of Respiratory and Critical Care Medicine 151(5):1644–1648.

Martinez FD, Wright AL, Taussig LM, Holberg CJ, Halonen M, Morgan WJ. 1995. Asthma and wheezing in the first six years of life. The Group Health Medical Associates. New England Journal of Medicine 332(3):133–138.

Menon PK, Stankus RP, Rando RJ, Salvaggio JE, Lehrer SB. 1991. Asthmatic responses to passive cigarette smoke: persistence of reactivity and effect of medications. Journal of Allergy Clinical Immunology 88(6):861–869.

Murphy S. 1997. Expert Panel Report 2: Guidelines for the Diagnosis and Management of Asthma. 97-4051, 1–86. National Institutes of Health, National Heart, Lung, and Blood Institute, Washington, DC.

Park ES, Golding J, Carswell F, Stewart-Brown S. 1986. Preschool wheezing and prognosis at 10. Archives of Disease in Childhood 61(7):642–646.

Pattemore PK, Holgate ST. 1993. Bronchial hyperresponsiveness and its relationship to asthma in childhood. Clinical and Experimental Allergy 23:886–900.

Pattemore PK, Johnston SL, Bardin PG. 1992. Viruses as precipitants of asthma symptoms. I. Epidemiology. Clinical and Experimental Allergy 22(3):325–336.

Richmond I, Booth H, Ward C, Walters EH. 1996. Intrasubject variability in airway inflammation in biopsies in mild to moderate stable asthma. American Journal of Respiratory and Critical Care Medicine 153(3):899–903.

Ross S, Godden DJ, Abdalla M, McMurray D, Douglas A, Oldman D, Friend JA, Legge JS, Douglas JG. 1995. Outcome of wheeze in childhood: the influence of atopy. European Respiratory Journal 8(12):2081–2087.

Rowe BH, Oxman AD. 1993. Performance of an asthma quality of life questionnaire in an outpatient setting. American Review of Respiratory Disease 148(3):675–681.

Samet JM. 1987. Epidemiologic approaches for the identification of asthma. Chest 91(6 Suppl):74S–78S.

Schwartz J, Gold D, Dockery DW, Weiss ST, Speizer FE. 1990. Predictors of asthma and persistent wheeze in a national sample of children in the United States. Association with social class, perinatal events, and race. American Review of Respiratory Disease 142(3):555–562.

Sherman CB, Tosteson TD, Tager IB, Speizer FE, Weiss ST. 1990. Early childhood predictors of asthma. American Journal of Epidemiology 132(1):83–95.

Toelle BG, Peat JK, Salome CM, Mellis CM, Woolcock AJ. 1992. Toward a definition of asthma for epidemiology. American Review of Respiratory Disease 146(3):633–637.

Toelle BG, Peat JK, van den Berg RH, Dermand J, Woolcock AJ. 1997. Comparison of three definitions of asthma: A longitudinal perspective. Journal of Asthma 34(2):161–167.

Vollmer WM, Osborne ML, Buist AS. 1998. 20-year trends in the prevalence of asthma and chronic airflow obstruction in an HMO. American Journal of Respiratory and Critical Care Medicine 157(4 Pt 1):1079–1084.

von Hertzen L, Toyryla M, Gimishanov A, Bloigu A, Leinonen M, Saikku P, Haahtela T. 1999. Asthma, atopy and *Chlamydia pneumoniae* antibodies in adults. Clinical and Experimental Allergy 29(4):522–528.

von Mutius E. 1996. Progression of allergy and asthma through childhood to adolescence. Thorax 51(Suppl 1):S3–S6.

Wever AMJ, Wever-Hess J. 1993. Testing for inhalant allergy in asthma. Clinical and Experimental Allergy 23:976–981.

Williams HB, McNicol KN. 1969. Prevalence, natural history, and relationship of wheezy bronchitis and asthma in children. An epidemiological study. British Medical Journal 4(679):321–325.

Woolley KL, Gibson PG, Carty K, Wilson AJ, Twaddell SH, Woolley MJ. 1996. Eosinophil apoptosis and the resolution of airway inflammation in asthma. American Journal of Respiratory and Critical Care Medicine 154(1):237–243.
Yunginger JW, Reed CE, O'Connell EJ, Melton LJ III, O'Fallon WM, Silverstein MD. 1992. A community-based study of the epidemiology of asthma. Incidence rates, 1964–1983. American Review of Respiratory Disease 146(4):888–894.

2
METHODOLOGICAL CONSIDERATIONS IN EVALUATING THE *Evidence*

The U.S. Environmental Protection Agency (EPA) charged the committee responsible for this report with two primary objectives:

1. To provide the scientific and technical basis for communications to the public on

- the health impacts of indoor pollutants related to asthma; and
- mitigation and prevention strategies to reduce these pollutants.

2. To help determine what research is needed in these areas.

To help operationalize the first objective, EPA posed several questions for the committee's consideration. The committee was asked to evaluate the strength of the scientific evidence associating exposure to indoor pollutants with asthma, to discuss what was known about how and in what way(s) various pollutants influence asthma, and to examine the risk for development or exacerbation of asthma associated with indoor exposures.

EPA asked for information on the characteristics of the individuals most at risk for these exposures and on the role of genetic and other environmental factors in the occurrence of asthma. It requested information about whether effective strategies to mitigate or prevent problematic exposures had been developed and tested, whether these strategies had been shown to decrease asthma as well, at what exposure levels such decreases had been shown to occur, and whether the strategies were reasonable and cost-effective for affected individuals to undertake.

The ensuing chapters of the report address these questions, to the extent permitted by currently available science. They also touch on issues identified by the committee as relevant to its charge.

EVALUATING THE EVIDENCE

The evaluation of evidence involves several stages: (1) assessing the quality and relevance of individual reports; (2) deciding on the possible influence of error, bias, or confounding on the reported results; (3) integrating the overall evidence within and across diverse areas of research; and (4) formulating the conclusions themselves. These aspects of a review require thoughtful consideration of both quantitative and qualitative information—they cannot be accomplished by adherence to a prescribed formula.

The approach applied by the committee to this task evolved throughout the process of review and was determined in important respects by the nature of the evidence, exposures, and outcomes at issue. Ultimately, the conclusions expressed in this report are based on the committee's collective judgment. The committee endeavored to express its judgments as clearly and precisely as the available information allowed.

This section describes more fully how the evidence was evaluated. It discusses the research approach used to develop information, the methodologic considerations underlying the evaluation, considerations in assessing the strength of the evidence, and the categories of evidence used to summarize the committee's conclusions. The section is based on similar discussions in the Institute of Medicine (IOM) reports characterizing scientific evidence regarding vaccine safety (IOM, 1991, 1993) and the health effects of herbicides used in Vietnam (IOM, 1994, 1996, 1999), adapted to the current task.

Research Approach

To answer the questions posed by the EPA, the committee undertook a wide-ranging evaluation of the research on asthma and indoor air. While it did not review all such literature—an undertaking beyond the scope of this report—the committee attempted to cover the work it believed to be influential in shaping scientific understanding at the time it completed its task in mid 1999.

The committee consulted several sources of information in the course of its work. For conclusions regarding asthma outcomes, the primary source was epidemiologic studies. Most of these studies examined general population exposures to indoor agents at home, reflecting the focus of researchers working in this field. A small number of studies of occupationally exposed individuals were also evaluated. Some clinical research—for example, that addressing challenge tests—and animal studies were considered where appropriate. Engineering, architecture, and physical sciences literature informed the discussions of building characteristics, exposure assessment and characterization, indoor dampness, pollutant transport, and related topics; public health and behavioral sciences research was consulted for data on the effectiveness of interventions to limit exposure to problematic indoor agents. The committee also benefited from presentations of cutting-edge research given during two workshops it held in early 1999. A listing of the participating researchers and their topics is given in Appendix B.

The committee attempted to fairly consider and weigh all relevant information in reaching its conclusions. The failure to cite a particular study or research effort, however, does not necessarily mean that the committee did not consider its results.

Methodologic Considerations in Evaluating the Evidence

Uncertainty and Confidence

All science is characterized by uncertainty. Scientific conclusions concerning the result of a particular analysis or set of analyses can range from highly uncertain to highly confident—the theoretical concept of "proof" does not apply in evaluating actual observations. In its review, the committee evaluated the degree of uncertainty associated with the results on which it had to base its conclusions.

Statistical significance is a quantitative measure of the extent to which chance—that is, sampling variation—might be responsible for the observed exposure–adverse event association. The magnitude of the probability value or the width of the confidence interval associated with an effect measure such as the relative risk or risk difference is generally used to estimate the role of chance in producing the observed association. This type of quantitative estimation is firmly founded in statistical theory on the basis of repeated sampling.

For individual studies, confidence intervals around estimated results such as relative risks represent a quantitative measure of uncertainty. Confidence intervals present a range of results that, with a predetermined level of certainty, is consistent with the observed data. The confidence interval, in other words, presents a statistically plausible range of possible values for the true relative risk. When it is possible to use meta-analysis to combine the results of different studies, a combined estimate of the relative risk and confidence interval may be obtained.

For an overall judgment about an association between an exposure and a disease outcome based on a whole body of evidence, no quantitative method exists to characterize the uncertainty of the conclusions. Thus, to assess the appropriate level of confidence to be placed in the ultimate conclusions, it is useful to consider qualitative as well as quantitative aspects.

Analytic Bias

Analytic bias is a systematic error in the estimate of association between the exposure and the adverse event. It can be categorized under four types: selection bias, information bias, confounding bias, and reverse causality bias. *Selection bias* refers to the way that the sample of subjects for a study has been selected (from a source population) and retained. If the subjects in whom the exposure–adverse event association has been analyzed differ from the source population in ways linked to *both* exposure *and* development of the adverse event, the resulting estimate of association will be biased. *Information bias* can result in a bias toward the null hypothesis (no association between the exposure and the adverse event), particularly when ascertainment of either exposure or outcome has been sloppy, or it may create a bias away from the null hypothesis through such mechanisms as recall bias or unequal surveillance in exposed versus unexposed subjects. *Confounding bias*—addressed in greater detail below—occurs when the exposure–adverse event association is biased as a result of a third factor that is both capable of causing the adverse event and is statistically associated with the exposure itself. Finally, *reverse causality bias* can be a concern where it is possible that the outcome in question might influence the probability of experiencing the exposure being studied. It is not generally possible to quantify the impact of such nonrandom errors in estimating the strength of the association.

Confounding

In any epidemiologic study comparing an exposed to an unexposed group, it is likely that characteristics other than exposure may differ between the two groups. For example, the group exposed to a particular indoor pollutant may be of lower socioeconomic status than the unexposed group. When the groups differ with respect to factors that are also associated with the risk of the outcome of interest, a simple comparison of the groups may either exaggerate or hide the true difference in disease rates that is due to the exposure of interest. In the example of socioeconomic status, a simple comparison of asthma rates among the exposed and unexposed would exaggerate an apparent difference in asthma rates, since socioeconomic status is also thought to influence asthma incidence. If exposed individuals were of higher socioeconomic status, the simple comparison would tend to mask any true association between exposure and asthma by spuriously elevating the risk of disease in the unexposed group. This phenomenon, known as confounding, represents a major challenge to researchers and those evaluating their work.

Publication Bias

An important aspect of the quality of a review is the extent to which all appropriate information is considered and any serious omission or inappropriate exclusion of evidence is avoided. A primary concern in this regard is the phenomenon known as publication bias. It is well documented (Begg and Berlin, 1989; Berlin et al., 1989; Callaham et al., 1998; Dickersin, 1990; Dickersin et al., 1992; Easterbrook et al., 1991) in the scientific literature that studies with a statistically significant finding are more likely to be published than studies with nonsignificant results. Where such bias is present, evaluations of disease–exposure associations based solely on published literature could be biased in favor of showing a positive association. Other forms of bias related to reporting and publication of results have also been suggested. These include multiple publications of positive results, slower publication of nonsignificant and negative results, and publication of nonsignificant and negative results in non-English-language and low-circulation journals (Sutton et al., 1998). Several researchers have addressed the specific topic of whether there is bias in the publication of studies regarding the health impacts of exposure to environmental tobacco smoke (Bero et al., 1994; Kawachi and Colditz, 1996; Lee, 1998; Misakian and Bero, 1998).

The committee did not in general consider the risk of publication bias to be high among studies of indoor air exposures and asthma because

1. there were numerous published studies showing no positive association;

2. the committee was aware of the results of some unpublished research; and

3. The committee felt that the interest of the research community, public health professionals, government, and the general public surrounding the issue of asthma is so intense that any studies showing no association would be unlikely to be viewed as unimportant by investigators. In short, there would also be pressure to publish "negative" findings.

Nonetheless, the committee was mindful of the possibility that studies showing a positive association might be overrepresented in the published literature.

Considerations in Assessing the Strength of Scientific Evidence

Causality Definitions

The question of causality is of cardinal importance in health research, clinical practice, and public health policy. Despite its importance, however, causality is not a concept that is easy to define or understand (Kramer and Lane, 1992). Consider, for example, the relation between a hypothetical exposure X and asthma. Does the statement "X causes asthma" mean that (1) all persons exposed to X will develop asthma, (2) all cases of asthma are caused by exposure to X, or (3) there is at least one person whose asthma was caused or will be caused by X?

The first interpretation corresponds to the notion of a *sufficient cause*; X is a sufficient cause of asthma if all individuals exposed to X develop the disease. X is a *necessary cause* of asthma if the disease occurs only among those exposed to X, the second interpretation above. The idea that a "proper" cause must be both necessary and sufficient underlies the postulates of causality articulated by Koch in the 1800s (Susser, 1973). However, it is now generally recognized that for most exposure–outcome relations, a particular exposure need not be necessary or sufficient in order to cause the outcome—the third interpretation above. In other words, most health outcomes of interest have multifactorial etiologies.

This third form of causality is what is meant when scientists say that cigarette smoking causes lung cancer. Not everyone who smokes will develop lung cancer and not everyone who develops lung cancer smokes. However, individuals who smoke are more likely to develop lung cancer than those who do not, and the more they smoke the more likely they are to develop it.

Types of Causal Questions

The causal relation between an exposure and a given adverse event can be considered in terms of three different questions (Kramer and Lane, 1992):

1. *Can It?* (potential causality): Can the exposure cause the adverse event, at least in certain people under certain circumstances?

2. *Did It?* (retrodictive causality): Given that an individual who was subjected to the exposure developed the adverse event, was the event caused by the exposure?

3. *Will It?* (predictive causality): Will the next person who is subjected to the exposure experience the adverse event because of the exposure? Equivalently, how frequently will those subjected to the exposure experience the adverse event as a result of the exposure?

The form of causality relevant to this report is the first of these—potential or "can it?" causality. In the section below, this form of causality is discussed with reference to how it relates to the committee's charges and how the committee attempted to answer it.

Evaluation Criteria

Much of the epidemiologic literature on causality has focused on potential causality, and a widely used set of criteria has evolved for its assessment (Bradford Hill, 1965; Bradford Hill and Hill, 1991; Susser, 1973; U.S. Public Health Service, 1964). These criteria are also often used to inform public health policy recommendations and decisions (Weed, 1997).

For each indoor air exposure for which evidence indicated the presence of an association with asthma, the committee assessed the applicability of each of five general considerations, based on these criteria:

1. *Strength of Association*: Strength of association is usually expressed in epidemiologic studies as the magnitude of the measure of effect, for example, relative risk or odds ratio. Generally, the higher the relative risk, the greater is the likelihood that the exposure-disease association is "real" or, in other words, the less likely it is to be due to undetected error, bias, or confounding. Small increases in relative risk that are consistent across a number of studies, however, may also provide evidence of an association.

2. *Biologic Gradient (Dose–Response Relationship)*: In general, potential causality is strengthened by evidence that the risk of occurrence of an outcome increases with higher doses or frequencies of exposure. In the case of asthma, however, this is complicated by the central roles that susceptibility and sensitization play in the disease. The same exposure may have very different effects in susceptible and nonsusceptible, sensitized and nonsensitized individuals. Thus, the absence of a dose–response effect might not constitute strong evidence against a causal relation.

3. *Consistency of Association*: Consistency of association requires that an association be found regularly in a variety of studies, for example, in more than one study population and with different study methods. The committee considered findings that were consistent across different categories of studies as being supportive of an association. Note that the committee did not interpret "consistency" to mean that one should expect to see exactly the same magnitude of association in different populations. Rather, consistency of a positive association was taken to mean that the results of most studies were positive and that the differences in measured effects were within the range expected on the basis of all types of error including sampling, selection bias, misclassification, confounding, and differences in actual exposure levels.

4. *Biologic Plausibility and Coherence*: Biologic plausibility is based on whether a possible association fits existing biologic or medical knowledge. The existence of a possible mechanism increases the likelihood that the exposure-disease association in a particular study reflects a true association. In addition, the committee considered factors such as evidence in humans of an association between the exposure in question and diseases known to have causal mechanisms similar to asthma and evidence that asthma outcomes are associated with occupational exposure levels.

Considerations of biologic plausibility informed the committee's decisions about how to categorize the association between various indoor exposures and asthma, but the committee recognized that research regarding mechanisms is still in its infancy and did not predicate decisions on the existence of specific evidence regarding biological plausibility. Chapter 4 addresses the state of the science on asthma mechanisms.

5. *Temporally Correct Association*: If an observed association is real, exposure must precede the onset or exacerbation of the disease by at least the duration of disease induction. Temporality can be difficult to evaluate for some indoor agents because exposure to them is recurrent and pervasive. If individuals are exposed to an agent almost every day and in an environment where they spend most of their time it can be difficult to discern a relationship between exposure and effect. The lack of an appropriate time sequence is thus evidence against association, but the lack of knowledge about the natural history and pathogenesis of asthma limits the utility of this consideration. The committee also considered whether the outcome being studied occurred within a time interval following exposure that was consistent with current understanding of its natural history.

Other Considerations   As noted above, it is important also to consider whether alternative explanations—error, bias, confounding, or chance—might account for the finding of an association. If an association could be sufficiently explained by one or more of these alternate considerations, there would be no need to invoke the several considerations listed above. Because these alternative explanations can rarely be excluded sufficiently, however, assessment of the applicable considerations listed above almost invariably remains appropriate. The final judgment is then a balance between the strength of support for the association and the degree of exclusion of alternatives.

SUMMARIZING CONCLUSIONS REGARDING THE EVIDENCE

Categories of Association

The committee summarized its conclusions using a common format, described below, categorizing the strength of the scientific evidence in two areas:

1. *health effects*: the association between exposure to an indoor agent and asthma development or exacerbation; and

2. *exposure reduction strategies*: the effectiveness of exposure mitigation and prevention measures.

The five categories described below were adapted by the committee from those used by the International Agency for Research on Cancer (IARC, 1977) to summarize the scientific evidence for the carcinogenicity of various agents. Similar sets of categories have been used in National Academies' reports characterizing scientific evidence regarding vaccine safety (IOM, 1991, 1993) and the health effects of herbicides used in Vietnam (IOM, 1994, 1996, 1999). The distinctions reflect the committee's judgment that an association would be found in a large, well-designed study of the outcome in question in which exposure was sufficiently high, well characterized, and appropriately measured on an individual basis.

For health effects, the categories relate to the association between exposure to the agent and asthma, not to the likelihood that any individual's health problem is associated with or caused by the exposure.

Each of the categories describes the strength of the scientific evidence regarding the relationship between an *action* and an *outcome* related to indoor exposures and asthma. Table 2-1 gives examples of these.

Sufficient Evidence of a Causal Relationship

Evidence is sufficient to conclude that a causal relationship exists between the action or agent and the outcome. That is, the

TABLE 2-1 Examples of Actions and Outcomes Used in Categories of Evidence

Category	Action	Outcome
Health effects	Exposure to an indoor agent	Asthma development or exacerbation
Exposure reduction strategies	Implementation of a strategy to avoid or reduce exposure to an indoor agent	Actual reduction of exposure or reduction of asthma incidence

evidence fulfills the criteria below for sufficient evidence of an association and, in addition, satisfies criteria used to assess causality:

- *Strength of Association*: Is the exposure or action associated with a high probability of the outcome?
- *Biologic Gradient (Dose–Response Effect)*: Is increased exposure to the possible cause associated with increased response? For consideration of exposure reduction strategies, does more effective implementation of the remediation strategy result in a greater reduction in exposure or in asthma incidence?
- *Consistency of Association*: Have similar results been observed in multiple studies?
- *Biologic Plausibility and Coherence*: Is there a reasonable postulated biologic mechanism linking the possible cause and the effect?
- *Temporally Correct Association*: Does the presumed cause plausibly precede the effect?

These criteria have been discussed in greater detail above.

The finding of sufficient evidence of a causal relationship between an exposure and an asthma outcome (development or exacerbation) does not mean that the exposure inevitably leads to that outcome. Rather it means that the exposure *can* cause the outcome, at least in certain people under certain circumstances.

Sufficient Evidence of an Association

Evidence is sufficient to conclude that there is an association. That is, an association between the action or agent and the outcome has been observed in studies in which chance, bias, and confounding could be ruled out with reasonable confidence. For example, if several small studies that are free from bias and confounding show an association that is consistent in magnitude and direction, there may be sufficient evidence of an association.

Limited or Suggestive Evidence of an Association

Evidence is suggestive of an association between the action or agent and the outcome but is limited because chance, bias, and confounding could not be ruled out with confidence. For example, at least one high-quality study shows a positive association, but the results of other studies are inconsistent.

Inadequate or Insufficient Evidence to Determine Whether or Not an Association Exists

The available studies are of insufficient quality, consistency, or statistical power to permit a conclusion regarding the presence or absence of an association; or no studies exist that examine the relationship. For example, available studies may have failed to control for confounding or may have inadequate exposure assessment.

Limited or Suggestive Evidence of No Association

Several adequate studies are mutually consistent in not showing an association between the action or agent and the outcome. A conclusion of "no association" is inevitably limited to the conditions, level of exposure, and length of observation covered by the available studies. *In addition, the possibility of a very small elevation in risk at the levels of exposure studied can never be excluded.*

ASSESSING EXPOSURES TO AGENTS IN INDOOR AIR

Assessments of exposure to environmental agents in indoor air play a central role in epidemiology studies seeking to characterize population risks, in screening studies aimed at identifying individuals at risk, and in interventions designed to reduce risk. Because of the central importance of exposure assessment, it is important to understand the strengths and limitations of the approaches that are available to assess exposures in these contexts.

Definitions of Exposure

Exposure has been defined in various ways in the past. For example, an Institute of Medicine report (IOM, 1994) defines exposure as "the concentration of an agent in the environment in close proximity to a study subject."

The National Research Council report *Human Exposure Assessment for Airborne Pollutants* (NRC, 1991) adds the concept of time to that of concentration:

> An exposure to a contaminant is defined as an event that occurs when there is contact at a boundary between a human and the environment with a contaminant of a specific concentration for an interval of time.

For the purposes of this report, it is convenient to distinguish between two classes of exposure measures: 1) the theoretically ideal (and typically unknown) *risk-relevant exposure metric* (E_{RR}) that represents the individual breathing zone concentration of an agent of interest over a time period that is relevant to the risk of developing the health outcome of interest; and 2) the practical and available *exposure surrogate* that correlates to a greater or lessor extent with the E_{RR}. The E_{RR} will differ depending on the health outcome under study (e.g., asthma development versus asthma exacerbation). An exposure surrogate is any available measure of exposure that is positively correlated with the E_{RR}. Better surrogates are those with higher correlations, and thus lower degrees of exposure misclassification. When used without qualification in this report, the term "exposure" refers to surrogate exposure measures.

The E_{RR} is that theoretical measure of exposure that best represents the risk of adverse health consequences. Researchers don't know enough about asthma pathogenesis to confidently identify the appropriate E_{RR}. One possible E_{RR} for development of asthma might be a long-term average of agent concentration in the breathing zone of the person at risk—with duration on the order of 6 months to 2 years—measured during a relevant period of lung or immune system development, such as during the early years of life. For this E_{RR} it would seem reasonable to assume that, to a first order approximation, the average exposure level best represents risk; however second order effects might relate to variation around the mean. In the case of asthma exacerbations, the E_{RR} might be a short-term average that captures peak agent exposures in the breathing zone immediately prior to the exacerbation. Relevant averaging times might range from approximately 20 minutes to 48 hours.

In this context, the quality of any particular exposure surrogate measure can be judged in terms of how close it comes (i.e., correlates) to the E_{RR} in terms of the following four components:

1. Does it represent the key agent; that is, the agent solely or primarily responsible for the outcome?
2. Does it represent breathing-zone concentration?
3. Does it average appropriately over time?
4. Does it capture exposure during the vulnerable period of life?

A wide range of exposure surrogates exists. Figure 2-1 summarizes the principal approaches that are available for developing exposure surrogates (adapted from NRC, 1991).

Direct exposure surrogates include personal monitoring—involving the measurement of agent concentrations using monitors carried by individual subjects, and biological markers—involving the measurement of the agent or its metabolite in biological samples such as urine and blood. These offer more proximal measures of individual exposure than do the indirect approaches, though usually at the expense of sample size or ability to characterize long-term exposures. Indirect measures include environmental area monitoring (e.g., room sampling), models (e.g., microenvironmental model), recall questionnaires and real-time diaries. These approaches tend to be more practical in large-scale

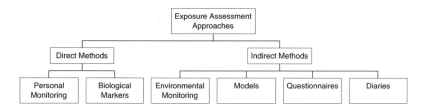

FIGURE 2-1. Classification of surrogate exposure assessment approaches. SOURCE: adapted from NRC (1991).

studies, and often are better suited to long-term exposure characterization than are direct measures.

The Microenvironmental Model

The microenvironmental model of total human exposure is a widely accepted tool for environmental exposure assessment (Sexton and Ryan, 1988). In this model, exposure of an individual to an airborne agent is defined as the time-weighted average of agent concentrations encountered as the individual passes through a series of microenvironments:

$$E_i = \sum_j^J C_j t_{ij} \, ,$$

where E_i is the time-weighted integrated exposure for person i over the specified time period; C_j is the is concentration of agent in microenvironment j; t_{ij} is the amount of time person i spent in microenvironment j, and J is the total number of microenvironments that person i moves through during the specified time period (24 hours, for example).

Sexton and Ryan (1988) define a microenvironment as "a 3-dimensional space with a volume in which contaminant concentration is spatially uniform during some specific interval." Examples include homes, offices, schools, and vehicles. An additional key assumption that is usually made is that the microenvironmental concentration is independent of time. These assumptions underlie the many attempts to use this model to reconstruct personal exposures to airborne contaminants based on environmental monitoring of microenvironments along with assessment of time-activity patterns. Unfortunately, the space and time continuity assumptions are rarely met in practice, severely limiting the utility of the microenvironmental modeling approach for estimating personal exposures. This is also likely to be the case for many of the indoor agents with plausible roles in asthma, such as particulate allergens. Exposures to such agents occur episodically due to the inadvertent disturbance and resuspension of allergen reservoirs by individual activities. These episodic exposure patterns are not likely to be accurately captured by environmental area samplers. Possible exceptions include ETS and airborne cat allergen, for which within-location spatial and temporal variations appear to be less pronounced (Samet, 1999; Chew et al., 1999).

Dose Concepts

Individual exposure to an airborne agent results in inhalation of the agent and deposition in the lungs. *Dose* is the amount of an agent that is absorbed or deposited in the body of an exposed organism for an increment of time (NRC, 1991). The NRC report further distinguishes between *internal dose* and *biologically effective dose*. Internal dose is defined as the amount of an agent that is absorbed into the body over a given time whereas biologically effective dose is the amount of an agent or its metabolites that has interacted with a target site over a given period of time.

For inhaled gases, the primary determinant of the lung deposition fraction (the proportion of inhaled mass that is deposited in the lungs) and the pattern of deposition in the lungs is the solubility of the gas. The primary determinant of deposition for particles is the aerodynamic particle diameter (d_{ae}). Aerodynamic particle diameter, as distinct from physical diameter, determines the motion of particles in air. The d_{ae} of a particle is defined as the diameter of the unit density sphere that has the same terminal settling velocity as the particle of interest (ICRP, 1994). Particles with d_{ae} larger than 15 μm are captured preferentially (though not exclusively) in the upper respiratory tract (i.e., nose and throat). Particles with d_{ae} between 2.5 and 15 μm enter the lungs but tend to deposit in the upper conducting airways where their mass and the high velocities favor inertial impaction. Because they lack inertia, smaller particles move with the inhaled airstream into the alveolar region, where they may or may not deposit. Deposition fraction in the deep lung increases with decreasing d_{ae} below 0.5 μm due to the high diffusion constants of very small particles.

The role of particle density in determining d_{ae} is critical. A spherical particle with a physical diameter of 16 μm but with a density of 0.1 will behave aerodynamically like a 5 μm water droplet. This property helps to explain the ability of large-diameter, low-density pollen grains to penetrate and deposit in the lung. Once deposited in the lungs, airborne agents may react with biomolecules, be absorbed into the blood, or be cleared from the lungs. From the viewpoint of asthma development or exacerbation, the relevant sites and nature of interactions between inhaled agents and human body remain uncertain, limiting our ability to define biologically effective dose in this context. In the case of allergens, however, the measurements of serum IgE and allergen skin test reactivity represent surrogates for biologically effective dose. It is important to note that all measures of dose, like those of exposure, can be viewed as surrogates for the theoretical risk-relevant dose measure.

Exposure Assessment for Specific Agents Considered in This Report

Table 2-2 lists the exposure and dose surrogates that have been used in past studies of agents with possible links to asthma development or exacerbation. The text that follows addresses issues related to assessing exposures to some of the agents addressed in this report. More detailed information on these agents and others evaluated in the report is given in Chapters 5 through 10.

House dust mite (HDM) exposure is associated primarily with inhalation of mite fecal pellets and aggregates (Chapman and Platts-Mills, 1980; Tovey et al., 1981). Most allergen-related particles are in the size range from 10–25 μm and are thought to become airborne primarily via active disturbance of allergen reservoirs in beds, soft furniture, and carpets (Tovey et al., 1981). Because of their large size, HDM allergen-related particles remain airborne for relatively short time periods (on the order of minutes). As such, area sampling of air concentrations has not proven a useful method of exposure assessment. Personal sampling is theoretically possible but requires further development. The currently accepted method for routine characterization of HDM exposures is to assay concentrations of group 1 allergens in dust samples collected by vacuuming, preferably in the bed or bedroom. Allergen concentrations are usually expressed in units of μg HDM/gram of dust collected. A theoretical advantage of dustborne allergen sampling is the presumed time-integration

TABLE 2-2 Exposure and Dose Surrogates Used in Asthma Research

Agent	Exposure Surrogates	Dose Surrogates
House dust mite (HDM) allergen	Dust mite count in bedroom dust HDM allergen in bedroom dust HDM allergen in other dust	HDM IgE HDM skin test
Cockroach (CR) allergen	CR counts by trapping CR allergen in bedroom dust CR allergen in kitchen dust	CR IgE CR skin test
Animal (dog, cat, etc.) allergen	Self-reported animal Pet allergen in dust Pet allergen in air	Pet-specific IgE Pet-specific skin test
Fungal allergen	Mold odor Moisture problems Visual evidence Culturable fungi Spore counts	Fungal-specific IgE Fungal-specific skin test
Pollens and plant allergen	Pollen counts in air Allergen concentration in air	Pollen-specific IgE Pollen-specific skin test
Environmental tobacco smoke (ETS)	Self-reported household smoking $PM_{2.5}$ sampling Airborne nicotine or other ETS markers	Cotinine in urine, blood, saliva
Nitrogen dioxide (NO_2)	Self-reported gas appliances Area monitoring for NO_2 in air Personal monitoring for NO_2 in air	None
Volatile organic compounds (VOCs)	Self-reported material presence (e.g., freshly painted surfaces) Area monitoring for VOCs in air Personal monitoring for VOCs in air	Exhaled-breath VOCs Blood VOCs concentrations
Formaldehyde	Self-reported material presence Area or personal air sampling	None
Pesticides	Self-reported use Concentrations in dust Concentrations in air	Blood concentrations

that occurs in the deposition of allergen particles on surfaces over time. It is worth noting one limitation of the common practice of reporting concentrations of allergen in settled dust samples in units of mass of allergen per gram of dust collected. By dividing by total dust collected, this expression of exposure does a poor job of characterizing the total allergen burden in a dwelling. For example, two homes A and B could have the same amount of allergen per gram of house dust by the conventional measure, whereas home A might have 10 times more house dust than home B, resulting in a 10-fold higher average exposure to occupants of home A. A high priority research need is the development of improved sampling methods that enable better standardization for area sampled than is possible using current methods.

Like HDM, **cockroach** allergens are thought to be associated primarily with larger particles that become airborne during and immediately after active disturbance of dust reservoirs. Thus, the same measurement issues apply here as for HDM. In contrast to HDM, which thrive primarily in bed, furniture and carpet materials, cockroach populations are usually concentrated in kitchens and bathrooms due to the availability of water and food sources. As a result, dust concentrations of cockroach allergen ($\mu g/g$) are often an order of magnitude higher in kitchen samples than in bedrooms (Sarpong et al., 1996). Even so, bedroom concentrations are generally thought to represent a better measure of human exposure to cockroach allergen for most individuals, due both to the duration of time spent in the bedroom and the likelihood of allergen disturbance there (Eggleston et al., 1998; Rosenstreich et al., 1997). For very young children who crawl or toddle on the floor, dustborne cockroach allergen on floors throughout the home may be relevant to exposures.

The principal **cat** allergen *Fel d* I is produced by salivary, sebaceous, and anal glands (De Andrade et al., 1996). *Fel d* I can be quantified in dust and air samples and also by specific IgE. A significant portion of airborne *Fel d* I is associated with particles less than 5 μm d_{ae} which remain airborne for extended periods and therefore tend to be distributed widely within interior spaces (Custovic et al., 1998). *Fel d* I also can be transported between locations via adherence to and resuspension from clothing, leading to measurable concentrations not only in homes with cats but also in schools, offices, vehicles, and homes with no cats. Because of this, home-based *Fel d* I concentrations often represent only one component of total human exposure. **Dog** and **other animal allergen** exposures have been studied less extensively; however, the basic parameters of exposure appear to be similar to those discussed above for cats (Custovic et al., 1997).

While several methods currently exist for measuring and characterizing **fungal** populations, methods for assessing human exposure to fungal allergens remain poorly developed at present, and represent a high priority research need. Part of the difficulty relates to the large number of fungal species that are measurable indoors, and the fact that fungal allergen content varies across species and across morphological forms within species (Cruz et al, 1997; Fadel et al., 1992). In addition, the most common methods for fungal assessment, counting cultured colonies and the identification and counting of spores, have variable and uncertain relationships to allergen content. Exposure surrogates based on questionnaire or inspection—such as water damage and visible fungal growth—also have very uncertain relationships with exposure to airborne fungal allergens. Although it is clear that individuals can be allergic to fungi, measurements of fungal allergen concentrations are very rarely included in epidemiological studies.

Indoor concentrations of airborne **pollens** occur via penetration of outdoor pollens into interior spaces, rather than emissions indoors. Penetration efficiency depends primarily on the size and shape of openings through which air enters the building—open windows versus small cracks, for example—and on aerodynamic particle diameter. Pollen grains often have large physical diameters but relatively small d_{ae} due to their low densities, favoring penetration. This allows pollens to remain airborne for long periods outdoors and be widely distributed by winds, as well as to penetrate indoors. Allergens from some plants, such as grass and birch, are located within particles that are much smaller than pollen grains (see Chapter 10). Indoor concentrations of particles from outdoors depend on the rate of depositional losses to indoor surfaces, the building ventilation rate, and the particle penetration efficiency.

Environmental tobacco smoke (ETS) is a complex mixture of submicron particles and gases produced by the combustion of tobacco products (Daisey et al., 1994; U.S. EPA, 1992). ETS remains airborne for long periods after emission, allowing time for dispersion and spread throughout interior spaces. Removal mechanisms include deposition onto interior surfaces and dilution by building ventilation. Significant ETS concentrations typically occur only indoors.

Questionnaire-based assessment of indoor smoking patterns has been used in many studies as an exposure surrogate. An important advantage of this approach is the potential it offers to capture long-term average indoor ETS emission patterns, which is less feasible to do with airborne measurements. A limitation of questionnaire-based exposure assessment is the potential for differential self-reporting of smoking patterns as a function of education and cultural attitudes.

Methods now exist for airborne measurements of chemical markers of the gaseous and particle phases of airborne ETS; however, the relationship between these marker compounds and the concentrations of the broader mixture of ETS constituents is still under investigation. While ETS is a major source of indoor $PM_{2.5}$ concentrations, $PM_{2.5}$ is not specific to ETS. For example, airborne nicotine concentrations are often used as a specific marker for the gaseous constituents of ETS. (Samet, 1999). Nicotine is a semi-volatile organic component of ETS.

Little data exist on the temporal variability of indoor ETS concentrations and it is not known what averaging time is adequate for characterizing long-term airborne ETS exposure of building occupants. Personal sampling represents an attractive approach in terms of sampling location; however, to be valid, the averaging time must be sufficiently long to estimate the long-term average exposure. Sampling with area monitors (i.e., nicotine sampler in bedroom or main activity room) enables more convenient and extensive sampling of long-term ETS exposures (e.g., over 1–2 weeks duration), and is thus recommended for epidemiology studies.

Biomarkers of ETS exposure also play an important role in research on ETS exposure and health. Cotinine, a biological metabolite of nicotine, can be measured in urine, blood, and saliva

(Benowitz, 1996). An important attribute of biomarkers such as cotinine is that they reflect exposures from all routes and locations—such as cotinine concentration in urine reflects not only ETS exposure in the home but also at work, in school or daycare, while shopping, in vehicles, and the like. This may be an advantage or disadvantage depending on the context. A limitation of cotinine for long-term exposure assessment is its relatively short biological half-life, on the order of several hours. Thus, cotinine measurements provide a measure of recent exposure, with significant modification by time since last exposure and individual metabolism rate. Because of these limitations, a single measurement of cotinine in urine or blood provides a good indication of whether recent exposure has occurred, but is generally not an accurate measure of long-term exposure levels.

Nitrogen dioxide (NO_2) is an irritant gas produced by high temperature combustion. Indoor sources include gas stoves and unvented space heaters. Indoor NO_2 levels are also influenced by the penetration of outdoor NO_2, which is elevated in urban areas where motor vehicles are the dominant source. Factors that influence concentrations of indoor NO_2 include frequency and duration of combustion appliance usage, emission rates of individual appliances, and home ventilation rate (Samet et al., 1987). High ventilation rates act to reduce NO_2 levels generated indoors, but conversely to increase penetration of NO_2 of outdoor origin. Available measures of indoor NO_2 exposures include questionnaire-based self-reporting of gas appliance presence or usage, environmental area sampling of airborne NO_2 levels, and personal sampling of airborne NO_2 levels. Questionnaire-based assessment is logistically simple but does not account well for variations in emission and ventilation rates. The sampling methods inherently account for these factors, as well has being specific for NO_2. However, as with most sampling methods, area and personal sampling characterizes only a snapshot in time, which may or may not be a good surrogate for long-term average indoor NO_2 levels.

Given the variety and complexity of indoor **volatile organic compound (VOC)** emission sources and rates, there is in general no reliable method for characterizing indoor VOC levels other than air sampling. A number of methods exist for both area and personal sampling of airborne VOCs. Of particular interest are recently developed passive diffusion badges that can measure VOC levels in the $\mu g/m^3$ range with sampling duration of 48 hours or more (Stock et al., 1996). Internal dose assessment is possible based on both exhaled breath and blood sampling; however, both methods address only recent exposures (e.g., within the previous 24 hours).

OTHER CONSIDERATIONS

The committee did not feel that there was sufficient evidence to generate confident quantitative estimates of the asthma risk associated with indoor air exposures. It is not possible to make general statements about the relative risk of various exposures because this is highly dependent on the characteristics of a particular environment and its occupants. House dust mites, for example, are a very common exposure in temperate and humid regions such as the southeastern United States but do not typically present a problem in cooler and drier climates such as northern Europe. Cockroaches, which also thrive in temperate and humid regions, are an important exposure in some urban environments. Fungi are ubiquitous and can be the primary source of allergen in some arid climates. Endotoxins may be found in humidifiers in urban settings or in organic dusts that infiltrate rural homes from outdoors. Occupant choice has a major role in determining indoor exposure to animals, plants, environmental tobacco smoke, indoor combustion sources, and chemicals used in cleaning and other activities. Indoor chemical exposures also result from outdoor infiltrates and certain building materials and furnishings.

Much of the literature regarding indoor exposures and asthma outcomes focuses on single agents, and the report thus has this same focus. Real indoor environments, however, are complex. They subject occupants to multiple exposures that may interact physically or chemically with one another and with the other characteristics of the environment like humidity, temperature, and ventilation levels. Synergistic effects—that is, interactions among agents that result in a combined effect greater than the sum of the individual effects—may also take place. Information on the combined effects of multiple exposures and on synergistic effects among agents is cited wherever possible. However, rather little data are available on this topic and it remains an area of active research interest.

Exposures in the indoor environment are not the only factor that may influence asthma outcomes, and interventions that consider only indoor factors may miss important opportunities to improve health. This report touches on the roles that genetics and socioeconomic status may play, although these subjects are not addressed in detail. Research has also examined the possible influence of several other factors, including antibiotic use (von Mutius et al., 1999), breastfeeding (Oddy et al., 1999) and other aspects of diet (Kimber, 1998; Weiss, 1999), low birth weight (Shaheen et al., 1999), number of siblings (Ponsonby et al., 1999), and obesity (Luder et al., 1998).

REFERENCES

Begg CB, Berlin JA. 1989. Publication bias and dissemination of clinical research. Journal of the National Cancer Institute 81(2):107–115.

Benowitz NL. 1996. Cotinine as a biomarker of environmental tobacco smoke exposure. Epidemiologic Reviews 18(2):188–204.

Berlin JA, Begg CB, Louis TA. 1989. An assessment of publication bias using a sample of published clinical trials. Journal of the American Statistical Association 84:381–392.

Bero LA, Glantz SA, Rennie D. 1994. Publication bias and public health policy on environmental tobacco smoke. Journal of the American Medical Association 272(2):133–136.

Bradford Hill A. 1965. The environment and disease: association or causation. Proceedings of the Royal Society of Medicine 58:295–300.

Bradford Hill A, Hill ID. 1991. Bradford Hill's Principles of Medical Statistics (Twelfth Edition). London: Hodder & Stoughton.

Callaham ML, Wears RL, Weber EJ, Barton C, Young G. 1998. Positive-outcome bias and other limitations in the outcome of research abstracts submitted to a scientific meeting. Journal of the American Medical Association 280(3):254–257. [Published erratum appears in JAMA 1998 280(14):1232.]

Chapman MD, Platts-Mills TA. 1980. Purification and characterization of the major allergen from *Dermatophagoides pteronyssinus*-antigen P1. Journal of Immunology 125(2):587–592.

Chew GL, Higgins KM, Gold DR, Muilenberg ML, Burge HA. 1999. Monthly measurements of indoor allergens and the influence of housing type in a northeastern US city. Allergy 54(10):1058–1066.

Cruz A, Saenz de Santamaria M, Martinez J, Martinez A, Guisantes J, Palacios R.

1997. Fungal allergens from important allergenic fungi imperfecti [Review]. Allergologia et Immunopathologia 25(3):153–158.

Custovic A, Green R, Fletcher A, Smith A, Pickering CA, Chapman MD, Woodcock A. 1997. Aerodynamic properties of the major dog allergen *Can f* I: distribution in homes, concentration, and particle size of allergen in the air. American Journal of Respiratory and Critical Care Medicine 155(1):94–98.

Custovic A, Simpson A, Pahdi H, Green RM, Chapman MD, Woodcock A. 1998. Distribution, aerodynamic characteristics, and removal of the major cat allergen *Fel d* I in British homes. Thorax 53(1):33–38.

Daisey JM, Mahanama KRR, Hodgson AT. 1994. Toxic volatile organic compounds in environmental tobacco smoke: Emission factors for modeling exposures of California populations. A133–186; California Air Resources Board, Sacramento, CA.

De Andrade AD, Birnbaum J, Magalon C, Magnol JP, Lanteaume A, Charpin D, Vervloet D. 1996. *Fel d* I levels in cat anal glands. Clinical and Experimental Allergy 26(2):178–180.

Dickersin K. 1990. The existence of publication bias and risk factors for its occurrence. Journal of the American Medical Association 263(10):1385–1389.

Dickersin K, Min YI, Meinert CL. 1992. Factors influencing publication of research results: follow-up of applications submitted to two institutional review boards. Journal of the American Medical Association 267(3):374–378.

Easterbrook PJ, Berlin JA, Gopalan R, Matthews DR. 1991. Publication bias in clinical research. Lancet 337(8746):867–872.

Eggleston PA, Rosenstreich D, Lynn H, Gergen P, Baker D, Kattan M, Mortimer KM, Mitchell H, Ownby D, Slavin R, Malveaux F. 1998. Relationship of indoor allergen exposure to skin test sensitivity in inner-city children with asthma. Journal of Allergy and Clinical Immunology 102(4 Pt 1):563–570.

Fadel R, David B, Paris S, Guesdon JL. 1992. *Alternaria* spore and mycelium sensitivity in allergic patients: in vivo and in vitro studies. Annals of Allergy 69(4):329–335.

IARC (International Agency for Research on Cancer). 1977. Some Fumigants, The Herbicides 2,4-D and 2,4,5-T, Chlorinated Dibenzodioxins and Miscellaneous Industrial Chemicals. IARC Monographs on the Evaluation of the Carcinogenic Risk of Chemicals to Man, Vol. 15. Lyon: IARC.

ICRP (International Commission on Radiological Protection). 1994. ICRP 66: Human Respiratory Tract Model for Radiological Protection. Annals of the ICRP 24(13).

IOM (Institute of Medicine). 1991. Adverse Effects of Pertussis and Rubella Vaccines. Washington, DC: National Academy Press.

IOM. 1993. Adverse Events Associated with Childhood Vaccines: Evidence Bearing on Causality. Stratton KR, Howe CJ, Johnston RB, eds. Washington, DC: National Academy Press.

IOM. 1994. Veterans and Agent Orange Health Effects of Herbicides Used in Vietnam. Washington, DC: National Academy Press.

IOM. 1996. Veterans and Agent Orange: Update 1996. Washington, DC: National Academy Press.

IOM. 1999. Veterans and Agent Orange: Update 1998. Washington, DC: National Academy Press.

Kawachi I, Colditz GA. 1996. Invited commentary: confounding, measurement error, and publication bias in studies of passive smoking. American Journal of Epidemiology 144(10):909–915.

Kimber I. 1998. Allergy, asthma and the environment: an introduction. Toxicology Letters 28(102–103):301–306.

Kramer MS, Lane DA. 1992. Causal propositions in clinical research and practice. Journal of Clinical Epidemiology 45(6):639–649.

Lee PN. 1998. Difficulties in assessing the relationship between passive smoking and lung cancer. Statistical Methods in Medical Research 7(2):137–163.

Luder E, Melnik TA, DiMaio M. 1998. Association of being overweight with greater asthma symptoms in inner city black and Hispanic children. Journal of Pediatrics 132(4):699–703.

Misakian AL, Bero LA. 1998. Publication bias and research on passive smoking: comparison of published and unpublished studies. Journal of the American Medical Association 280(3):250–253.

National Research Council (NRC). 1991. Human Exposure Assessment for Airborne Pollutants. Washington, DC: National Academy Press.

Oddy WH, Holt PG, Sly PD, Read AW, Landau LI, Stanley FJ, Kendall GE, Burton PR. 1999. Association between breast feeding and asthma in 6 year old children: findings of a prospective birth cohort study. British Medical Journal 319(7213):815–819.

Ponsonby AL, Couper D, Dwyer T, Carmichael A, Kemp A. 1999. Relationship between early life respiratory illness, family size over time, and the development of asthma and hay fever: a seven year follow up study. Thorax 54(8):664–669.

Rosenstreich DL, Eggleston P, Kattan M, Baker D, Slavin RG, Gergen P, Mitchell H, McNiff-Mortimer K, Lynn H, Ownby D, Malveaux F. 1997. The role of cockroach allergy and exposure to cockroach allergen in causing morbidity among inner-city children with asthma. New England Journal of Medicine 336(19):1356–1363. [Comment in N Engl J Med 1997. 336(19):1382–1384 and 337(11):791–792.]

Samet JM, Marbury MC, Spengler JD. 1987. Health effects and sources of indoor air pollution. Part I. American Review of Respiratory Disease 136(6):1486–1508.

Samet JM. 1999. Workshop summary: assessing exposure to environmental tobacco smoke in the workplace. Environmental Health Perspectives 107(Suppl 2):309–312.

Sarpong SB, Wood RA, Eggleston PA. 1996. Short-term effect of extermination and cleaning on cockroach allergen *Bla g* II in settled dust. Annals of Allergy, Asthma, and Immunology 76(3):257–260.

Sexton K and Ryan PB. 1988. Assessment of Human Exposure to Air Pollution: Methods, Measurements, and Models. In Air Pollution, the Automobile, and

Public Health, AY Watson, RR Bates, and D Kennedy, editors, pp. 207–238. Sponsored by the Health Effects Institute, Cambridge, MA. Washington, DC: National Academy Press.

Shaheen SO, Sterne JA, Montgomery SM, Azima H. 1999. Birth weight, body mass index and asthma in young adults. Thorax 54(5):396–402.

Stock TH, Morandi MT, Afshar M. 1996. A Modified Diffusion Sampler for Measuring 24-Hour VOC Concentrations in Personal, Indoor and Community Air, in *Proceedings of the 7th International Conference on Indoor Air Quality and Climate*, Vol. 2, S Yoshizawa, K Kimura, K Ikeda, S Tanabe, T Iwata, Eds., Indoor Air '96, Nagoya, Japan, pp. 73–77.

Susser M. 1973. Causal Thinking in the Health Sciences: Concepts and Strategies in Epidemiology. New York: Oxford University Press.

Sutton AJ, Abrams KR, Jones DR, Sheldon TA, Song F. 1998. Systematic reviews of trials and other studies. Health Technology Assessment 2(19):1–276.

Tovey ER, Chapman MD, Platts-Mills TA. 1981. Mite faeces are a major source of house dust allergens. Nature 289(5798):592–593.

U.S. EPA (U.S. Environmental Protection Agency). 1992. Respiratory Health Effects of Passive Smoking: Lung Cancer and Other Disorders. EPA/600/6-90/006F. Washington, DC.

U.S. Public Health Service, U.S. Department of Health, Education, and Welfare. 1964. Assessing Causes of Adverse Drug Reactions with Special Reference to Standardized Methods. Venulet J, Ed. London: Academic Press.

von Mutius E, Illi S, Hirsch T, Leupold W, Keil U, Weiland SK. 1999. Frequency of infections and risk of asthma, atopy and airway hyperresponsiveness in children. European Respiratory Journal 14(1):4–11.

Weed DL. 1997. On the use of causal criteria. International Journal of Epidemiology 26(6):1137–1141.

Weiss ST. 1999. Gene by environment interaction and asthma. Clinical and Experimental Allergy 29(Suppl 2):96–99.

3
PATTERNS OF ASTHMA
MORBIDITY AND Mortality

*E*pidemiology is the study of the distribution of a disease and its risk factors in a population. Epidemiologists attempt to understand the causes of a disease by studying its occurrence across time, place, and persons. Studying the observed differences will identify risk factors contributing to the development and activity of a disease. Once these risk factors are understood, interventions can be developed to ameliorate their effects. As we see in this chapter, asthma is not homogeneously spread throughout the population, nor has it remained static in terms of its burden on the U.S. population. These differences are reviewed to help the reader gain a better understanding of the importance of and interplay between the environment and genetics in asthma. Because the epidemiology of asthma is difficult to cover in such a brief space, the reader seeking further information is directed to a number of more extensive reviews (Mannino et al., 1998; Weiss et al., 1993).

THE BURDEN OF ASTHMA

Asthma is an important health problem in the United States. In the 1995 National Health Interview Survey (NHIS), approximately 15 million individuals identified themselves as asthmatics, with approximately 5 million being under age 18 (Benson and

Marano, 1998). The Centers for Disease Control and Prevention (CDC) estimated that there were about 17.3 million people with the illness in 1998 (Rappaport and Boodram, 1998). Despite these high numbers, asthma is a relatively minor cause of mortality in the United States. Asthma was listed as the underlying cause of death in 5,667 out of 2.3 million deaths in the United States during 1996 (U.S. Vital Statistics, 1996). In contrast, the morbidity burden is much greater. During 1996 there were 474,000 asthma hospitalizations, of which 195,000 occurred in children less than 15 years of age (Graves and Kozak, 1998). In addition, there were approximately 11.9 million medical visits for asthma, of which 9.1 million occurred in physicians' offices, 0.9 million in hospital outpatient departments, and 1.9 million in emergency rooms (Schappert, 1998). The economic burden of asthma was estimated to range from $5.8 billion to $6.2 billion in the early 1990s (Smith et al., 1997; Weiss et al., 1992). Asthma was estimated to account for approximately 1% of U.S. health care costs in the mid-1980s (Weiss et al., 1992).

MORTALITY

Asthma mortality data are readily available from the U.S. Vital Statistics System. The U.S. asthma mortality rate of 2.13 deaths per 100,000 population in 1996 is low compared to as many as 7 to 9 deaths per 100,000 population in other parts of the world—the highest rates being noted in New Zealand, West Germany, and Norway (Sears, 1991). Although the number of deaths from asthma is not large, an understanding of their causes is important since asthma mortality is considered preventable. Asthma mortality primarily affects adults, with approximately 67% occurring at or after 45 years of age. Males tend to have higher asthma death rates than females until about age 25, after which females have the higher rates for the rest of the life span. African Americans have consistently higher asthma death rates than whites. The difference is greatest in the younger group—approximately 10 times higher in 1- to 4-year-olds—and decreases with age until the ratio is only 1.2 times greater at age 85 and over (Figure 3-1). Little difference in asthma mortality rates is seen when the United States is divided into four census regions (U.S. Vital Statistics, 1995,

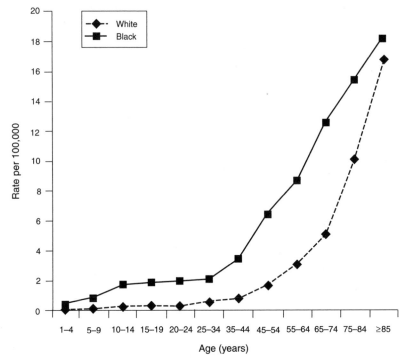

FIGURE 3-1 Asthma mortality by race, 1995. SOURCE: Anderson et al., 1997.

1996). However, numerous studies conducted in urban areas have much higher asthma mortality in areas characterized by high levels of poverty and minority populations (Carr et al., 1992; Lang and Polansky, 1994). Asthma mortality has been increasing in the United States since the late 1970s, although in the past several years the increase appears to have slowed down. No group has been spared this increase.

UTILIZATION OF HEALTH CARE SERVICES

Data on asthma health care utilization patterns in the United States are available from two national surveys—the National Hospital Discharge Survey and the National Health Care Survey. In contrast to mortality, approximately 37% of asthma hospitalizations occur before 15 years of age. Rates are highest in the young (less than 5 years of age), with a second, lower peak in the group age 65 or more (Graves and Kozak, 1998). In the young, hospitalization rates are higher for males, where as among adults, females have higher rates (Skobeloff et al., 1992). Hospitalization rates are elevated in urban areas with high levels of poverty and/or minority populations (Carr et al., 1992; Gergen and Weiss, 1990; Gottlieb et al., 1995). Readmissions can account for up to approximately 20% of hospitalizations, with the readmission rate being higher in the younger age group (Goldring et al., 1997). Hospitalizations show a regional variation in the United States, with the highest rates in the Northeast and the lowest in the West. However, this variation is true of hospitalizations in general—not just for asthma (Graves and Kozak, 1998). Asthma hospitalizations have primarily been rising in the United States in the under-15 age group, especially among those under 5 years of age. African-American children appear to have a higher rate of increase than white children do (Gergen and Weiss, 1990).

Somewhat less detail is available on outpatient care. Emergency room (ER) visits have been tracked in the United States only since 1992. Similar to hospitalizations, African Americans and women have higher rates. The rates for ER visits are the highest in the very young but, unlike hospitalizations, do not increase in the older age groups. Ambulatory care visits for asthma in physicians' offices are only slightly higher in the young versus older ages. The most striking difference is the low number of visits in the 15- to 34-year age range. African Americans and women have slightly higher rates. The number of office visits for asthma increased from 4.6 million in 1975 to 10.4 million in 1993–1994 (Mannino et al., 1998).

PREVALENCE

The prevalence of asthma is highly dependent on the definition chosen. The prevalence of asthma varied almost threefold, from 3.6 to 9.5%, when various combinations of physician diagnosis and wheezing were used to define asthma among children age 3 to 17 who participated in the second National Health and Nutrition Examination Survey (NHANES II) from 1976 to 1980

TABLE 3-1 Prevalence of Asthma by Type of Diagnosis

Type of Diagnosis	Percentage
Ever diagnosed with asthma by physician	7.0
Current physician diagnosis of asthma	3.6
Self-reported wheezing	5.3
Ever diagnosed with wheeze by physician	9.5
Current physician diagnosis of wheeze	6.7

SOURCE: Gergen et al., 1988. Adapted with permission from Pediatrics, 81:1–7, 1988.

(see Table 3-1) (Gergen et al., 1988). Despite this dependence on definition, certain patterns appear consistently in the age distribution of asthma. Most childhood asthma begins before age 5 (Gergen et al., 1988). A review of medical records from the Mayo Clinic found that the highest incidence of "asthma" occurred in the first year of life, with incidence rates falling throughout childhood and reaching their lowest levels among adults (Yunginger et al., 1992). More males develop asthma during childhood, while the prevalence in females surpasses that in males during adolescence. After adolescence, males and females have equivalent rates when a broad range of ages is examined (Figure 3-2) (Benson and Marano, 1998; Turkeltaub and Gergen, 1991).

Asthma prevalence is found to vary with race or ethnicity and urban location. In national surveys, African Americans reported higher rates of asthma and a higher rate of increase over the past decade (see Figure 3-3) (Benson and Marano, 1998; Gergen et al., 1988; Turkeltaub and Gergen, 1991). Hispanics reported wide discrepancies in prevalence: Mexican-American children in the Southwest reported some of the lowest rates of asthma in the United States while Puerto Rican children living on the East Coast of the United States reported some of the highest rates (Carter-Pokras and Gergen, 1993). Other data suggest that racial differences may reflect, at least in part, diagnostic acquisition rather than true differences in disease prevalence. A survey of 9- to 11-year-old school children in Philadelphia, Pennsylvania, found that white and African-American children reported the same

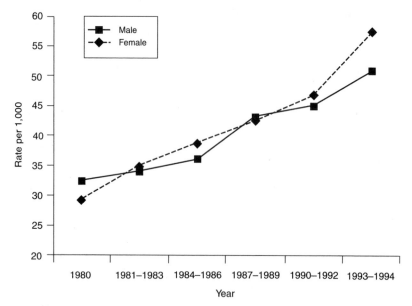

FIGURE 3-2   Asthma prevalence by gender, 1980–1994. SOURCE: Mannino et al., 1998.

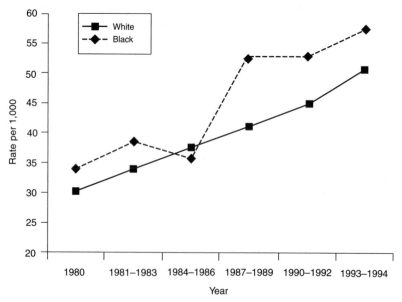

FIGURE 3-3   Asthma prevalence by race, 1980–1994. SOURCE: Mannino et al., 1998.

prevalence of persistent wheezing, while among children with wheeze, the African-American children were more likely to have received the diagnosis of asthma (Cunningham et al., 1996).

The NHIS has reported the prevalence of asthma to be higher in urban areas. More focused studies have shown that the prevalence of asthma can be just as high in less populated areas of the United States—for example, Los Alamos, New Mexico (Sporik et al., 1995)—compared to highly populated areas such as the Bronx, New York (Crain et al., 1994). Regional data from the 1996 NHIS show asthma prevalence to be highest in the Northeast and lowest in the Midwest and South (Benson and Marano, 1998). State-specific asthma prevalence rates have been estimated by multiplying age-, race-, sex-, and region-specific asthma prevalence rates calculated from the NHIS with the state's demographic composition. The 1998 state-level estimates are provided in Table 3-2. California, New York, and Texas were found to have the greatest number of asthmatics. The state-specific prevalence rates ranged from 5.8 to 7.2%, leading the authors to conclude that the demographic composition of states had minimal influence on estimated asthma prevalences (Rappaport and Boodram, 1998).

Based on data from the NHIS, the self-reported prevalence of asthma increased 75% between 1980 and 1994 (Mannino et al., 1998). All groups are thought to be affected by the increase in asthma prevalence. However, a study of the Mayo Clinic patient population reported that between 1964 and 1983, the incidence of asthma increased only in the 1- to 14-year age group, with no change in other age groups (Yunginger et al., 1992).

Epidemiologic investigations of asthma prevalence have demonstrated wide variability both between and within different countries. The International Study of Asthma and Allergy in Childhood (ISAAC) has shown that asthma-like symptom prevalence varies almost thirtyfold, from 1.6% in Indonesia to 36.8% in the United Kingdom. Generally, more developed Western countries have higher rates of asthma, while less developed countries have lower rates (ISAAC, 1998). The ISAAC project found that 12-month prevalence rates of wheeze were greater than 25% in the United Kingdom, New Zealand, Australia, Ireland, Canada, Peru, Costa Rica, Brazil and the United States, while Indonesia, Albania, Romania, Georgia, Greece, China, and Russia had

TABLE 3-2 Forecasted Estimates of Self-Reported Asthma Prevalence by State, 1998

Region or State	Number of Cases	Estimated Prevalence (%)	95% Confidence Interval (%)	Standard Error (%)
Northeast				
Connecticut	215,900	6.6	5.6–7.5	7.2
Maine	80,300	6.4	5.4–7.4	7.8
Massachusetts	401,000	6.5	5.6–7.5	7.2
New Hampshire	78,500	6.6	5.5–7.6	7.8
New Jersey	540,400	6.7	5.7–7.6	7.2
New York	1,236,200	6.8	5.8–7.8	7.3
Pennsylvania	800,900	6.6	5.6–7.5	7.2
Rhode Island	64,400	6.5	5.5–7.4	7.3
Vermont	39,500	6.5	5.5–7.6	7.8
Total	3,241,200	6.7	5.7–7.6	7.3
Midwest				
Iowa	190,100	6.6	5.6–7.6	7.5
Illinois	795,200	6.7	5.7–7.6	7.5
Indiana	398,400	6.7	5.7–7.7	7.3
Kansas	174,900	6.7	5.7–7.6	7.3
Michigan	642,300	6.7	5.7–7.7	7.5
Minnesota	318,600	6.7	5.8–7.7	7.1
Missouri	362,300	6.1	4.7–7.4	11.3
Nebraska	112,100	6.7	5.7–7.7	7.4
North Dakota	43,600	6.7	5.7–7.6	7.3
Ohio	748,200	6.7	5.7–7.6	7.4
South Dakota	51,000	6.7	5.8–7.7	7.3
Wisconsin	350,800	6.7	5.7–7.7	7.2
Total	4,187,600	6.6	5.6–7.6	7.4
South				
Alabama	280,500	6.0	4.8–7.1	9.5
Arkansas	162,600	5.9	4.9–6.9	6.9
District of Columbia	31,400	5.9	3.6–8.2	19.7
Delaware	44,300	5.9	4.9–6.9	8.5
Florida	863,900	5.8	4.9–6.8	8.0
Georgia	458,700	6.0	4.9–7.2	9.7
Kentucky	232,800	5.9	4.9–6.9	8.2
Louisiana	265,500	6.1	4.8–7.3	10.5
Maryland	307,300	6.5	5.6–7.5	7.2
Mississippi	167,900	6.1	4.7–7.4	11.3
North Carolina	447,200	5.9	4.9–7.0	8.9

TABLE 3-2 Continued

Region or State	Number of Cases	Estimated Prevalence (%)	95% Confidence Interval (%)	Standard Error (%)
Oklahoma	191,700	5.8	4.8–6.7	7.9
South Carolina	228,600	6.0	4.8–7.2	10.1
Tennessee	328,300	5.9	4.9–6.9	8.3
Texas	1,175,100	6.0	5.0–7.0	8.2
Virginia	403,400	5.9	4.9–6.9	8.6
West Virginia	108,600	5.8	4.9–6.8	8.2
Total	5,697,800	5.9	4.9–7.0	8.8
West				
Alaska	42,500	6.7	5.7–7.7	7.7
Arizona	316,200	6.9	6.0–7.9	6.9
California	2,268,300	7.1	6.1–8.0	6.8
Colorado	283,700	7.1	6.1–8.0	6.8
Hawaii	73,100	6.0	4.1–7.8	15.3
Idaho	86,100	6.7	5.7–7.8	7.6
Montana	61,600	6.6	5.7–7.6	7.4
Nevada	125,700	7.2	6.3–8.1	6.4
New Mexico	121,800	6.8	5.8–7.8	7.2
Oregon	225,900	6.9	5.9–7.8	6.9
Utah	141,200	6.7	5.6–7.8	8.1
Washington	391,900	6.9	5.9–7.8	6.8
Total	4,172,400	7.0	6.0–8.0	7.0
U.S. total	17,299,000	6.4	5.5–7.5	7.8

NOTE: Persons were considered to have asthma if asthma had been diagnosed by a physician at some time in their life and they had reported symptoms of asthma during the preceding 12 months.

SOURCE: Rappaport and Boodram, 1998.

12-month prevalence rates of less than 5%. Figure 3-4 illustrates the range of country-specific rates.

SEVERITY

Based on the data shown above, one could conclude that the severity of asthma has increased. However, asthma mortality,

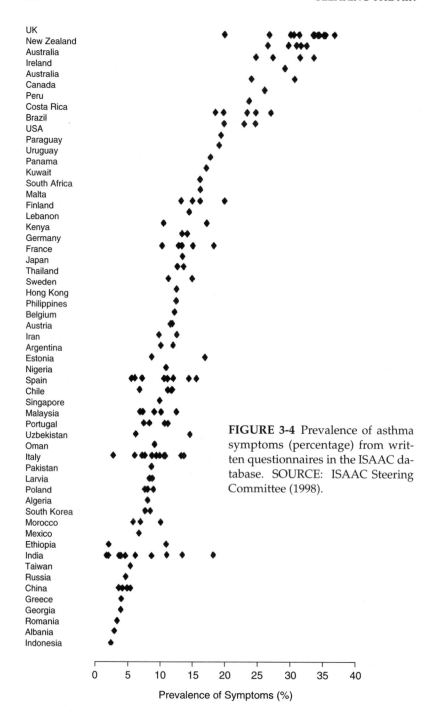

FIGURE 3-4 Prevalence of asthma symptoms (percentage) from written questionnaires in the ISAAC database. SOURCE: ISAAC Steering Committee (1998).

health care utilization, and the like are as much a reflection of asthma control as intrinsic severity (Cockcroft and Swystun, 1996). Asthma control reflects the access to appropriate health care, the use of medications, and environmental controls. Data from a number of sources imply that the increase in asthma is not due solely to an increase in its severity. Data collected on self-reported asthmatics participating in the Child Health Supplements of the NHIS—a nationally representative health survey—found that, in comparing 1988 with 1981, a number of markers of asthma impact had decreased: the percentage of asthmatics reporting fair or poor health status, the number reporting 30 or more days in bed over the past year, the rate of extreme behavior problem scores, and the number of school days missed. The reductions were similar in white and African-American children (Weitzman et al., 1992). Repeat surveys among 7- and 8-year-olds in the London borough of Croydon during 1978 and 1991 reported significant increases in the 12-month prevalence of attacks of wheezing or asthma, the one-month prevalence of wheezing episodes, and night waking, but not the prevalence of individuals reporting five or more attacks per year. In addition, there was a decrease in the number reporting 10 or more school days missed per year due to wheezing and any days in bed or restrictions in activities (Anderson et al., 1994).

TRENDS IN RISK FACTORS

The changing epidemiology of asthma provides the opportunity to compare the changes in the reported asthma risk factors to determine the role these are playing. Data are available on a number of asthma risk factors during the period of the observed increase in asthma. Outdoor air pollution does not appear to play an important role in the increase in asthma in the United States since pollutant levels have been decreasing during the time of asthma increase. Between 1987 and 1996, the ambient concentrations of a number of pollutants decreased: nitrogen dioxide by 10%, ozone by 15%, and sulfur dioxide by 37%; also, between 1988 and 1996, ambient particulate matter (PM_{10}) decreased by 25% (U.S. EPA, 1998). Similarly, cigarette smoking does not appear to be driving the increase in U.S. asthma rates since cigarette smoking decreased during the time of the well-documented asthma increase. Between 1979 and 1994, the percentage of active smokers dropped from 33.5 to 25.5% among individuals 18 years of age or more (Fingerhut and Warner, 1997). As discussed elsewhere in this report, there is little good information on allergen levels in the environment. A longitudinal study reporting that exposure to house dust mites was associated with the development of asthma found no change in the household levels of mites over the 10-year period of the study (Sporik et al., 1990). The rate of breast feeding—reported by some (Oddy et al., 1999) to protect against the development of asthma—has remained stable in the United States since the early 1980s. In 1981–1983, 58.1% of babies in the United States were reported to be breast-fed in the National Survey of Family Growth. Although the rate fluctuated a few percentage points during the intervening years, during 1993–1994 the rate was still 58.1% (Fingerhut and Warner, 1997).

There is some evidence that infections early in life could be protective for asthma and allergies (Martinez, 1994). However, this runs counter to data suggesting an increase in childhood asthma incidence at a time when the use of day care increased (from 13% in 1994 to 29.4% in 1997 according to a 1998 U.S. Census Bureau estimate [1998]), given that day care attendance is associated with an increase in infections (Nafstad et al., 1999). The relationship of infections to asthma is clearly complex. Although they may exacerbate asthma (Johnston et al., 1995), their role in the development of or protection from asthma is controversial (Kramer et al., 1999; Sporik et al., 1990). The role of infectious agents in asthma exacerbation and development is addressed in greater detail in Chapter 5.

TWIN, ADOPTION, AND MIGRANT STUDIES

Twin studies have long been used to determine the relative importance of environment and heredity in disease development. A study of 13,888 Finnish twin pairs age 18 to 70+ years reported a heritability estimate of 35.6% (Nieminen et al., 1991). Surprisingly, when examined by gender, the heritability estimate was 67.8% for females and 0% for males. In contrast, a report on 5,864 Norwegian twins studied at 18–25 years of age reported a 75%

heritability estimate (Harris et al., 1997). No separate estimates were given by gender. In a preliminary report on the development of asthma among adoptees, the presence of asthma in the adoptive parents was found to be a risk factor, implying that the home environment was important in the development of asthma (Smith et al., 1998).

Other empirical evidence for the role of the environment in the development of asthma comes from the study of migrant populations, which allows one to compare the development of asthma between two relatively homogeneous groups of individuals living in different environments. For example, a study of children living on Tokelau, an island in the Pacific Ocean, and children of Tokelau origin living in New Zealand found that the children living in New Zealand had about double the asthma prevalence of those living in Tokelau. This was true even when the children were new immigrants to New Zealand (Waite et al., 1980).

SOCIOECONOMIC STATUS VERSUS RACE OR ETHNICITY

Poverty and minority status are consistently reported to be related to increased asthma morbidity and mortality (Carr et al., 1992; Gottlieb et al., 1995). Thus far, research has not provided evidence that asthma is intrinsically a different disease among the various racial or ethnic groups in the United States. However, is asthma more prevalent in poor communities? Although asthma prevalence rates have been reported to be very high in poor urban areas with a high percentage of minority residents, such as the Bronx, New York (Crain et al., 1994), equally high rates have been reported in less urban areas with low levels of poverty and minority residents, such as Los Alamos, New Mexico (Sporik et al., 1995). According to asthma prevalence data for poor and non-poor children participating in the 1988 Child Health Supplement of the NHIS, there was only approximately 10% more asthma among poor children than non-poor children (Halfon and Newacheck, 1993). The previously mentioned differences in utilization and mortality are much greater. Thus, the differences in utilization appear to more clearly reflect differences in the control of asthma. It is clear that access to high-quality medical care can greatly decrease the impact of asthma: the variation in hospital rates for asthma among three cities in the United States was found to be related to the quality of primary care received (Homer et al., 1996); almost 80% of ER visits for asthma were found to be potentially preventable with better self-management training (Wasilewski et al., 1996). In a managed care setting, the use of an action plan was associated with reduced asthma hospitalizations and emergency room visits (Lieu et al., 1997). The quality of asthma care received by some poor and minority individuals is clearly inadequate (Finkelstein et al., 1995; Kattan et al., 1997).

ASTHMA RATES IN GERMANY—A NATURAL EXPERIMENT

The reunification of Germany in the early 1990s offered a chance to compare allergic diseases between two genetically similar populations living under vastly different and changing environmental conditions. A cross-sectional study found that 9- to 11-year-old children in West Germany reported more asthma, and allergic rhinitis, and had higher rates of allergen skin test reactivity and bronchial hyperreactivity (BHR). The East German children reported higher rates of bronchitis (von Mutius et al., 1994). A second cross-sectional study was done in the same East Germany city four years later, during which time pollution levels had dropped and prosperity had increased. The allergic skin test reactivity and reported rates of hay fever were found to have increased, but the prevalence of asthma and BHR had not (von Mutius et al., 1998). Somewhat different results were reported when two cross-sectional surveys (1992–1993 and 1995–1996) were carried out among 5- to 7-year-olds in East Germany. Self-reported allergies increased and bronchitis decreased after unification, while the level of allergen sensitivity as measured by RAST (radioallergosorbent test) did not change. The authors concluded that the differences were due to changes in physician diagnostic practices and/or reporting patterns by parents (Heinrich, 1999). The lack of change in asthma in both of these reports could represent either the lack of importance of these environmental changes to the development of asthma or an indication that there is a relatively brief period of vulnerability early in life when these types of environmental influences can have an effect.

REFLECTIONS ON THE TRENDS

There is no question that the prevalence of, health care utilization for, and morbidity from asthma have been reported to be increasing in the United States. The crucial issue is whether the observed changes represent a true increase or merely a better recognition of previously undiagnosed asthma. Better recognition has the potential to greatly increase the reported prevalence of asthma, since numerous population-based studies have shown that asthma is underdiagnosed in all age groups (Banerjee et al., 1987; Speight et al., 1983). One review of the reported increase in asthma prevalence concluded that much of the increase could be explained by changes in labeling and information bias, since few studies have employed objective measures to diagnose asthma (Magnus and Jaakkola, 1997). However, one of the few studies that employed an objective measure (peak flow) to define asthma reported an increase in prevalence over a 15-year time span (Burr et al., 1989). Markers of utilizations are subject to their own set of biases. Practice patterns have been found to vary greatly from one region to another for many diseases—not just asthma (Wennberg and Gittelsohn, 1982). Asthma hospital admission rates are influenced by many factors other than disease prevalence or severity. For example, higher hospitalization rates are found in areas with increased numbers of available hospital beds (Goodman et al., 1994), and it has been suggested that the majority of children hospitalized for asthma could have been treated in outpatient settings (McConnochie et al., 1999).

Lower socioeconomic status is clearly linked with increased levels of asthma morbidity and mortality. Asthma is closely associated with other manifestations of atopy such as immunoglobulin E (IgE) levels and allergen skin test reactivity (Burrows et al., 1989). However, in the United States, allergen skin test reactivity increases with increasing socioeconomic status (Gergen et al., 1987).

Even mortality statistics should be interpreted with caution. The accuracy of the coding of death certificates for asthma decreases with age—dropping from an accuracy of more than 97% in those under age 35 to 50% in the oldest age groups. (Sears et al., 1986). Chronic obstructive pulmonary disease (COPD) can be mistaken for asthma. Since deaths from COPD outnumber asthma deaths by a ratio of approximately 14 to 1, even a slight shift in coding practices that assigns more deaths to asthma can cause a large increase in asthma deaths with a much smaller decrease in COPD.

Thus, many questions about asthma remain unanswered. Although we have now accumulated a wealth of data regarding changes over time, variation among geographic areas and population groups, the cause(s) of asthma continue to elude scientists. It is clear that asthma is a disease with both genetic and environmental components. A better understanding of the complex interaction between genetics and environment awaits further study.

REFERENCES

Anderson HR, Butland BK, Strachan DP. 1994. Trends in prevalence and severity of childhood asthma. British Medical Journal 308(6944):1600–1604.

Anderson RN, Kochanek MA, Murphy SL. 1997. Report of final mortality statistics, 1995. Monthly Vital Statistics Report; vol. 45 no. 11, supplement 2. Hyattsville, MD, National Center for Health Statistics.

Banerjee DK, Lee GS, Malik SK, Daly S. 1987. Underdiagnosis of asthma in the elderly. British Journal of Diseases of the Chest 81(1):23–29.

Benson V, Marano M. 1998. Current estimates from the National Health Interview Survey, 1995. National Center for Health Statistics. Vital and Health Statistics. Series 10: Data from the National Health Survey (199):1–428.

Burr ML, Butland BK, King S, Vaughan-Williams E. 1989. Changes in asthma prevalence: two surveys 15 years apart. Archives of Disease in Childhood 64(1):1452–1456.

Burrows B, Martinez FD, Halonen M, Barbee RA, Cline MG. 1989. Association of asthma with serum IgE levels and skin-test reactivity to allergens. New England Journal of Medicine 320(5):271–277.

Carr W, Zeitel L, Weiss K. 1992. Variations in asthma hospitalizations and deaths in New York City. American Journal of Public Health 82(1):59–65.

Carter-Pokras OD, Gergen PJ. 1993. Reported asthma among Puerto Rican, Mexican-American, and Cuban children, 1982 through 1984. American Journal of Public Health 83(4):580–582.

Cockcroft DW, Swystun VA. 1996. Asthma control versus asthma severity. Journal of Allergy and Clinical Immunology 98(6 Pt 1):1016–1018.

Crain EF, Weiss KB, Bijur PE, Hersh M, Westbrook L, Stein RE. 1994. An estimate of the prevalence of asthma and wheezing among inner-city children. Pediatrics 94(3):356–362.

Cunningham J, Dockery DW, Speizer FE. 1996. Race, asthma, and persistent wheeze in Philadelphia schoolchildren. American Journal of Public Health 86(10):1406–1409.

Fingerhut L, Warner M. 1997. Injury chartbook. Health, United States, 1996–97. Hyattsville, Maryland: National Center for Health Statistics.

Finkelstein JA, Brown RW, Schneider LC, Weiss ST, Quintana JM, Goldmann DA, Homer CJ. 1995. Quality of care for preschool children with asthma: the role of social factors and practice setting. Pediatrics 95(3):389–394.

Gergen PJ, Turkeltaub PC, Kovar MG. 1987. The prevalence of allergic skin test reactivity to eight common aeroallergens in the U.S. population: results from the second National Health and Nutrition Examination Survey. Journal of Allergy and Clinical Immunology 80(5):669–679.

Gergen PJ, Mullally DI, Evans R III. 1988. National survey of prevalence of asthma among children in the United States, 1976 to 1980. Pediatrics 81(1):1–7.

Gergen PJ, Weiss KB. 1990. Changing patterns of asthma hospitalization among children: 1979 to 1987. Journal of the American Medical Association 264(13): 1688–1692.

Goldring J, Hanrahan L, Anderson H, Remington P. 1997. Asthma hospitalizations and readmissions among children and young adults—Wisconsin, 1991–1995. Morbidity and Mortality Weekly Report 46:726–729.

Goodman DC, Fisher ES, Gittelsohn A, Chang CH, Fleming C. 1994. Why are children hospitalized? The role of non-clinical factors in pediatric hospitalizations. Pediatrics 93(6 Pt 1):896–902.

Gottlieb DJ, Beiser AS, O'Connor GT. 1995. Poverty, race, and medication use are correlates of asthma hospitalization rates. A small area analysis in Boston. Chest 108(1):28–35.

Graves E, Kozak L. 1998. Detailed diagnoses and procedures. National Hospital Discharge Survey, 1996. National Center for Health Statistics. Vital and Health Statistics. Series 13: Data from the National Health Survey (138):i–iii,1–151.

Halfon N, Newacheck PW. 1993. Childhood asthma and poverty: differential impacts and utilization of health services. Pediatrics 91(1):56–61.

Harris JR, Magnus P, Samuelsen SO, Tambs K. 1997. No evidence for effects of family environment on asthma. A retrospective study of Norwegian twins. American Journal of Respiratory and Critical Care Medicine 156(1):43–49.

Heinrich J, Hoelscher B, Jacob B, Wjst M, Wichmann HE. 1999. Trends in allergies among children in a region of former East Germany between 1992–1993 and 1995–1996. European Journal of Medical Research 4(3):107–113.

Homer CJ, Szilagyi P, Rodewald L, Bloom SR, Greenspan P, Yazdgerdi S, Leventhal JM, Finkelstein D, Perrin JM. 1996. Does quality of care affect rates of hospitalization for childhood asthma? Pediatrics 98(1):18–23.

ISAAC (The International Study of Asthma and Allergies in Childhood) Steering Committee. 1998. Worldwide variation in prevalence of symptoms of asthma, allergic rhinoconjunctivitis, and atopic eczema: ISAAC. Lancet 351(9111):1225–1232.

Johnston SL, Pattemore PK, Sanderson G, Smith S, Lampe F, Josephs L, Symington P, O'Toole S, Myint SH, Tyrrell DA, et al. 1995. Community study of role of viral infections in exacerbations of asthma in 9–11 year old children. British Medical Journal 310(6989):1225–1229.

Kattan M, Mitchell H, Eggleston P, Gergen P, Crain E, Redline S, Weiss K, Evans R III, Kaslow R, Kercsmar C, Leickly F, Malveaux F, Wedner HJ. 1997. Characteristics of inner-city children with asthma: the National Cooperative Inner-City Asthma Study. Pediatric Pulmonology 24(4):253–262.

Kramer U, Heinrich J, Wjst M, Wichmann HE. 1999. Age of entry to day nursery and allergy in later childhood. Lancet 353(9151):450–454.

Lang DM, Polansky M. 1994. Patterns of asthma mortality in Philadelphia from 1969 to 1991. New England Journal of Medicine 331(23):1542–1546.

Lieu TA, Quesenberry CP Jr, Capra AM, Sorel ME, Martin KE, Mendoza GR. 1997. Outpatient management practices associated with reduced risk of pediatric asthma hospitalization and emergency department visits. Pediatrics 100(3):334–341.

Magnus P, Jaakkola JJ. 1997. Secular trend in the occurrence of asthma among children and young adults: critical appraisal of repeated cross sectional surveys. British Medical Journal 314(7097):1795–1799.

Mannino DM, Homa DM, Pertowski CA, Ashizawa A, Nixon LL, Johnson CA, Ball LB, Jack E, Kang DS. 1998. Centers for Disease Control and Prevention. Surveillance for Asthma Prevalence—United States, 1960–1995. Morbidity and Mortality Weekly Report 47(no. SS–1):1–28.

Martinez, FD. 1994. Role of viral infections in the inception of asthma and allergies during childhood: Could they be protective? Thorax 49(12):1189–1191.

McConnochie KM, Russo MJ, McBride JT, Szilagyi PG, Brooks AM, Roghmann KJ. 1999. How commonly are children hospitalized for asthma eligible for care in alternative settings? Archives of Pediatric and Adolescent Medicine 153(1):49–55.

Nafstad P, Hagen JA, Øie L, Magnus P, Jaakkola JJ. 1999. Day care centers and respiratory health. Pediatrics 103(4 Pt 1):753–758.

Nieminen MM, Kaprio J, Koskenvuo M. 1991. A population-based study of bronchial asthma in adult twin pairs. Chest 100(1):70–75.

Oddy WH, Holt PG, Sly PD, Read AW, Landau LI, Stanley FJ, Kendall GE, Burton PR. 1999. Association between breast feeding and asthma in 6 year old children: findings of a prospective birth cohort study. British Medical Journal 319(7213):815–819.

Rappaport S, Boodram B. 1998. Forecasted state-specific estimates of self-reported asthma prevalence—United States, 1998. Morbidity and Mortality Weekly Report 47(47):1022–1025.

Schappert S. 1998. Ambulatory care visits to physician offices, hospital outpatient departments, and emergency departments: United States, 1996. National Center for Health Statistics. Vital and Health Statistics. Series 13: Data from the National Health Survey (134):1–37.

Sears MR. 1991. Worldwide trends in asthma mortality. Bulletin of the International Union Against Tuberculosis and Lung Disease 66(2–3):79–83.

Sears MR, Rea HH, De Boer G, Beaglehole R, Gillies AJ, Holst PE, O'Donnell TV, Rothwell RP. 1986. Accuracy of certification of deaths due to asthma. A national study. American Journal of Epidemiology 124(6):1004–1011.

Skobeloff EM, Spivey WH, St. Clair SS, Schoffstall JM. 1992. The influence of age and sex on asthma admissions. Journal of the American Medical Association 268(24):3437–3440.

Smith DH, Malone DC, Lawson KA, Okamoto LJ, Battista C, Saunders WB. 1997. A national estimate of the economic costs of asthma. American Journal of Respiratory and Critical Care Medicine 156(3 Pt 1):787–793.

Smith JM, Cadoret RJ, Burns TL, Troughton EP. 1998. Asthma and allergic rhinitis in adoptees and their adoptive parents. Annals of Allergy, Asthma, and Immunology 81(2):135–139.

Speight AN, Lee DA, Hey EN. 1983. Underdiagnosis and undertreatment of asthma in childhood. British Medical Journal (Clinical Research Edition) 286(6373):1253–1256.

Sporik R, Holgate ST, Platts-Mills TAE, Cogswell JJ. 1990. Exposure to house-dust mite allergen (*Der p* I) and the development of asthma in childhood. A prospective study. New England Journal of Medicine 323(8):502–507.

Sporik R, Ingram JM, Price W, Sussman JH, Honsinger RW, Platts-Mills TA. 1995. Association of asthma with serum IgE and skin test reactivity to allergens among children living at high altitude. Tickling the dragon's breath. American Journal of Respiratory and Critical Care Medicine 151(5):1388–1392.

Turkeltaub PC, Gergen PJ. 1991. Prevalence of upper and lower respiratory conditions in the U.S. population by social and environmental factors: data from the second National Health and Nutrition Examination Survey, 1976–80 (NHANES II). Annals of Allergy 67(2 Pt 1):147–154.

U.S. Bureau of the Census. 1998. Primary Child Care Arrangements used for preschoolers by families with employed mothers. Internet Release data: January 14, 1998.

U.S. EPA (U.S. Environmental Protection Agency). 1998. National air quality and emissions trends report, 1996. Office of Air Quality Planning and Standards. Vol. EPA-454/R-97-013. Research Triangle Park, United States Environmental Protection Agency.

U.S. Vital Statistics. 1995. National Center for Health Statistics. Hyattsville, MD: Public Health Service.

U.S. Vital Statistics. 1996. National Center for Health Statistics. Hyattsville, MD: Public Health Service.

von Mutius E, Martinez FD, Fritzsch C, Nicolai T, Roell G, Thiemann HH. 1994. Prevalence of asthma and atopy in two areas of West and East Germany. American Journal of Respiratory and Critical Care Medicine 149(2 Pt 1):358–364.

von Mutius E, Weiland SK, Fritzsch C, Duhme H, Keil U. 1998. Increasing prevalence of hay fever and atopy among children in Leipzig, East Germany. Lancet 351(9106):862–866.

Waite DA, Eyles EF, Tonkin SL, O'Donnell TV. 1980. Asthma prevalence in Tokelauan children in two environments. Clinical Allergy 10(1):71–75.

Wasilewski Y, Clark NM, Evans D, Levison MJ, Levin B, Mellins RB. 1996. Factors associated with emergency department visits by children with asthma: implications for health education. American Journal of Public Health 86(1):1410–1415.

Weiss KB, Gergen PJ, Hodgson TA. 1992. An economic evaluation of asthma in the United States. New England Journal of Medicine 326(13):862–866.

Weiss KB, Gergen PJ, Wagener DK. 1993. Breathing better or wheezing worse? The changing epidemiology of asthma morbidity and mortality. Annual Review of Public Health 14:491–513.

Weitzman M, Gortmaker SL, Sobol AM, Perrin JM. 1992. Recent trends in the prevalence and severity of childhood asthma. Journal of the American Medical Association 268(19):2673–2677.

Wennberg J, Gittelsohn A. 1982. Variations in medical care among small areas. Scientific American 246(4):120–134.

Yunginger JW, Reed CE, O'Connell EJ, Melton LJ III , O'Fallon WM, Silverstein MD. 1992. A community-based study of the epidemiology of asthma. Incidence rates, 1964–84. American Review of Respiratory Disease 146(4):888–894.

4
PATHOPHYSIOLOGICAL
BASIS OF *Asthma*

In the current working definition of asthma, provided in the 1997 National Institutes of Health, Expert Panel Report 2: *Guidelines for the Diagnosis and Management of Asthma* (Murphy, 1997; see Chapter 1), the disease is characterized by two fundamental features, (1) an excessive sensitivity of the airways to a variety of endogenous and/or exogenous bronchoconstrictor agents, a feature referred to as "bronchial hyperresponsiveness"; and (2) a pathophysiological link between bronchial hyperresponsiveness and the presence of inflammation of the airways. A major focus of research in the past two decades has been to identify the physiological, cellular, and molecular mechanisms that underlie the association between bronchial hyperresponsiveness and airways inflammation in establishing the asthmatic condition. This research has led to key advances in elucidating the roles of specific inflammatory cells and other processes in the pathobiology of asthma. While there is a wealth of information indicating or suggesting an association between environmental exposures and asthma outcomes, very little is known about the means by which the exposures bring about the changes that manifest as asthma. A better understanding of the molecular mechanism(s) regulating the recognition of and response to environmental exposures may lead to more safe and effective asthma interventions. This chapter summarizes the state of the science regarding research on these mechanisms.

AIRWAY INFLAMMATION IN ASTHMA

Role of Mast Cells

In the mid-1960s, immunoglobulin E (IgE) was first clearly identified as the principal mediating agent of the allergic proasthmatic response (Ishizaka et al., 1966). Accordingly, in response to specific allergens bound to IgE molecules present on the surface of mast cells, basophils, and other cell types, a host of preformed cellular bronchoactive mediators are acutely released. In this manner, at least with respect to allergic asthma, IgE has been implicated as the fundamental inherent determinant of the "immediate" hypersensitivity airway response to allergen exposure. In accordance with this concept, allergic asthmatic individuals classically present with elevated serum concentrations of IgE, which defines the atopic state. Moreover, the degree of elevation in serum IgE concentration has been linked to the severity of asthma and has been identified as an important risk factor in the development of the disease (Sears et al., 1991).

Antigen coupled to cell-bound IgE is now known to activate a number of proinflammatory cells, principally including the airway mast cells. Upon activation of their high-affinity receptors for IgE, mast cells release a variety of preformed mediators including histamine, leukotrienes, various cytokines, and other proinflammatory molecules. This array of mast cell-derived mediators largely serves to elicit the immediate airway hypersensitivity response to allergen exposure, which is characterized by acute constriction of the airways, airway mucosal gland secretion, and airway edema secondary to increased airway microvascular permeability. In addition to this "acute-phase" response, various mast cell-derived chemotactic mediators have been implicated further in the development of a subsequent "late-phase" response, hours after allergen exposure, which is characterized by prolonged or sustained bronchoconstriction associated with infiltration of the airways by a variety of inflammatory cell types (Lemanske and Kaliner, 1981–1982; Robertson et al., 1974). Among the mast cell-derived mediators importantly implicated in the development of the late-phase response are the cysteinyl leukotrienes, previously identified as slow-reacting substance of anaphylaxis (SRS-A). Other important proinflammatory mediators of airway inflammation include eosinophil chemotactic factor (ECF), neutrophil chemotactic factor (NCF), eotaxin, and others. The orchestrated release of these mediators apparently serves to propagate the airway inflammatory response and the associated sustained constriction of the airways (Metzger et al., 1985; Strek and Leff, 1997).

As part of the proinflammatory late-phase response to allergen exposure, it has been demonstrated that the airways display nonspecific bronchial hyperresponsiveness to a variety of biologic and chemical agents, a feature that represents the pathognomonic functional disturbance in asthma. Moreover, particularly severe late-phase responses have also been associated with recurrent episodes of exacerbation of asthma (Cartier et al., 1982). Given this evidence, together with that stemming from the recent application of flexible bronchoscopy to obtain bronchoalveolar lavage (BAL) fluid for analysis of lung cellular infiltrates (Riedler et al., 1995), the recruitment of eosinophils and T lymphocytes in the lung has been identified as a key feature in the trafficking of inflammatory cells in the airways and in the establishment of bronchial hyperresponsiveness.

Role of Eosinophils

Peripheral blood and airway tissue eosinophilia have long been recognized in association with asthma. In more recent years, considerable insight has been gained into the role of the airway eosinophilic infiltration in the pathobiology of the disease. Accordingly, activation of airway eosinophils, resulting in release from their granules of preformed mediators, has been implicated in producing constriction of airway smooth muscle, bronchial hyperresponsiveness, recruitment of other inflammatory cell types, and airway tissue (e.g., epithelium) damage. In mediating these diverse actions, eosinophils release a variety of cationic proteins including major basic protein (MBP), eosinophil cationic protein (ECP), eosinophil derived neurotoxin (EDN), eosinophil peroxidase (EPO), and unique to eosinophils, lysophopholipase protein, which forms the Charcot-Leyden crystals that are characteristically found in asthmatic sputum specimens. In addition, eosinophils also secrete such enzymes as collagenase, ß-glucuronidase, acid phosphatase, and others (Strek and Leff, 1997). Among these secreted products, MBP has been found to produce damage to the airway epithelial cell lining, inhibit airway ciliary beat activity, stimulate eicosanoid production, and enhance histamine release from mast cells (Gleich et al., 1974). ECP is neurotoxic, has ribonuclease activity, and also causes damage to the airway epithelium (Motojima et al., 1989), whereas EPO has been associated with inducing increases in lung microvascular permeability (Yoshikawa et al., 1993). Collectively, these eosinophil functions, together with the generation of toxic oxygen radicals, have been implicated in establishing a number of the histological and physiological perturbations that characterize the asthmatic airway. In accordance with this evidence, an association between airway eosinophilia and the clinical presentation of asthma severity and bronchial hyperresponsiveness has been well documented (Strek and Leff, 1997). It remains to be established, however, whether this associative relationship is also mechanistically causative.

Role of T lymphocytes

T helper (TH) lymphocytes have also been implicated importantly in the regulation of various immune functions, including the development of allergic inflammation of the airways. In this regard, TH cells have been phenotypically partitioned into two profiles of differentiated cell function. These are represented by cells expressing either a TH1 or a TH2 profile of cytokine release upon activation. TH cells expressing the TH1 phenotype generate cytokines, including interleukin-2 (IL-2), IL-12, and interferon gamma (IFN-γ), which, although generally associated with host defense against infection, also act to modulate airway function. Indeed, in this regard, it has been demonstrated that the TH1-type cytokines, notably IFN-γ, largely play a protective role in countering the IgE-dependent expression of allergic responses and atopic asthma (Coffman and Carty, 1986; Lack and Gelfand,

1996; Pene et al., 1988). In contrast, lymphocytes of the TH2 phenotype release cytokines (e.g., IL-4 and IL-5) that have been implicated in orchestrating various proinflammatory humoral and cellular immune responses, including IgE synthesis and eosinophil recruitment and activation, both of which are characteristic features of the inflammatory state in asthmatic airways (Koning et al., 1997; Romagnani, 1995). In this connection, it has been recently demonstrated that both TH1- and TH2-type cytokines may, independent of the presence of inflammatory cells, directly exert potent opposing actions on the airway smooth muscle itself (Hakonarson et al., 1999). Accordingly, TH2-type cytokines have been shown to facilitate expression of the proasthmatic phenotype of altered airway smooth muscle responsiveness, whereas TH1 cytokines were found to act directly on the airway smooth muscle to attenuate its proasthmatic phenotype (Hakonarson et al., 1999).

In light of the above information pertaining to the roles of TH1 and TH2 lymphocytes, a popular contemporary paradigm states that the expression of the asthmatic state reflects a relative imbalance between TH1- and TH2-type cytokine production and action. Thus, an induced upregulated TH2 cytokine response, together with a relatively downregulated TH1 cytokine response, is considered to underlie the cellular and humoral airway inflammatory responses and bronchoconstrictor responsiveness in asthma (Ackerman et al., 1994; Corrigan et al., 1995; Robinson et al., 1992). There exists substantial evidence in support of this concept, based largely on recent clinical studies conducted in children and adults. These studies have reported that relative to nonallergic or nonasthmatic individuals, both serum and BAL fluid samples isolated from atopic asthmatic patients reveal significantly increased levels of the TH2 cytokines IL-4 and IL-5, in association with relatively decreased levels of the TH1-type cytokine IFN-γ (Hamid et al., 1991; Umetsu and DeKruyff, 1997; Ying et al., 1995). Moreover, it has been demonstrated that mononuclear cells isolated from serum or BAL fluid samples from atopic asthmatic patients also display a similar altered TH1- versus TH2-type profile of cytokine release when the cells are stimulated with antigen (i.e., favoring the TH2-type cytokine response). Finally, in extended support of the paradigm of altered TH1- versus TH2-type cytokine expression in asthma, it has been demonstrated that treatment of asthmatic patients with corticosteroids reduces their airway constrictor hyperresponsiveness and BAL fluid levels of IL-4 and IL-5, as well as the number of cells expressing these cytokines, while IFN-γ levels and cells expressing IFN-γ in the lung are increased (Bentley et al., 1996; Leung et al., 1995; Robinson et al., 1993).

In view of the above compelling body of evidence, current research into the pathobiology of asthma is largely directed at elucidating those mechanisms that regulate the expression of the TH1 and TH2 profiles of cytokine expression in the lung. In this regard, among the principal areas of research pursuit are studies directed at identifying the genetic basis for development of the TH1 or TH2 phenotype, as well as the influence of allergic and other environmental factors in modulating the TH1–TH2 cytokine balance.

Role of Cell Adhesion Molecules

The localized accumulation of inflammatory cells, particularly eosinophils and lymphocytes, in the asthmatic airway is, in large part, regulated by the actions of cell adhesion molecules. Together with their sequential interaction with cytokines or chemokines and other chemoattractants, cell adhesion molecules contribute importantly to the process of recruitment and activation of specific inflammatory cells at the primary inflammatory focus. The cell adhesion molecules have been classified into three families that include the selectins, integrins, and immunoglobulin supergene family (Albelda et al., 1994; Springer, 1990). Members of all of these families play critical roles in regulating leukocyte–endothelial cell interactions and other functions. Accordingly, in the initial phase of inflammatory cell recruitment from the tissue microvasculature in response to specific chemotactic stimuli, the tethering and rolling behavior (i.e., margination) of circulating leukocytes toward the affected site is mediated by the actions of E- and P-selectins on the vascular endothelium and by the action of L-selectin on the leukocyte surface. Thereafter, the integrin family of adhesion molecules, when bound to their respective counterreceptors in endothelial and other cell types, contributes to enhanced adhesion of the selected leukocytes. Finally, firm leukocyte adhesion is followed by transmigration of the inflammatory cells through the endothelial cell junctions (diapedesis) and their directed movement along a chemotactic gradient to the tissue inflammatory site.

Given the above sequence of events, in recent years a host of studies have examined the roles of cell adhesion molecules in the pathogenesis of the airway inflammatory response in asthma. Although many mechanistic processes remain unidentified, the accumulated data to date support the general notion that mast cell and TH2 lymphocyte activation, occurring following exposure to a sensitizing antigen, elicits the release of a host of soluble mediators, which in turn induce airway endothelial cells to upregulate their expression of E-selectin, intercellular adhesion molecule-1 (ICAM-1), and vascular cell adhesion molecule-1 (VCAM-1). This effect, together with the stimulated release of specific chemoattractants, subsequently mediates the recruitment of specific leukocytes, most notably eosinophils and lymphocytes, into the airway tissue. In accordance with this concept, Wegner and colleagues (1990) demonstrated that ICAM-1 expression in bronchial endothelium and epithelium is increased after antigen challenge in *Ascaris*-sensitized monkeys. Moreover, this effect was associated with airway eosinophil recruitment and the manifestation of bronchial hyperresponsiveness; and both these phenomena were inhibited by pretreatment of the animals with a monoclonal blocking antibody to ICAM-1 (Wegner et al., 1990). Comparably, upregulated VCAM-1 expression has also been correlated with increased IL-4 and IL-13 expression, in association with infiltration of eosinophils, macrophages, and T lymphocytes in allergen-induced late phase cutaneous reactions in atopic individuals (Ying et al., 1997). Thus, these findings, together with those from a series of related studies, have lent extended support to the above concept of cell adhesion molecule-dependent regulation of allergic inflammatory reactions and bronchial hyperresponsiveness in asthmatic individuals following inhaled antigen challenge (Georas et al., 1992; Montefort et al., 1994; Ohkawara et al., 1995; Takahashi et al., 1994). Collectively, this evidence underscores the need to further identify the mechanistic interplay between specific inflammatory cells, cell adhesion molecules, and the changes in airway function that characterize the asthmatic condition.

Current understanding of the above proposed mechanisms related to the role of inflammation in the pathophysiology of allergic asthma is summarized schematically in Figure 4–1. In the development of the immune and inflammatory responses in the airways, the inhaled sensitizing antigen is initially processed by antigen-presenting cells and then the antigen protein is bound to a complex of intercellular co-stimulatory molecules that includes major histocompatibility complex (MHC) class II, T cell receptor, and B7/CD28 molecules. This interaction leads to CD4+ T helper cell activation (Banchereau and Steinman, 1998). The latter results

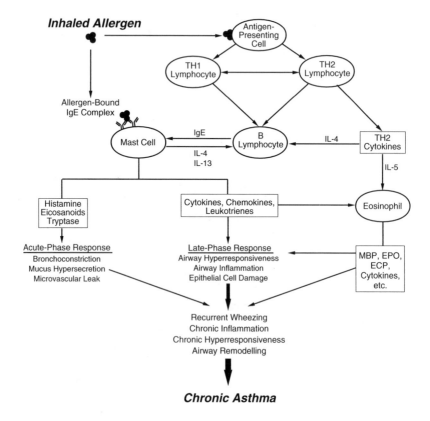

FIGURE 4-1 Proinflammatory mechanisms in allergic asthma.

in the subsequent differentation of T cells into those expressing either a TH1- or TH2-type profile of cytokine release. The TH2 phenotype is generally proinflammatory in nature in the airways, as represented by the release of the cytokines IL-4, IL-13, and IL-5. Among other functions, these cytokines act to direct IgE synthesis and to recruit and activate eosinophils. In the presence of IgE bound to an inhaled antigen, the high-affinity surface receptors for IgE (i.e., FcεRI) found on the surface of mast cells (also basophils) are activated, which leads to the release of various preformed mediators. Among these, histamine, tryptase, and certain eicosanoids are key mast cell-derived mediators that are largely responsible for eliciting the acute-phase response to the inhaled antigen. Other mast cell-derived mediators including leukotrienes and specific cytokines (e.g., IL-4, IL-5, IL-6, and IL-13) act cooperatively to orchestrate the subsequent late-phase proinflammatory response (Shimizu and Schwartz, 1997). These events, together with eosinophil activation and the release of various eosinophil-derived mediators, as well as activation of other cell types (e.g., basophils and mononuclear cells), serve to further perpetuate the airway inflammatory response and produce the state of chronic airway inflammation, perturbed airway function, and structural remodeling of the airway that characterizes the atopic asthmatic phenotype.

Apart from the above contemporary view related to the role of airway inflammation in the pathophysiology of allergic asthma, it is well recognized that respiratory inhalation of nonbiological (nonallergenic) agents (e.g., certain particulates, noxious gases, tobacco smoke) can also trigger acute asthma symptoms, potentially in the absence of any concomitant airway inflammation. Under these circumstances, it is generally believed that the mechanism underlying such asthmatic reactions is related to "nonspecific" irritant effects of the offending inhaled agent in the lung, which are attributed to the activation of specific bronchoactive reflexes. These reflexes are characteristically mediated by airway irritant receptors and/or receptors associated with small pulmonary c-type neural fibers that release specific neuropeptides in the lung (e.g., substance P) (Undem and Riccio, 1997).

THE AIRWAY SMOOTH MUSCLE IN ASTHMA

Role in Altered Airway Responsiveness

The characteristic functional perturbations of the asthmatic airway are its heightened contractile responsiveness to broncho-constrictor agents (e.g., mediators, neurotransmitters) and impaired relaxation responsiveness to bronchodilatory agents (i.e., beta-adrenergic drugs, prostaglandin E_2). Although substantial evidence exists in support of an important association between airway inflammation and altered airway function, implicating a complex interplay between activated inflammatory cells, airway epithelial cells, and airway smooth muscle (ASM), the cellular and molecular mechanisms underlying the functional perturbations in ASM responsiveness in asthma remain to be identified. In this regard, it is relevant to note that based on studies using isolated asthmatic and antigen-sensitized airways, the impaired relaxation responsiveness to beta-adrenergic receptor agonists does not appear to be related to reductions in either the density or the affinity of beta-adrenergic receptors in asthmatic ASM (Bai et al., 1992; Sharma and Jeffery, 1990; Spina et al., 1989; van Koppen et al., 1989). Rather, the changes in responsiveness of asthmatic ASM appear to be fundamentally linked to perturbations in certain key postreceptor-coupled transmembrane signal transduction mechanisms that regulate ASM contraction and relaxation. In recent years, this concept has received considerable attention and evidence has been accumulated demonstrating that, under atopic asthmatic conditions, the sensitized ASM displays attenuated beta-adrenoceptor-induced accumulation of cyclic adenosine monophosphate (cAMP), the key intracellular second messenger that mediates ASM relaxation. Moreover, in related studies, it has been shown that both the impaired beta-adrenoceptor-coupled accumulation of cAMP and the attenuated relaxation responsiveness in atopic asthmatic-sensitized ASM are largely attributable to an induced increased expression and action of the receptor-coupled G protein G_1, which inhibits adenylate cyclase and, hence, agonist-mediated cAMP accumulation (Hakonarson et al., 1995).

The changes identified in transmembrane signaling and tissue responsiveness in isolated atopic asthmatic ASM has raised the possibility that mechanisms intrinsic to the ASM itself contribute importantly to its autocrine induction of the proasthmatic phenotype. In support of this new concept, recent studies have demonstrated that under in vitro conditions of passive sensitization of isolated ASM with atopic asthmatic serum, the sensitized ASM itself is induced to release various proinflammatory cytokines, including IL-1ß (Hakonarson et al., 1997), as well as both TH1- and TH2-type cytokines (Hakonarson et al., 1999). In turn, these cytokines apparently act in an autocrine fashion to elicit proasthmatic changes in ASM responsiveness. Moreover, in extending this concept, the release of such proinflammatory cytokines, as well as the potential release of certain chemokines (Elias et al., 1997; Ghaffar et al., 1999; John et al., 1997) by the sensitized ASM itself, may further facilitate the recruitment of inflammatory cells into the airway tissue and thereby propagate the local inflammatory reaction in the asthmatic airway.

Role in Airway Remodeling

An additional important characteristic feature of asthmatic airways, particularly in the setting of chronic severe asthma, is the presence of an increase in ASM tissue mass, reflecting ASM cell hyperplasia and/or hypertrophy. This remodeling of the airways, together with a disruption of the airway epithelium and altered airway tissue extracellular matrix, may contribute importantly to the presence of the fixed (i.e., acutely nonreversible) airway narrowing that is often seen in long-standing severely asthmatic individuals. Although the precise mechanisms regulating airway remodeling remain to be identified, in recent years there has been considerable progress in our understanding of certain processes that control ASM cell growth. Accordingly, in concert with the effects of inflammatory cell-derived mediators and growth factors, ASM cells have also been shown to intrinsically express various cell adhesion molecules, extracellular matrix proteins, and as noted above, various cytokines or chemokines. The localized release of such a diverse collection of extracellular autocrine and paracrine stimuli appears to induce ASM cell proliferation, at least in part, by stimulating certain common intracellular signaling pathways (Panettieri and Grunstein, 1997). Moreover, the complex interaction between these signaling pathways likely determines the ultimate manifestation of airway remodeling via a coordinated regulation of promitogenic and antipro-liferative (i.e., apoptosis) intracellular signals.

THE GENETICS OF ASTHMA

There exists a substantial long-standing body of evidence that predisposition to asthma represents an inheritable phenomenon. Epidemiologic and immunologic studies have demonstrated that there is an increased prevalence of asthma within families and that monozygotic twins depict greater concordance than do dizygotic twins (Duffy et al., 1990; Edfors-Lubs, 1971). An inheritable basis for atopy has also been reported with respect to the expression of serum IgE (Pirson et al., 1991; Sibbald et al., 1980). Despite this strongly suggestive evidence, the gene(s) responsible for asthma remains unidentified. This deficiency in our current understanding of the genetic basis of asthma is largely reflective of the notion that like many other common diseases, asthma represents a polygenic disorder in which the phenotypic manifestation of the disease is greatly influenced by environmental factors.

Different approaches have been used to identify and map the genes causing asthma. One approach involves complex segregation analysis, followed by linkage analysis using the most compatible genetic paradigm identified by the segregation analysis. In addition, linkage analysis also has been applied in analyzing affected pairs of relatives without a predefined specific genetic model. Another approach, referred to as the candidate gene approach, is based on testing for simple associations with specific polymorphisms of potentially relevant genes in affected and unaffected individuals. Use of the candidate gene approach in asthma is dependent on information about potential mechanisms related to the development of the disease process. Finally, an approach involving a genome-wide search, wherein polymorphic DNA markers are measured throughout the human genome, followed by linkage analysis, has been applied to asthmatic individuals and their familial relations.

To date, the collection of evidence based on the above analyses of potential genetic determinants of asthma has identified a variety of candidate genes related to the disease. Accordingly, studies have reported that genes contained within the cytokine cluster on chromosome 5 (encoding IL-3, IL-4, IL-5, IL-9, and IL-13), chromosome 11 (encoding the high-affinity receptor for IgE, FcεRI), chromosome 12 (encoding insulin-like growth factor, stem cell factor, IFN-γ, and Stat 6), and chromosome 16 (encoding the IL-3 receptor) may possibly contribute to the development of asthma and atopy (Borish, 1999; Daniels et al., 1996; Marsh et al., 1994). Moreover, there is mounting evidence in support of the involvement of genes that regulate antigen presentation (i.e., MHC class II genes), as well as T lymphocyte responses (i.e., T cell receptor gene) (Blumenthal et al., 1992; Marsh et al., 1981). Finally, polymorphisms have been reported in genes encoding the β-adrenergic receptor, 5'-lipoxygenase, and leukotriene C_4 synthase (Borish, 1999; Chandrasekharappa et al., 1990). Collectively, this information highlights the complexity of the molecular genetics of asthma. Moreover, it emphasizes that much more research, based on combining data from genetic analyses with those identifying pathophysiological processes involved in asthma, is needed to ultimately determine the genetic basis of asthma, as well as the potential development of new strategies for therapeutic intervention.

CONCLUSION

In the past, the quest to understand asthma was a process much like that of blind men trying to understand the elephant — impressions of the beast were based largely on the part of the animal that was touched. In recent years, however, the synthesis of evidence stemming from the diversity of basic and clinical research studies on asthma has led to major advances in our overall understanding of the pathobiology of this disease. While we know that asthma is a genetically predisposed condition that is associated with chronic inflammation of the airways, ongoing research continues to uncover a multiplicity of cellular and molecular mechanisms involved in regulating the phenotypic expression of the disease. Importantly, these mechanisms are activated largely in response to a variety of environmental factors, including exposure to biologic and nonbiologic agents. In this regard, exposure to various allergens, specific viruses, and certain air pollutants have been importantly linked to pulmonary complications that include the clinical manifestation of asthma.

REFERENCES

Ackerman V, Marini M, Vittori E, Bellini A, Vassali G, Mattoli S. 1994. Detection of cytokines and their cell sources in bronchial biopsy specimens from asthmatic patients. Relationship to atopic status, symptoms, and level of airway hyperresponsiveness. Chest 105(3):687–696.

Albelda SM, Smith CW, Ward PA. 1994. Adhesion molecules and inflammatory injury. FASEB Journal 8(8): 504–512.

Bai TR, Mak JC, Barnes PJ. 1992. A comparison of beta-adrenergic receptors and *in vitro* relaxant responses to isoproterenol in asthmatic airway smooth muscle. American Journal of Respiratory Cell and Molecular Biology 6(6):647–651.

Banchereau J, Steinman RM. 1998. Dendritic cells and the control of immunity. Nature 392(6673):245–252.

Bentley AM, Hamid Q, Robinson DS, Schotman E, Meng Q, Assoufi B, Kay AB, Durham SR. 1996. Prednisolone treatment in asthma. Reduction in the numbers of eosinophils, T cells, tryptase-only positive mast cells, and modulation of IL-4, IL-5, and interferon-gamma cytokine gene expression within the bronchial mucosa. American Journal of Respiratory and Critical Care Medicine 153(2):551–556.

Blumenthal M, Marcus-Bagley D, Awdeh Z, Johnson B, Yunis EJ, Alper CA. 1992. HLA-DR2 [HLA-B7, SC31, DR2], and [HLA-B8, SC01, DR3] haplotypes distinguish subjects with asthma from those with rhinitis only in ragweed pollen allergy. Journal of Immunology 148(2):411–416.

Borish L. 1999. Genetics of allergy and asthma. Annals of Allergy, Asthma, and Immunology 82(5):413–426.

Cartier A, Thomson NC, Frith PA, Roberts R, Hargreave FE. 1982. Allergen-induced increase in bronchial responsiveness to histamine: relationship to the late asthmatic response and change in airway caliber. Journal of Allergy and Clinical Immunology 70(3):170–177.

Chandrasekharappa SC, Rebelsky MS, Firak TA, Le Beau MM, Westbrook CA. 1990. A long-range restriction map of the interleukin-4 and interleukin-5 linkage group on chromosome 5. Genomics 6(1):94–99.

Coffman RL, Carty J. 1986. A T cell activity that enhances polyclonal IgE production and its inhibition by interferon-gamma. Journal of Immunology 136(3):949–954.

Corrigan CJ, Hamid Q, North J, Barkans J, Moqbel R, Durham S, Gemou-Engesaeth V, Kay AB. 1995. Peripheral blood CD4 but not CD8 t-lymphocytes in patients with exacerbation of asthma transcribe and translate messenger RNA encoding cytokines which prolong eosinophil survival in the context of a Th2-type pattern: effect of glucocorticoid therapy. American Journal of Respiratory Cell and Molecular Biology 12(5):567–578.

Daniels SE, Bhattacharrya S, James A, Leaves NI, Young A, Hill MR, Faux JA, Ryan GF, le Souef PN, Lathrop GM, Musk AW, Cookson WO. 1996. A genome-wide search for quantitative trait loci underlying asthma. Nature 383(6597):247–250.

Duffy DL, Martin NG, Battistutta D, Hopper JL, Mathews JD. 1990. Genetics of asthma and hay fever in Australian twins. American Review of Respiratory Disease 142(6 Pt 1):1351–1358.

Edfors-Lubs ML. 1971. Allergy in 7000 twin pairs. Acta Allergologica 26(4):249–285.

Elias JA, Wu Y, Zheng T, Panettieri R. 1997. Cytokine- and virus-stimulated airway smooth muscle cells produce IL-11 and other IL-6-type cytokines. American Journal of Physiology 273(3 Pt 1): L648–L655.

Georas SN, Liu MC, Newman W, Beall LD, Stealey BA, Bochner BS. 1992. Altered adhesion molecule expression and endothelial cell activation accompany the recruitment of human granulocytes to the lung after segmental antigen challenge. American Journal of Respiratory Cell and Molecular Biology 7(3):261–269.

Ghaffar O, Hamid Q, Renzi PM, Allakhverdi Z, Molet S, Hogg JC, Shore SA, Luster AD, Lamkhioued B. 1999. Constitutive and cytokine-stimulated expression of eotaxin by human airway smooth muscle cells. American Journal of Critical Care Medicine 159(6):1933–1942.

Gleich GJ, Loegering DA, Kueppers F, Bajaj SP, Mann KG. 1974. Physiochemical and biological properties of the major basic protein from guinea pig eosinophil granules. Journal of Experimental Medicine 140(2):313–332.

Hakonarson H, Herrick DJ, Grunstein MM. 1995. Mechanism of impaired beta-adrenoceptor responsiveness in atopic sensitized airway smooth muscle. American Journal of Physiology 269 (5 Pt 1): L645–L652.

Hakonarson H, Herrick DJ, Serrano PG, Grunstein MM. 1997. Autocrine role of interleukin 1beta in altered responsiveness of atopic asthmatic sensitized airway smooth muscle. Journal of Clinical Investigation 99(1):117–124.

Hakonarson H, Maskeri N, Carter C, Grunstein MM. 1999. Regulation of TH1- and TH2-type cytokine expression and action in atopic asthmatic sensitized airway smooth muscle. Journal of Clinical Investigation 103(7):1077–1087.

Hamid Q, Azzawi M, Ying S, Moqbel R, Wardlaw AJ, Corrigan CJ, Bradley B, Durham SR, Collins JV, Jeffrey PK, et al. 1991. Expression of mRNA for interleukin-5 in mucosal bronchial biopsies from asthma. Journal of Clinical Investigation 87(5):1541–1546.

Ishizaka K, Ishizaka T, Hornbrook, MM. 1966. Physicochemical properties of reaginic antibody. V. Correlation of reaginic activity with gamma-E-globulin antibody. Journal of Immunology 97(6):840–853.

John M, Hirst SJ, Jose PJ, Robichaud A, Berkman N, Witt C, Twort CH, Barnes PJ, Chung KF. 1997. Human airway smooth muscle cells express and release RANTES in response to T helper 1 cytokines: regulation by T helper 2 cytokines and corticosteroids. Journal of Immunology 158(4):1841–1847.

Koning H, Neijens HJ, Baert MR, Oranje AP, Savelkoul HF. 1997. T cells subsets and cytokines in allergic and non-allergic children. II. Analysis of IL-5 and IL-10 mRNA expression and protein production. Cytokine 9(6):427–436.

Lack G, Gelfand EW. 1996. The role of nebulized IFN-gamma in the modulation of allergic responses. Advances in Experimental Medicine and Biology 409:17–23.

Lemanske RF Jr, Kaliner M. 1981–1982. Mast cell-dependent late phase reactions. Clinical and Immunology Review 1(4):547–580.

Leung DY, Martin RJ, Szefler SJ, Sher ER, Ying S, Kay AB, Hamid Q. 1995. Dysregulation of interleukin 4, interleukin 5, and interferon gamma gene expression in steroid-resistant asthma. Journal of Experimental Medicine 181(1):33–40.

Marsh DG, Meyers DA, Bias WB. 1981. The epidemiology and genetics of atopic allergy. New England Journal of Medicine 305(26):1551–1559.

Marsh DG, Neely JD, Breazeale DR, Ghosh B, Friedhoff LR, Ehrlich-Kautzky E, Shou C, Krishnaswamy G, Beaty TH. 1994. Linkage analysis of IL4 and other chromosome 5q31.1 markers and total serum immunoglobulin E concentrations. Science 264(5162):1152–1156.

Metzger WJ, Hunninghake GW, Richerson HB. 1985. Late asthmatic responses: inquiry into mechanisms and significance. Clinical Reviews in Allergy 3(2):145–165.

Montefort S, Lai CK, Kapahi P, Leung J, Lai KN, Chan HS, Haskard DO, Howarth PH, Holgate ST. 1994. Circulating adhesion molecules in asthma. American Journal of Respiratory and Critical Care Medicine 149(5):1149–1152.

Motojima S, Frigas E, Loegering DA, Gleich GJ. 1989. Toxicity of eosinophil cationic proteins for guinea pig tracheal epithelium in vitro. American Review of Respiratory Disease 139(3):801–805.

Murphy S. 1997. Expert Panel Report 2: Guidelines for the Diagnosis and Management of Asthma. 97-4051, 1-86. National Institutes of Health, National Heart, Lung, and Blood Institute, Washington, DC.

Ohkawara Y, Yamauchi K, Maruyama N, Hoshi H, Ohno I, Honma M, Tanno Y, Tamura G, Shirato K, Ohtani H. 1995. In situ expression of the cell adhesion molecules in bronchial tissues from asthmatics with air flow limitation: in vivo evidence of VCAM-1/VLA-4 interaction in selective eosinophil infiltration. American Journal of Respiratory Cell and Molecular Biology 12(1):4–12.

Panettieri RA, Grunstein MM. 1997. Airway smooth muscle hyperplasia and hypertrophy. In: Asthma. Barnes PJ, Grunstein MM, Leff AR, Woolcock AJ, Eds. Philadelphia; Lippincott-Raven Publishers. Chapter 59, pp. 823–842.

Pene J, Rousset F, Briere F, Chretien I, Bonnefoy JY, Spits H, Yokota T, Arai N, Arai K, Banchereau J, et al. 1988. IgE production by normal human lymphocytes is induced by interleukin 4 and suppressed by interferons gamma and alpha and prostaglandin E2. Proceedings of the National Academy of Sciences of the United States of America 85(18):6880–6884.

Pirson F, Charpin D, Sansonetti M, Lanteaume A, Kulling G, Charpin J, Vervloet D. 1991. Is intrinsic asthma a hereditary disease? Allergy 46(5):367–371.

Riedler J, Grigg J, Stone C, Tauro G, Robertson CF. 1995. Bronchoalveolar lavage cellularity in healthy children. American Journal of Respiratory and Critical Care Medicine 152(1):163–168.

Robertson DG, Kerigan AT, Hargreave FE, Chalmers R, Dolovich J. 1974. Late asthmatic responses induced by ragweed pollen allergen. Journal of Allergy and Clinical Immunology 54(4):244–254.

Robinson DS, Hamid Q, Ying S, Tsicopoulos A, Barkans J, Bentley AM, Corrigan C, Durham SR, Kay AB. 1992. Predominant TH2-like bronchoalveolar T-lymphocyte population in atopic asthma. New England Journal of Medicine 326(5):298–304.

Robinson D, Hamid Q, Ying S, Bentley A, Assoufi B, Durham S, Kay AB. 1993. Prednisolone treatment in asthma is associated with modulation of broncho-alveolar lavage cell interleukin-4, interleukin-5, and interferon-gamma cytokine gene expression. American Review of Respiratory Disease 148(2):401–406.

Romagnani S. 1995. Biology of human TH1 and TH2 cells. Journal of Clinical Immunology 15(3):121–129.

Sears MR, Burrows B, Flannery EM, Herbison GP, Hewitt CJ, Holdaway MD. 1991. Relation between airway responsiveness and serum IgE in children with asthma and in apparently normal children. New England Journal of Medicine 325(15):1067–1071.

Sharma RK, Jeffery PK. 1990. Airway beta-adrenoceptor number in cystic fibrosis and asthma. Clinical Science 78(4):409–417.

Shimizu Y, Schwartz LB. 1997. Mast cell involvement in asthma. In: Asthma. Barnes PJ, Grunstein MM, Leff AR, Woolcock AJ, Eds. Philadelphia, PA: Lippincott-Raven Publishers, pp. 353–366.

Sibbald B, Horn ME, Brain EA, Gregg I. 1980. Genetic factors in childhood asthma. Thorax 35(9):671–674.

Spina D, Rigby PJ, Paterson JW, Goldie RG. 1989. Autoradiographic localization of beta-adrenoceptors in asthmatic human lung. American Review of Respiratory Disease 140(5):1410–1415.

Springer TA. 1990. Adhesion receptors of the immune system. Nature 346(6283):425–434.

Strek ME, Leff AR. 1997. Eosinophils. In: Asthma. Barnes PJ, Grunstein MM, Leff AR,Woolcock AJ, Eds. Philadelphia; Lippincott-Raven Publishers, pp. 399–418.

Takahashi N, Liu MC, Proud D, Yu XY, Hasegawa S, Spannhake EW. 1994. Soluble intracellular adhesion molecule 1 in bronchoalveolar lavage fluid of allergic subjects following segmental antigen challenge. American Journal of Respiratory and Critical Care Medicine 150(3):704–709.

Umetsu DT, DeKruyff RH. 1997. Th1 and Th2 CD4+ cells in the pathogenesis of allergic diseases. Proceedings of the Society for Experimental Biology and Medicine 215(1):11–20.

Undem BJ, Riccio MM. 1997. Activation of airway afferent nerves. In: Asthma. Barnes PJ, Grunstein MM, Leff AR, Woolcock AJ, Eds. Philadelphia; Lippincott-Raven Publishers, pp. 1009–1026.

van Koppen CJ, de Miranda JF, Beld AJ, van Herwaarden CL, Lammers JW, van Ginneken CA. 1989. Beta adrenoceptor binding and induced relaxation in airway smooth muscle from patients with chronic airflow obstruction. Thorax 44(1):28–35.

Wegner CD, Gundel RH, Reilly P, Haynes N, Letts LG, Rothlein R. 1990. Intercellular adhesion molecule-1 (ICAM-1) in the pathogenesis of asthma. Science 247(4941):456–459.

Ying S, Durham SR, Corrigan CJ, Hamid Q, Kay AB. 1995. Phenotype of cells expressing mRNA for TH2-type (interleukin 4 and interleukin 5) and TH1-type (interleukin 2 and interferon gamma) cytokines in bronchoalveolar lavage and bronchial biopsies from atopic asthmatic and normal control subjects. American Journal of Respiratory Cell and Molecular Biology 12(5):477–487.

Ying S, Meng Q, Barata LT, Robinson DS, Durham SR, Kay AB. 1997. Associations between IL-13 and IL-4 (mRNA and protein), vascular cell adhesion molecule-1 expression, and the infiltration of eosinophils, macrophages, and T cells in allergen-induced late-phase cutaneous reactions in atopic subjects. Journal of Immunology 158(10):5050–5057.

Yoshikawa S, Kayes SG, Parker JC. 1993. Eosinophils increase lung microvascular permeability via the peroxidase–hydrogen peroxide-halide system. Bronchoconstriction and vasoconstriction unaffected by eosinophil peroxidase inhibition. American Review of Respiratory Disease 147(4):914–920.

5
INDOOR BIOLOGIC *Exposures*

The allergic constituents of indoor air are predominantly biologic in origin (Becher, 1996). As early as the sixteenth century, associations between a number of these exposures and asthma were suspected; however, the scientific data available were unable to confirm such an association. Concern in recent years regarding the potential health effects of indoor air, as well as the marked increase in the prevalence of asthma in industrialized countries, has prompted an influx of scientific data on exposure to airborne biologic agents and asthma.

The committee was charged with the task of evaluating the strength of the scientific evidence concerning the possible association between these agents and asthma prevalence and severity. The committee was also tasked with examining possible means of mitigating or preventing exposure to these agents. In this chapter the committee evaluates indoor exposure to biologic agents, addressing the following to the extent permitted by available research:

1. which factors influence exposures to the agent;
2. whether a relationship exists between the agent and asthma prevalence or severity, taking into account the strength of the scientific evidence and the appropriateness of the methods used to detect the relationship;

3. what type of relationship exists between the agent and asthma;

4. whether there are special considerations regarding the agent (e.g., subpopulations at risk and interactions with other exposures);

5. which strategies effectively mitigate or prevent exposure to the agent;

6. whether these strategies only reduce exposures or decrease the occurrence or exacerbation of asthma; and

7. whether these strategies are reasonable for use by the target populations.

Each section begins by providing a definition of the agent and a summary of the factors that influence exposure. The evidence concerning the possible association between the agent and asthma is discussed, followed by the committee's conclusions regarding the health impacts. Where information is available, evidence regarding possible means of mitigating or preventing exposure to the agent is addressed. Each section concludes with any committee recommendations for areas for which additional research is needed with respect to the agent. Because there are great differences in the amount and type of information available on specific agents, the sections vary in their depth and focus.

ANIMALS

Cats

Definition of the Agent and Means of Exposure

Cats are kept as pets in 27% of U.S. households. The major cat allergen, *Fel d* I, is a glycoprotein structured as a heterodimer with two chains of amino acids, which have been defined by polymerase chain reaction (PCR) and subsequent DNA sequencing (Griffith et al., 1992; Morgenstern et al., 1991; Schou, 1993). It is found on cat hair and is produced in cat sebaceous, salivary, and anal glands (De Andrade et al., 1996). In male cats, *Fel d* I glandular production is under hormonal control and decreases after castration (Zielonka et al., 1994). As discussed in the Third International Workshop report (Platts-Mills et al., 1997), the clinical significance of this decrease in allergen production is not certain, since many symptomatic cat-allergic asthmatics in the United States have neutered cats. Further investigation of hormonal and genetic control of *Fel d* I production could be relevant to the control of allergen levels in homes with cats. Although 90% of patients allergic to cats make immunoglobulin E (IgE) to *Fel d* I (de Groot et al., 1988; Schou, 1993), making *Fel d* I a marker for the immune response to cat allergens, at least eight other cat allergens have been identified (Duffort et al., 1987), suggesting that protection from *Fel d* I exposure may not be the equivalent of protection from cat allergen exposure. This conclusion is supported by the findings that 66% of the histamine-releasing activity of cat hair and dander extract, and about 60% of the cat dander radioallergosorbent test (RAST) activity, was carried by *Fel d* I (Schou, 1993).

Touching the cat is only one mode of contact that may result in airborne suspension of allergen and potential direct hand-to-nose deposition of allergen-associated particles. In contrast with cockroach allergen, which is airborne only transiently during the disturbance of household dust, cat allergen can remain airborne for long periods of time, in part because *Fel d* I is associated to a significant extent with smaller particles of less than 5 µm (Custovic et al., 1998b). Particles on which cat allergen is carried, coming primarily from cat dander, are also very adherent. Consequently cat allergen is spread easily throughout a house, even when cats are kept out of certain rooms. Moreover, cat allergen is easily carried from home to home, office, school, or day care center by those who touch cats or visit households with cats (Custovic et al., 1998a; Dybendal and Elsayed, 1994; Warner, 1992). At trace or small amounts that may be significant for sensitization or exacerbation of disease in sensitized individuals, *Fel d* I in settled dust is found in most homes without cats (Bollinger et al., 1996; Chew et al., 1998), although allergen levels are generally higher in homes with cats. A Baltimore, Maryland, study found measurable levels of airborne *Fel d* I in 37 homes with cats (range, 1.8–578 ng/m^3; median, 45.9 ng/m^3) and in 10 of 40 homes without cats (range for detectable samples, 2.8–88.5 ng/m^3; median, 17 ng/m^3) (Bollinger et al., 1996). In 38 of 40 homes without cats,

Fel d I was present in the settled dust (range, 39–3,750 ng/g; median, 258 ng/g); dust levels were weakly correlated with airborne levels. Carpeting, bedding, and upholstered furniture can be reservoirs for deposited cat allergen (Wood et al., 1989); shaking the bedding in rooms with cats resuspends cat allergen in the air.

As well as being detected in homes without cats, cat allergen has also been detected in public places such as hospitals and schools. In one British study that measured both settled and airborne cat allergen in hospitals, the amount of cat allergen in settled dust in upholstered chairs was as high as in homes with cats (geometric mean, 23 µg/g dust), but airborne levels were low (0.22 ng/m^3) (Custovic et al., 1998a).

Evidence Regarding Asthma Exacerbation

In cat-sensitized asthmatics, cat allergen can induce allergic symptoms, asthmatic symptoms, and decrements in lung function. Exposure to inhaled cat allergen in an experimental cat room led to significant decreases in forced expiratory volume in one second (FEV$_1$ range, 6–57%; mean, 25%) in a study of 13 adults with cat allergy. The percentage decrease in FEV$_1$ did not correlate significantly with either the intradermal titration end point with cat allergen or the magnitude of the RAST response with cat allergen. Those cat-allergic subjects classified as asthmatic by methacholine challenge testing experienced almost identical responses to environmental allergen challenge in an experimental cat room and inhalation challenge with cat allergen (Sicherer et al., 1997). Norman and colleagues documented progressive increases in both nasal and lung symptom scores during a 60-minute period in a cat room (Norman et al., 1996).

While initial cat room studies involved very high airborne cat allergen levels, a later experimental cat room exposure study by Bollinger and colleagues (1996) evaluated symptom and lung function responses of cat-sensitive subjects to low-level airborne cat allergens. They demonstrated that cat-sensitive individuals can have increases in upper- (congestion, rhinorrhea, pruritus) and lower- (chest tightness, wheezing) respiratory symptoms and decrements in lung function at levels of cat allergen occurring in homes without cats. The median FEV_1 change was 15% in the seven challenges with a *Fel d* I level less than 100 ng/m^3.

In a Delaware case-control study, Gelber and colleagues (1993) studied allergen predictors of emergency room visits for asthma. This study compared 93 patients, 15 to 55 years of age, who presented with breathlessness and airway obstruction, to 93 patients presenting without breathlessness. For cat and cockroach, the combination of sensitization and the presence of allergen in the house was associated with asthma presenting to hospitals (14/93 asthmatics versus 1/93 controls). Whether other exposures potentiate the response of cat-allergic asthmatics to cat exposure is unknown. In a cross-sectional study of children in New Mexico with asthma or bronchial hyperresponsiveness, cat sensitization and exposure to cat allergen were common (Sporik et al., 1995). Among children with asthma (defined as symptomatic bronchial reactivity), 13/19 were sensitized to cat. Numbers were too small to compare symptoms in cat-sensitized asthmatics with and without significant home exposure to cat allergen (Ingram et al., 1995).

Evidence Regarding Asthma Development

Insufficient data are available to assess whether exposure to cats influences the development of asthma. Cross-sectional studies of children suggest an association between sensitization to cats and home exposure in the first six months of life (Suoniemi et al., 1981; Warner et al., 1991). These studies may be subject to recall bias. A longitudinal birth cohort study from the Isle of Wight found that the presence of a cat in the home predicted a greater risk of skin test reactivity to cat and a greater risk of any skin test reactivity by 2 years of age (Hide et al., 1994). Sensitization to cat predicts asthma development, but this may simply be a confirmation of the well-documented fact that atopic individuals are more likely to develop asthma than nonatopic individuals. In a New Zealand birth cohort study, the development of asthma by 13 years of age was associated with sensitization to cats and dogs at age 13 (Sears et al., 1989). Sensitization to cats predicted the development of bronchial hyperresponsiveness in a longitudinal study of adults in Boston, Massachusetts (Litonjua et al., 1997). However, neither of these studies provides evidence of whether exposure to cats predicts either asthma or bronchial hyperresponsiveness. Since cat allergen exposure can potentially take place either at home or in schools and public places, the relative importance of home versus community-wide exposure to cat allergen in the risk of specific sensitization to cats is unknown.

Although the field of the genetics of asthma is in its infancy, preliminary studies suggest that certain genetic phenotypes are associated with allergy to specific insects or animals (Fukuda et al., 1995; Hizawa et al., 1998; Young et al., 1992, 1993). Understanding the genetics of allergy and asthma, including understanding the phenotypes associated with allergy specificity, may eventually prove useful in understanding gene-by-environment interaction in the development of asthma.

Conclusions: Asthma Exacerbation and Development

In cat-sensitive asthmatics, cat allergen exposure leads to worsening of respiratory symptoms and to a decline in lung function. Although sensitization to cats is a prerequisite to reactivity to cat exposure, the level of airborne cat allergen that exacerbates asthma varies by individual and is not necessarily predictable by the size of the skin test reaction to cat or the titer of IgE antibody. However, specific sensitive subgroups have not been defined. The relationship between cat allergen in the home and asthma development is uncertain. Because cat allergen is frequently found outside the home and in households without cats, the assessment of individual exposures to cats is difficult, making evaluation of the association between cat allergen exposure and asthma development difficult as well. In summary:

• There is sufficient evidence of a causal relationship between cat allergen exposure and exacerbation of asthma in individuals specifically sensitized to cats.

• There is inadequate or insufficient evidence to determine whether or not an association exists between cat allergen exposure and the development of asthma.

Evidence Regarding Exposure Mitigation and Prevention:
Homes

Removal of the cat from the home will decrease exposure to cat allergen and is widely recommended for symptomatic cat-sensitive asthmatics. However even when the owner removes the cat, cat allergen levels may remain elevated for 20 weeks or more (Wood et al., 1989). Removal of carpets and upholstery, with encasement of mattresses and pillows, may be required for diminishing cat allergen to levels commonly measured in homes without cats (Wood et al., 1989).

No studies are available that evaluate symptoms or lung function in cat-sensitive asthmatics before and after removal of the cat from the home. Nor are there studies of change in symptoms or lung function after moving from a home with a cat to a home without a cat. However, exposure studies suggest that in cat-sensitive subjects the decline in lung function associated with low-level airborne exposure to cat allergen, which can be present in a home with no cats, tends to be less extreme than the decline in lung function associated with higher-level airborne exposure (Bollinger et al., 1996). The experiments demonstrating entry into a room with a cat as a source of exacerbation of asthma in cat-sensitive individuals also suggest that removal of the cat from the household may decrease symptoms in cat-allergic asthmatic.

Because of the reluctance of cat-allergic symptomatic asthmatics to get rid of their cats, a number of studies have focused on the potential for lowering cat allergen levels by washing the cat. In eight households with cats, de Blay and colleagues (1991a) found that the combination of washing the cat weekly, reducing furnishings, vacuum cleaning, and air filtration reduced airborne cat allergen levels. In a second study, however, Avner and colleagues (1997) from the Platts-Mills group found that while washing cats by immersion will transiently remove significant allergen from the cat and reduce the quantity of airborne *Fel d* I, this reduction in allergen is not maintained by one week. Klucka and colleagues (1995) found no significant reduction of *Fel d* I by washing, use of Allerpet-C (a widely advertised topical spray), or acepromazine, a tranquilizer advocated as efficacious in subsedating doses. While removal of the cat from the living room and bedroom areas of the home and use of a High-Efficiency Particulate Air (HEPA) filter reduced airborne levels of cat allergen in homes with cats, the reduction was not evenly spread across the particle size range (Custovic et al., 1998b). It is unclear that the level of reduction of allergen obtained in this study is sufficient to influence symptoms in cat-sensitized asthmatics. No studies are available to assess the efficacy of recommendations to wash cats in reducing symptoms in cat allergic-asthmatics. Although the combination of HEPA filter use, mattress and pillow covers, and exclusion of cats from the bedroom reduced airborne cat allergen levels, a Maryland study detected no improvement in daily symptom scores, peak flow rates, medication use, monthly spirometry, pre- and post-study cat-specific IgE levels, and methacholine challenge studies in cat-allergic subjects (Wood et al., 1998). On the other hand, in a double-blind, placebo-controlled, cross-over study of twenty asthmatic children sensitized to cat or dog allergens, and living in homes with these animals, airway hyperresponsiveness was improved and peak flow variation was decreased during the use of air cleaners in the living room and bedroom of the child (van der Heide et al., 1999). The authors report that substantial amounts of cat and dog allergen were captured by the air cleaners; floor cat and dog allergen levels were unchanged by air cleaner use.

Evidence Regarding Exposure Mitigation and Prevention: Schools and Hospitals

Even if the cat is removed from the home, continued low-grade exposures may occur in public places or via clothes from cat owners. Since cat allergen is everywhere, there is little potential for absolute avoidance (Dybendal and Elsayed, 1994; Warner, 1992). Norwegian investigators have demonstrated the presence of *Fel d* 1 in schools on both smooth and carpeted floors, with approximately 11 times more allergen on the carpeted floors (Dybendal et al., 1991, 1989a, 1989b). The frequency of cleaning floors and furniture was believed to influence the level of cat and dog allergen, which were higher on chairs than in floor dust (Warner, 1992). Upholstered chairs and mattresses in hospitals are also demonstrated reservoirs for cat and dog allergen (Custovic et al., 1998a).

To lessen the risk of exacerbation of asthma in cat- or dog-sensitized asthmatics in public buildings, Warner and others (1992) have recommended the use of smooth floors and frequently cleaned wooden or plastic chairs. In their study demonstrating the presence of significant cat and dog allergen levels in a hospital in Manchester, England, Custovic and colleagues (1998a) questioned the introduction of soft furnishings and carpets into hospitals where highly cat- or dog-allergic asthmatics may come for care. Where upholstered chairs were present, they demonstrated that vacuuming three times a week significantly reduced allergen levels.

Conclusions: Exposure Mitigation and Prevention

Cat allergen levels can be reduced to levels found in homes without cats by removal of the cat from the home, but the reduction in allergen levels may require a prolonged period of time. The combination of HEPA filter use, mattress and pillow covers, and exclusion of cats from the bedroom may not reduce airborne cat allergen levels sufficiently to improve symptoms in cat-sensitive asthmatics. The absence of carpet, the use of plastic or wooden rather than upholstered chairs, and of frequent vacuuming in schools and hospitals may decrease the levels of cat allergen in public places. No studies are available to evaluate whether these measures improve symptoms or lung function in cat-sensitized asthmatics or whether they decrease the potential for sensitization in nonsensitized individuals. In summary:

• There is sufficient evidence of an association between removal of a cat from the home and a decrease in levels of cat allergen in the home; this decrease in levels of allergen may be slow if reservoirs of cat allergen are not simultaneously removed from the home.

• There is limited or suggestive evidence of an association between removal of a cat from the home and improvement of symptoms or lung function in cat-allergic asthmatics.

• There is limited or suggestive evidence of an association

between measures short of removal of a cat from the home (e.g., washing the cat, HEPA filter use) and some transient reduction in cat allergen levels in the home.

• There is inadequate or insufficient evidence to determine whether or not an association exists between measures short of removal of a cat from the home (e.g., washing the cat, HEPA filter use) and improvement in symptoms in cat-allergic asthmatics.

Dogs

Definition of the Agent and Means of Exposure

Dogs are present in 31% of U.S. households and are also sources of allergens (Schou, 1993). Allergy to cats is reported to be about twice as common as allergy to dogs, despite the fact that dogs are as common in U.S. households as cats (Bollinger et al., 1996). $Can f$ I and $Can f$ II are purified dog allergens that have been identified (Schou, 1993). $Can f$ I is a polypeptide whose molecular weight and structure have been partially but not fully defined (Schou, 1993). It is present in dander, pelt, hair, and saliva, but not in the urine or feces of dogs (Schou, 1993). Though there are likely to be other dog allergens, no others have been found to have clinical importance. $Can f$ I is considered a major allergen, because it accounts for at least half of the allergenic activity in dog hair and dander. In addition, 92% of dog-allergic patients had a positive skin prick test to $Can f$ I (Schou, 1993; Yman et al., 1973). It is still a controversial matter whether true breed-specific dog allergens exist or whether the differences observed between breeds are quantitative rather than qualitative (Schou, 1993). Hair is not the only source of dog allergen, and it is not known whether short-haired dogs are less allergenic. Cross-reactivity can be found between dog and cat allergen (Vanto and Koivikko, 1983).

Dog allergen, like cat allergen and unlike cockroach, is easily aerosolized and widely disseminated throughout the community (Custovic et al., 1997). In a Baltimore study of 42 homes, dog antigen was demonstrated in more than half of households (Lind et al., 1987; Schou, 1993). In Sweden, dog allergen has been measured in homes that have never had dogs (Munir et al., 1992). Like cat allergen, dog allergen has been found in significant amounts in public buildings such as schools (Berge et al., 1998; Dybendal et al., 1989a; Schou, 1993; Warner, 1992) and hospitals (Custovic et al., 1998a). In dust from upholstered English hospital chairs, *Can f* I levels (geometric mean = 22 µg/g, range, 4–63) were as high as levels in settled dust from households with dogs (Custovic et al., 1998a; Munir et al., 1994). Hospital airborne *Can f* I levels were detectable in 7 of 10 testing days but were lower (range 0.09–0.22 ng/m³) than those often found in homes with dogs (range 0–100 ng/m³ *Can f* I) (Custovic et al., 1998a; Hodson et al., 1999).

Evidence Regarding Asthma Exacerbation

The asthmatic response to bronchial provocation test (PT) with dog allergen was evaluated in a cross-sectional Finnish study of 203 asthmatic children selected from the Children's Asthma Registry (Vanto and Koivikko, 1983). Of those with a positive PT, 64% had kept dogs, whereas only 36% with a negative PT had kept dogs. A positive PT was correlated with a positive skin prick test to dog (correlation coefficient = 0.8), but not with the frequency of reported asthma symptoms. In immunotherapy trials, positive response to bronchial provocation with dog allergen has also been associated with elevated levels of IgE to dog allergen in asthmatic subjects (Hedlin et al., 1995; Valovirta et al., 1984; Vanto et al., 1980). Some investigators consider symptomatic and bronchial response to animal allergen in an experimental animal room to be more definitive proof that animal allergen triggers asthma than allergen bronchial PT. They question whether the airway response to bronchial provocation with an allergen is always an allergic rather than an irritant response. The committee could find no published studies of the response of dog-sensitized asthmatics to exposure to dogs in an experimental dog room analogous to the cat room set up by the Hopkins group (Sicherer et al., 1997).

Evidence Regarding Asthma Development

There is insufficient evidence regarding the role of dog allergen in the development of asthma. In keeping with the ecology of Los Alamos, New Mexico, which is high and dry, with less dust mite, the allergens to which asthmatics are sensitized tend to be the predominant allergens in the indoor and outdoor environment. These allergens include dog and *Alternaria*, as well as cat (Ingram et al., 1995; Sporik et al., 1995). A cross-sectional retrospective Finnish study suggested that dog allergy was more prevalent in children from homes where dogs were present in the first few months of life, compared to homes where the dog was introduced after the child had reached the age of 1 (Vanto and Koivikko, 1983). As mentioned above, in a New Zealand birth cohort study the development of asthma by 13 years of age was associated with sensitization to dogs at age 13 (Sears et al., 1989). In contrast, the presence of a dog in the home in childhood was negatively associated with asthma (OR = 0.85, 95% CI = 0.78–0.92) among adults reporting no parental allergy, in the cross-sectional European Community Respiratory Health Survey of 13,932 20- to 44-year-old subjects from 36 areas in Europe, New Zealand, Australia, and the United States (Svanes et al., 1999). Among 1,649 Swedish school children aged 7–13 years, a report of keeping a cat or a dog in the first year of life was also negatively associated with asthma and allergic rhinitis (Hesselmar et al., 1999).

Conclusions: Asthma Exacerbation and Development

• There is sufficient evidence of an association between dog allergen exposure and exacerbation of asthma in individuals specifically sensitized to dogs.
• There is inadequate or insufficient evidence to determine whether or not an association exists between dog allergen exposure and development of asthma.

Evidence Regarding Exposure Mitigation and Prevention

Because the aerodynamic properties, carrier material, and chemical composition of dog and cat allergens are similar, issues related to mitigation of dog allergen are likely to be similar. However fewer studies are available related to the mitigation of dog allergen exposure and its health consequences. In a cross-sectional study of 203 asthmatic children listed in the Finnish Asthma Register, 68 of 203 had kept a dog, but 59 of 68 (87%) had removed the dogs from their homes. Parents tended to report that removal of the dog had improved asthma symptoms and had not been detrimental because of emotional deprivation (Vanto and Koivikko, 1983). The Finnish study also demonstrated that homes with dogs had higher dog antigen levels in dust than homes without dogs, some of which had a past history of a dog in the home. Homes in which occupants had indirect contact with dogs had more dog allergen than homes in which no one reported contact with dogs. Despite reported avoidance of dogs, a rising or steadily high level of dog-specific IgE was observed in follow-up of 24 dog-allergic subjects. The authors hypothesized that dog allergen encountered outside the home might be sufficient to boost IgE synthesis in most sensitive subjects (Vanto and Koivikko, 1983).

In a study of 25 homes with dogs, Custovic found that dogs had to be washed at least twice a week to maintain a reduction in recoverable $Can\,f\,I$ from the hair (Hodson et al., 1999). Airborne dog allergen levels were not significantly affected by washing the dog. No studies are available regarding the effect of this mitigation measure on symptoms or lung function in dog-sensitive asthmatics.

Conclusions: Exposure Mitigation and Prevention

• There is limited or suggestive evidence of an association between removal of the dog from the home and reduction of dog allergen levels. This evidence comes from an association between the absence of a dog in the home and the measurement of low dog allergen levels in a study including homes that had a history of keeping dogs (Vanto and Koivikko, 1983).

• There is inadequate or insufficient evidence to determine whether or not an association exists between removal of a dog from the home and improvement in symptoms or lung function in dog-sensitized asthmatics. The one epidemiologic study reporting an association between dog removal and symptom improvement in asthmatic children relies on retrospective parental reporting without measures of sensitization at the time of dog removal or measures of symptoms or lung function before and after dog removal (Vanto and Koivikko, 1983).

Rodents

Definition of the Agent and Means of Exposure

Exposure to rodents can come either from keeping pets or from their presence as pests in the home. Rodents (mouse, rat, and guinea pig) can also be found in school settings. They have been studied as sources of allergens, particularly because of their extensive use as laboratory animals (Schou, 1993). Hair and epithelial fragments carry allergenic molecules, the allergens measured are believed to be from urine, saliva, or skin. The relative importance of these allergen sources has been debated (Karn, 1994; Longbottom and Austwick, 1987; Walls and Longbottom, 1985). As described by Schou (1993), rodents have permanent proteinuria; allergenic protein from sprayed urine dries up and becomes airborne on dust particles. Airborne rodent allergen has been measured in laboratory facilities (Sakaguchi et al., 1990a; Swanson et al., 1985; Twiggs et al., 1982) and, in one study, in inner city apartments (Swanson et al., 1985). Clinically important allergens that have been identified include *Mus m* I and *Mus m* II for mouse, *Rat n* I for rat, and *Cav p* I and II for guinea pig (Schou, 1993). Assessment of rodent exposure in the home has been limited, to some extent, by limitations in the ability to measure allergens from all species of wild mice potentially present in the home.

Evidence Regarding Asthma Exacerbation

A number of cross-sectional studies document the association between handling animals in a laboratory setting and allergy (Beeson et al., 1983; Cockcroft et al., 1981; Cullinan et al., 1994; Davies and McArdle, 1981; Gross, 1980; Newman-Taylor, 1982; Schumacher et al., 1981; Venables et al., 1988). Hollander and colleagues (1996) conducted a prospective panel study of self-reported symptoms and peak flow in Dutch laboratory animal workers. Workers who reported asthmatic symptoms (chest tightness) due to working with rats had significant decreases in peak expiratory flow on days they worked with the animals; 86% of them were sensitized to rat allergens (Hollander et al., 1996). A cross-sectional study of British laboratory workers demonstrated an association between positive skin tests to animal extracts and asthmatic symptoms (Cockcroft et al., 1981). Whereas IgE antibody to rat allergen was present in only 2 of 135 laboratory workers without asthmatic symptoms, it was present in 12 of 18 laboratory workers with symptoms (Platts-Mills et al., 1987).

In the U.S. National Cooperative Inner City Asthma Study, 19% of asthmatic children were allergic to rats and 15% were allergic to mice, suggesting exposure to rat or mouse allergens in the home (Kattan et al., 1997). No data are available on rat or mouse allergen levels in the home and the exacerbation of asthma among rodent-sensitized asthmatics.

Evidence Regarding Asthma Development

Although there are retrospective reports of the incidence of asthma symptoms after beginning laboratory work with rodents (Platts-Mills et al., 1987), the committee could find no relevant studies on rodent allergen exposure and the development of asthma.

Conclusions: Asthma Exacerbation and Development

• There is sufficient evidence of an association between exposure to rodents in a laboratory setting and exacerbation of symptoms or lung function in rodent-sensitized asthmatics.

• There is inadequate or insufficient evidence to determine whether or not an association exists between exposure to rodents (wild or as pets) in the home and exacerbation of symptoms or lung function in rodent-sensitized asthmatics.

• There is inadequate or insufficient evidence to determine whether or not an association exists between exposure to rodents and the development of asthma.

Evidence and Conclusions: Exposure Mitigation and Prevention

No studies are available to document the influence of rodent eradication on the level of rodent allergen or on the severity of asthma in sensitized asthmatics.

• There is inadequate or insufficient evidence to determine whether or not an association exists between removal of rodents from the home and the reduction of rodent allergen levels.

• There is inadequate or insufficient evidence to determine whether or not an association exists between removal of rodents from the home and improvement in symptoms or lung function in rodent-sensitized asthmatics.

Cow and Horse

Definition of the Agent and Means of Exposure

Relatively few people in the United States live on farms where exposure to cows, horses, or pigs can be significant; only 1.5% of U.S. households keep horses as pets. Cow hair and dander contain at least 17 antigens, 4 of which have been identified as allergens and 3 of which have been purified (*Bos d* I, II, and III). Horse hair and dander have been demonstrated to contain three important allergens (*Equ c* I, II, and III) (Schou, 1993).

Evidence and Conclusions: Asthma Exacerbation and Development

Allergies to cows and horses are considered occupational diseases of farm workers and veterinarians (Prahl and Roedpetersen, 1979; Schou, 1993). Data on the effect of non-occupational exposures is lacking. In summary:

• There is inadequate or insufficient evidence to determine whether or not an association exists between cow or horse allergen in the home and the exacerbation of asthma in sensitive children or the development of asthma.

Living on a Farm and Development of Asthma

The epidemiologic literature on allergy in farmers and children from farming families was reviewed in a 1999 article demonstrating a lower prevalence of hay fever and allergic sensitization in farmer's children compared to peers from nonfarming families living in the same rural Swiss community (Braun-Fahrlander et al., 1999). While farm animals can be allergenic, lower rates of sensitization to pollen and animal dander have been reported in adult farmers compared to other occupational groups (Iversen and Pedersen, 1990; Kohler et al., 1983; Rautalahi et al., 1987; Sigsgaard et al., 1996). Swedish conscripts and Finnish university students raised on farms have reported fewer allergic symptoms than students from nonagricultural backgrounds (Åberg, 1989; Kilpeläinen et al., 1997). This may be the result of self-selection out of the farming community by individuals and families with a genetic predisposition toward allergy (von Mutius et al., 1994).

These findings have led investigators to question whether the farm environment itself might play a protective role in the development of allergy and allergic asthma. One hypothesis is that contact with farm animals and their bacterial products (including endotoxin, which is discussed later in this chapter) may be protective against allergy or asthma through early-life stimulation of TH1 immunity, particularly for individuals with specific genetic characteristics (Baldini et al., 1999). However, no additional evidence is available to confirm or reject the hypothesis that for some children, exposure to farm animals early in life might be protective against either allergy or asthma. The farm is a complex environment that varies by country and culture, and many other aspects of farm living may contribute to the observed epidemiologic differences between asthma or atopy in children from farming versus nonfarming families.

Birds

Definition of the Agent and Means of Exposure

Birds are kept in 5% of U.S. households. While it is clear that hypersensitivity pneumonitis can be associated with antigens from bird excreta, serum protein, and proteinaceous material in dispersed dust from birds (Christensen et al., 1975; Hendrick et al., 1978), specific bird antigens associated with allergy and asthma have not been defined with certainty. Tauer-Reich and colleagues (1994) studied five bird fanciers who complained of asthmatic symptoms during contact with their birds and also had documented bronchial hyperreactivity to acetylcholine. These individuals had positive IgE antibody reactions to bird sera as well as to extracts of feathers. Skin prick tests to mites, molds, pollen, and domestic animals other than birds were negative in all five patients.

Evidence Regarding Asthma Exacerbation and Development

A few additional case reports describe bird handlers with the combination of asthmatic symptoms in the presence of birds, symptoms of egg hypersensitivity, and specific IgE antibodies against blood serum proteins of chicken, parrot, budgerigar, or pigeon serum (de Blay et al., 1991b).

A portion of what is called bird allergy may be an expression of allergy to dust mites. Bird feathers can harbor mites (Kemp et al., 1996). The material used in skin prick testing for bird allergy may also be contaminated with mite allergen. Although clinicians have traditionally advised asthmatic patients not to use feather pillows, an epidemiologic study found an increased risk of wheeze in children using foam pillows compared to children using feather pillows (Strachan and Carey, 1995). This association could occur if parents of symptomatic children tend to provide their children with synthetic rather than feather pillows because of advice from health professionals. On the other hand, this association may relate to increase mite exposure from synthetic pillows. New Zealand researchers found that after four months of use, dust mite (*Der p* I) levels were significantly higher in synthetic than in feather pillows (Crane et al., 1997; Rains et al., 1999). To prevent the feathers from coming out, feather pillows may have more impermeable covers than synthetic pillows; which could result in less dust mite infestation.

Conclusions: Asthma Exacerbation and Development

• There is limited or suggestive evidence of an association between bird exposure and exacerbation of symptoms in bird-sensitized asthmatics. This association may be confounded by the allergic asthmatic response to mites harbored by birds.

- There is inadequate or insufficient evidence to determine whether or not an association exists between bird allergen exposure and the development of asthma.
- There is inadequate or insufficient evidence to determine whether or not an association exists between down pillows and exacerbation of symptoms or lung function in asthmatics. Pillows are believed to be a risk factor for asthma because of their documented mite content, rather than because of the presence of bird allergen.

Evidence and Conclusions: Exposure Mitigation and Prevention

There is inadequate or insufficient evidence to determine whether or not an association exists between removal of a bird from the home and reduction in bird allergen levels or improvement in symptoms or lung function in bird-sensitized asthmatics.

Research Needs

The associations between dust mite allergen, asthma exacerbation, and asthma development are much more well defined than the associations between larger animals and asthma. This is only partly a function of the number of years and intensity of efforts to investigate the health effects of dust mites. Compared with dust mite allergen, once the allergen source is present, cat and dog allergens are more easily dispersed throughout the household. Cat and dog allergens remain airborne for much longer than dust mite. The potential for exposure to allergens outside the home is markedly greater for cats and dogs than for dust mites. The potential for home exposure to cat or dog allergen in homes without cats or dogs has also been underestimated. The absence of adequate information regarding allergen exposure may, in part, account for contradictory data regarding the effects of cats or dogs in the home on the development of asthma.

Research is needed to assess whether removal of the cat or dog from the home results in sufficient reduction in overall allergen exposure to reduce symptoms and improve bronchial reactivity in specifically sensitized asthmatics. Further research is needed to assess the level of animal allergen exposure (cat and dog) in day care centers and schools. When significant levels are noted, the potential for lowering exposure should be investigated. Since so many cat- or dog-allergic asthmatics are emotionally attached to their pets, investigators should explore the success of efforts that recommend the removal of the pet for sensitized symptomatic child and adult asthmatics. Further research is also needed to evaluate the effect of mitigation measures short of animal removal on asthma symptoms, lung function, or bronchial responsiveness in specifically sensitized asthmatics. Although frequent animal washing and HEPA filter use are widely recommended, their efficacy in reducing asthma severity has not been proven.

Two retrospective cross-sectional studies suggest that exposure to cat or dog in early life may actually be protective against asthma development in some subsets of children (Hesselmar et al., 1999; Svanes et al., 1999). The relationship between cat or dog allergen exposure in early childhood, the development of sensitization, and the development of asthma merits further investigation. This investigation will require better assessment of exposure. It is likely that the genetic phenotype will modify the response to cat or dog allergen at different levels of exposure, but gene-by-environment interactions cannot be effectively explored until the genetics of asthma is better understood.

Further research is needed to evaluate rodent allergen exposure in the home as a potential factor in the exacerbation of asthma in rodent-sensitized asthmatics. Particularly in socially disadvantaged populations, research should focus on effective reduction of rodent allergen and its effect on symptoms or lung function in specifically sensitized asthmatics.

Researchers should also consider the possibility that animal (or animal allergen) exposure may be either protective or allergenic. The effects may depend on the mode of exposure, the genetic characteristics of the populations, the timing in the life cycle when exposure occurs, and many other cofactors (e.g., early-life viral, bacterial, and parasitic infection experience).

COCKROACH

Many insects have been identified as sources of inhalant allergens in case reports or small outbreaks; these include moths, crickets, locusts, beetles, "green nimitti" midges, lake flies, and houseflies (IOM, 1993). Cockroaches, however, are the only insect that has been repeatedly recognized as a common source of indoor allergens.

Definition of the Agent and Means of Exposure

Agent Definition and Biology

Cockroach is an important source of indoor allergen worldwide. Although more than 60 species have been identified, the most common indoor species in North America are the German (*Blattella germanica*), American (*Periplaneta americana*), and Oriental (*Blatta orientalis*). There are multiple allergens from cockroach that have been identified (*Bla g* I or *Per a* I, *Bla g* II, IV, and V) and cloned (*Bla g* II, IV, and V). Schou and colleagues (1990) purified an allergen from *B. germanica* and *P. americana* extract, which resulted in positive reaction to skin tests in 50% of patients who were allergic to cockroaches and was designated *Bla g* I (or *Per a* I). Monoclonal antibodies have been produced against extracts of both cockroach species and used for cockroach allergen identification and purification. Enzyme-linked immunosorbent assays have been developed that can be used to estimate exposure to some of these cockroach allergens (*Bla g* I, *Per a* I, and *Bla g* II) in the environment. *Bla g* I is a 25-kD (kilodaltons, a unit of molecular weight; also abbreviated kDa) cross-reacting antigen from both the German and the American cockroach. *Bla g* II is a 36-kD species-specific antigen derived from the German cockroach. Specific allergens have not been purified from Oriental cockroaches.

Neither the antigenic relationships between cockroach species nor the precise source of cockroach allergens are well understood. The source of the allergens derived from cockroaches is not known, but there are speculations to suggest that they may come from feces, parts of the body, or other sources in the body. Sequence homology searches are useful tools for investigation of the biologic function of cockroach allergens, and as more sequences become available it will be possible to make comparisons of biologic function and allergenicity, to compare allergen expression in different species, and to localize the source of the allergens in cockroach tissues.

At present, cockroach extracts are not standardized; however, the Food and Drug Administration's (FDA's) Center for Biologics Evaluation and Research has embarked on a program to standardize *B. germanica* extracts on the basis of skin testing, protein content, RAST inhibition, and specific allergen assays. Most patients in the United States appear to be sensitized to *B. germanica*, although there is a cross-reactivity between *B. germanica* and *P. americana* on skin tests and serum IgE antibody assay. The prevalences of IgE antibodies to *Bla g* I and *Bla g* II in patients allergic to cockroaches are 40% and 60–80%, respectively (Pollart et al., 1991a). Approximately 20% of patients lack detectable IgE antibody to either allergen, which suggests that cockroaches produce other important allergens (Chapman, 1993). However, the levels of the two allergens have been found to be highly correlated (r = 0.92; $p < .01$) in dust samples (Pollart et al., 1991a).

Factors Influencing Exposure

Studies have suggested that cockroach sensitization is an affliction of the inner city poor, but the complex interrelationship of race, poverty, and residence has been difficult to unravel (Bernton and Brown, 1964; Call et al., 1992; Garcia et al., 1994; Gelber et al., 1993; Koehler et al., 1987). Sarpong and colleagues (1996a) examined race and socioeconomic status (SES) as risk factors for cockroach allergen exposure and sensitization. In their cohort of 48 white and 39 African-American children, they found that both factors were independent predictors of cockroach sensitization. Among low-SES subjects, sensitization was common with 50% (4 of 8) of white participants and 75% (15 of 20) of African-American participants exhibiting a positive skin test. Cockroach antigens are widely distributed in homes and schools, and the kitchen is the most common source of cockroach allergen (Rosenstreich et al., 1997; Sarpong et al., 1996b, 1997). The cockroach allergen in school dust was similar to that reported in homes. The level of antigen reported in school dust is of concern because it may constitute a very important occupational risk to students, teachers, and other school workers. The source of the allergen is not known, but it is likely that the schools may have been infested with cockroaches since there was evidence of dead cockroaches in the schools examined (Sarpong et al., 1997).

In kitchens, food and water sources may be important factors for the proliferation of cockroaches. The humidity in the home may be an important factor for increased cockroach allergen in infested homes. However, in a study evaluating allergen levels in schools, the presence of air conditioners, which may lower the humidity, did not affect the distribution of allergen levels (Sarpong et al., 1997). Despite the evidence that cockroach allergen levels are higher in the kitchens of both asthmatic and nonasthmatic individuals, the concentration of cockroach allergen in the bedroom has been used as the surrogate marker of exposure (Eggleston et al., 1998; Rosenstreich et al., 1997; Sarpong and Han, 1999; Sarpong et al., 1996a, 1996b).

Exposure to cockroach allergens are dependent on their aerodynamic properties and the characteristics of the surface on which they are deposited. In an Ohio study, there was no difference between dust mite allergen concentrations in low-pile carpet and smooth floors, but allergen levels were significantly higher in high-pile carpets (Arlian et al., 1982). In a school study in the Baltimore metropolitan area, there was no difference in cockroach allergen, *Bla g* I between low-pile carpet and uncarpeted floors (Sarpong et al., 1997). However, kitchen areas are generally noncarpeted, and this may have distorted the level of allergen in uncarpeted compared to carpeted areas.

Cockroach allergens may behave like the dust mite antigen; that is, they are carried on large particles that become airborne for short periods of time during active disturbance. High concentrations of cockroach allergen are found in the kitchen, compared to the living room or bedroom. However, some studies have reported similar levels of cockroach allergen in all these sites. Taken together, the cockroach allergen may be more relevant in the bedroom than the kitchen or the living room because of close contact with the pillow while in bed. Interestingly, about 20% of homes with no evidence of cockroach infestation have significant levels of cockroach allergen in settled dust (Pollart et al., 1991b). Although the possibility of reporting bias cannot be excluded, it is likely that cockroach allergen may be present in a home long after the infestation has been controlled.

Cockroach hypersensitivity is a unique risk factor for asthma among the urban poor. Many case-control studies have documented that cockroach allergen exposure and sensitization are significantly more common in patients with asthma living in urban homes compared to those living in suburban homes (Bernton et al., 1972; Kang et al., 1993; Sarpong et al., 1996a). However, despite higher rates of exposure and sensitization among urban subjects, place of residence was not independently associated with cockroach exposure after correction for socioeconomic status; families of lower SES were likely to be exposed wherever they lived (Sarpong et al., 1996a). This agrees with the observations of Gelber and colleagues (1993), who used lack of health insurance as a marker of poverty and showed that this was a more important correlate of sensitization than urban residence per se as well as with others who have categorized patients on the basis of access to private medical care (Garcia et al., 1994). A 1999 report from Morgantown, West Virginia, documented the role of sensitization in infantile asthma (Wilson et al., 1999). These children who were at least 3 years of age with documented wheezing episodes demonstrated cockroach hypersensitivity at a rate of 25%. The rate of sensitivity in this predominantly white population was not related to low socioeconomic status.

Evidence Regarding Asthma
Exacerbation and Development

Biologic Plausibility

Bernton and colleagues (1972) reported that inhalation of cockroach extract could induce an immediate asthmatic reaction in cockroach-sensitive asthmatic subjects. Subsequently, controlled inhalation challenge confirmed the production of antigen-specific, acute and late bronchospasm in cockroach-sensitive asthmatics (Kang, 1976). These antigen-induced asthmatic reactions were blocked by premedication with disodium cromoglycate. Challenge of these subjects with inhaled cockroach extract produced a significant antigen-specific peripheral eosinophilia, which progressed to its peak 24–36 hours after exposure and often more than doubled the baseline value. In animal studies there is evidence to suggest that cockroach antigen can induce lung eosinophilia (Campbell et al., 1998). Previous data in guinea pigs have shown that aerosolized cockroach antigen can be utilized to induce airway inflammation and alter airway physiology (Kang et al., 1996). The development of these models will allow the evaluation of mediators involved in both stages of cockroach allergen challenge, as well as the testing of specific therapeutic modalities.

Consistency

There is now good evidence from epidemiologic studies in several parts of the world which demonstrates that the development of immediate sensitivity to cockroach allergens is associated with asthma morbidity and that sensitization is related to the degree of allergen exposure (Eggleston et al., 1998; Rosenstreich et al., 1997; Sarpong and Han, 1999; Sarpong et al., 1996a; Sastre et al., 1996). In the original description of cockroach sensitivity in 1967, Bernton and Brown (1967) found higher rates of sensitization among Puerto Ricans and African Americans than among Jews and Italians in New York City and believed that the difference was related to the improved economic conditions of the latter groups. Shulaner (1970) also found a relationship to poverty and overcrowding, and a report by Gelber and colleagues (1993) relates exposure and sensitization to urban housing, race, and poverty as indicated by lack of medical insurance.

Strength of Association

Sensitization, with production of specific IgE antibodies to cockroach, is a strong risk factor for acute severe asthma, especially when sensitized persons are exposed to high concentration of allergen in their homes. R.P. Nelson and colleagues (1996) compared 29 children with acute asthma, ages 3–16 years, first seen in a Florida emergency department, to 25 control subjects. They found that sensitization to cockroach allergens was associated with acute asthma that required emergency treatment. Gelber and colleagues (1993) and Call and colleagues (1992) showed that cockroach sensitization was a risk factor for acute asthma in patients visiting emergency departments in Wilmington, Delaware, and Atlanta, Georgia, respectively. The multicenter National Cooperative Inner City Asthma Study (NCICAS) of the home environment of asthmatic children aged 4–9 years living in seven major urban areas and eight centers reported that cockroach allergen was the predominant indoor allergen in these homes. In the NCICAS, it was suggested that sensitization to cockroach allergen and exposure to high levels of cockroach allergen may explain the asthma-related health problems of inner city children (Rosenstreich et al., 1997).

Dose–Response

A dose–response relationship between cockroach allergen exposure and sensitization has been established in asthmatic children. Sarpong and colleagues (1996a) report that children who were exposed to *Bla g* I or *Bla g* II of 1 unit per gram (U/g) or higher demonstrated skin sensitivity to cockroach allergen. For *Bla g* I, all the children who were exposed to more than 10 U/g were sensitized to cockroach allergen. Similarly, 100% of the children who were exposed to more than 5 U/g of *Bla g* II were sensitized to cockroach allergen. Following this report, the NCICAS has suggested that cockroach allergen (*Bla g* I) exposure is related to sensitization in a dose–response manner (Eggleston et al., 1998). Preliminary data from a population of pregnant women have suggested that total cockroach allergen (*Bla g* I and *Bla g* II) exposure and sensitization were related in a dose–response fashion (Sarpong and Han, 1999). This relationship was demonstrated in both asthmatic and nonasthmatic controls. However, the dose of cockroach allergen concentration required to induce sensitization rates in the asthmatic population was at least a factor of ten lower than in the nonasthmatic population. A dose–response relationship was also demonstrated in the inner city asthma study between cockroach allergen (*Bla g* I) exposure and morbidity due to asthma. Children who were exposed to more than 8 U/g of *Bla g* I were more likely to be hospitalized for asthma symptoms than those who were exposed to less. In a prospective study by Gold and colleagues (1999), infants who were born in homes with *Bla g* I levels greater than 2 U/g reported increased risk of wheezing by the age of 1 year. A study by Litonjua and colleagues available as a conference abstract (1998) evaluated the relationship between home allergen levels and the prevalence and incidence of asthma 16 months later among 215 children younger than 5 years old. The researchers found that, among the children with new onset of asthma, 7 (87.5%) lived in homes with high *Bla g* I or II levels while 1 (12.5%) lived in a home with low levels ($p = .028$).

Although it seems clear that exposure to cockroach allergen precedes disease, it has not been shown that sensitization consistently precedes disease. However, it is the combination of strong association, biological plausibility, dose–response, and provocation experiments that creates the strength of the argument.

Conclusions: Asthma Exacerbation and Development

Cockroach antigen exposure can elicit a strong IgE immune response to induce sensitization. Sensitization to cockroach antigen has been linked to the season of birth (Sarpong and Karrison, 1998a), although this study does not necessarily support cockroach antigen as a disease promoter, it at least gives some insight that exposure during the perinatal period may be critical. Because of the likely aerodynamic nature of the cockroach allergen, it is difficult to predict whether cockroach allergen per se is an initiator of asthmatic symptoms. However, data suggest that infants who are exposed to cockroach allergen are more likely to wheeze during infancy. Moreover, researchers have shown that cockroach sensitization is a risk factor for acute asthma in patients visiting emergency departments (Call et al., 1992; Gelber et al., 1993; R.P. Nelson et al., 1996). In a retrospective study of asthmatic children in Chicago, it was suggested that children with combined sensitization to cat, dog, dust mite, and cockroach allergens were at increased risk of having more severe asthma (Sarpong and Karrison, 1998b). Researchers have also found that there is an increased risk of sensitization to cockroach allergen in asthmatic children born in the winter months (Sarpong and Karrison, 1998a). Because the risk of respiratory syncytial virus (RSV) infection is high in winter months and RSV is also known to give rise to asthma-like symptoms, others have even suggested that it may promote IgE sensitization. The level of cockroach allergen in settled dust may not demonstrate seasonality. It is therefore attractive to postulate an interaction between RSV bronchiolitis and cockroach allergen-induced sensitization. Thus, there may be virus–antigen interaction in initiating sensitization and possibly inducing asthma.

Every individual is at risk for cockroach allergen exposure and sensitization. However sensitization risk is higher among the asthmatic population. Understanding the genetics of allergy and asthma, including understanding the phenotypes associated with cockroach allergy, may eventually prove useful in the understanding of gene–environment interaction in the development of asthma.

There are no prospective data demonstrating the association of exposure to cockroach allergen with subsequent development of sensitization to cockroach allergen and asthma. However, a number of case-control studies demonstrate a link between cockroach allergen exposure in sensitized individuals and asthma morbidity, as well as a dose–response relationship between cockroach exposure and sensitization (Gelber et al., 1993; Rosenstreich et al., 1997; Sarpong et al., 1996a).

In summary:

• There is sufficient evidence of a causal relationship between cockroach allergen exposure and exacerbation of asthma in individuals specifically sensitized to cockroaches.
• There is limited or suggestive evidence of an association between cockroach allergen exposure and the development of asthma in preschool-aged children.
• Inadequate or insufficient evidence exists to determine whether or not an association exists between cockroach allergen exposure and the development of asthma in older children and adults.

Evidence Regarding Exposure Mitigation and Prevention

In principle, control of exposure to cockroach allergens combines three approaches directed through an indoor environment:

1. control of reservoirs of allergen in beds, carpets, furnishings, and clothing, which are the main sources of exposure;
2. control of the sources of new allergen (e.g., cockroach reinfestation); and
3. direct control of cockroach airborne allergens.

There are several potential strategies for achieving these forms of control. These are noted below, along with some of their practical limitations.

The bed is probably the most important site of cockroach allergen exposure because of the high level of exposure during sleep, the proximity of the subject to the source, the proportion of time spent indoors at this one site, and the large amounts of dust present in the bed. The relative success and simplicity of interventions directed at this site make it a key target for allergen control. Strategies known to reduce allergen exposure include encasing mattresses and pillows and laundering bedding in hot (>130°F, or 55°C) water. (This strategy is also effective for dust mite allergen.) However, there may be problems of subject compliance with regular washing of some items, such as blankets, even in the setting of a clinical trial.

Carpets serve as a major reservoir of cockroach and many other indoor allergens and may serve as an additional primary source of allergen. Replacement of fitted carpets with smooth flooring has been shown to reduce dust mite allergen levels (Hayden et al., 1997) and is often recommended as a cockroach allergen management strategy. However, this intervention is sometimes unpopular and may be impractical in rental units, public housing, and other environments where occupants do not have control over floor coverings. Dry vacuum cleaners are useful for picking up excess dust and reducing reservoirs and allergen concentration. Whether this achieves significant reductions in exposure remains to be demonstrated. Further, while wet vacuuming and steam cleaning may reduce allergen levels, they also may enhance the environment for dust mites and other potential allergens if the surface does not dry properly or promptly.

Extermination is a primary approach to control new sources.

The use of avermectin or hydramethylnon has, for example, been shown to reduce cockroach populations (Cochran, 1995, 1996).

Finally, housekeeping measures aimed at limiting open food-stuffs (using sealed containers, cleaning under refrigerators and stoves, washing dishes promptly after meals, and the like) and water sources (fixing leaking faucets and pipes; eliminating pet water dishes) are often suggested as part of an overall cockroach management strategy.

As noted in Chapter 10, ventilation rates are not likely to have an appreciable direct impact on indoor concentrations of the larger particles associated with cockroach allergens.

Attempts to eliminate cockroach allergen from the environment have had limited success. A short-term effect of extermination and cleaning on cockroach antigen levels was noted in a study conducted in an urban college dormitory (Sarpong et al., 1996b). Other studies have found that the number of cockroaches seen in a dwelling can be drastically reduced with roach insecticides such as hydramethylnon. However, the elimination of cockroach sightings did not decrease the levels of *Bla g* I or *Bla g* II in vacuum dust over the next six months in single-family dwellings in North Carolina. No extra cleaning was performed in these homes (Williams et al., 1999). Eggleston and colleagues (1999) evaluated the effect of professional pest control and home cleaning on infestation and allergen concentrations in 13 inner-city homes with active infestations. Occupants were instructed on how to conduct follow-up cleaning and pest control measures, and study personnel monitored compliance over the eight months of the study. The researchers reported decreases in infestation and in allergen concentrations in settled dust. While mean allergen concentrations were reduced between 74% and 93% in various rooms of the homes, they still exceeded 20 U/g, a level associated with increased morbidity in the NCICAS study. In the NCICAS, integrated intervention to reduce cockroach allergen in homes was conducted. Despite a significant but short-lived reduction, cockroach allergen levels remained well above those previously found to be clinically significant (Gergen et al., 1999). Insufficient evidence is available to determine whether or not reduction of cockroach allergen levels in the home reduces asthma severity in cockroach-sensitized asthmatics. Additional studies are needed to evaluate, within both multi-unit and single family housing, the effectiveness of various proposed methods of reducing cockroach allergen to levels below those associated with symptoms in asthmatic subjects. Until more effective methods of reduction of cockroach allergen levels are developed, it may prove difficult to assess the effectiveness of cockroach allergen reduction programs in reducing asthma morbidity. Ongoing intervention studies may provide further data as to whether, for some subsets of sensitized asthmatics, moderate reductions in allergen levels or in allergen load influence asthma morbidity. Given the evidence that decreasing exposure to dust mites can help control the symptoms of dust mite-allergic asthmatics, it is prudent to identify patients who are allergic to cockroaches and educate them to reduce allergen exposure indoors.

Conclusions: Exposure Mitigation and Prevention

There is preliminary evidence to suggest that reduction in cockroach allergen in settled dust can be achieved with extermination on a short-term basis; there are ongoing studies to evaluate whether long-term reduction of cockroach allergen in settled dust can be achieved with extermination. It appears that the use of abatement strategies combining extermination and cleaning can reduce cockroach allergen exposure. However, research has not to date demonstrated that these strategies result in an improvement in symptoms or lung function in cockroach-sensitized asthmatics. It is important to remember that the absence of evidence does not mean an absence of effect. The science regarding indoor environmental interventions, exposure limitation, and effects on asthma outcomes is not nearly as well developed as that regarding the health effects of exposures. Given the evidence that decreasing exposure to dust mites can help control the symptoms of dust mite-allergic asthmatics, it is prudent to identify patients who are allergic to cockroaches and educate them to reduce allergen exposure indoors.

In summary:

• There is sufficient evidence of an association between the implementation of intensive cockroach allergen exposure mitigation strategies and short-term reduction of cockroach allergen levels. Such strategies must include both removal of the allergen from reservoirs and control of sources (i.e., abatement and prevention of reinfestation) to be effective.

• There is inadequate or insufficient information to determine whether or not an association exists between cockroach reduction interventions and improvement in symptoms or lung function in cockroach-sensitized asthmatics.

Research Needs

The preceding discussion suggests that there is still a need for fundamental research on cockroach allergens and asthma outcomes. Future research should focus on the efficacy of cockroach allergen reduction in the homes of asthmatic patients, the aerodynamic properties of cockroach allergen, the efficacy of cockroach immunotherapy, and B and T cell reactive epitopes. Further studies are also needed to better elucidate any relationship between cockroach allergen exposure and asthma development; explore the interaction of cockroach allergen with infectious agents, irritants, and other allergens in causing asthma; and examine the influences of genetics, socioeconomic status, and location on exposure and sensitization.

HOUSE DUST MITES

In 1967, Voorhorst and colleagues identified dust mites of the genus *Dermatophagoides* as the most important source of allergens in house dust (Voorhorst et al., 1969). Using dust mite extracts, they demonstrated that sensitization was common among children with asthma and later developed techniques for growing mites. By 1974, extracts were widely available for skin testing, and mite sensitization became recognized in many countries. Indeed, mite sensitization was found to be strongly associated with asthma in all countries that had reported increases in asthma up to 1980 (i.e., the United Kingdom, Australia, New Zealand, Japan) (Clarke and Aldons, 1979; Miyamoto et al., 1968; Smith et al., 1969).

Since the original research by Voorhurst and colleagues (1969), most of the studies on allergen avoidance have focused on mite-allergic patients. Studies have included moving children from Holland to Davos (Kerrebijn, 1970), keeping mite-allergic adults in special hospital rooms (Platts-Mills et al., 1982), and controlled trials of dust mite reduction in patients' rooms (Murray and Ferguson, 1983). In each of these models, change in environment with reduction in exposure to mite allergen was associated with decreased symptoms of asthma and decreased bronchial hyper-reactivity (BHR) (Platts-Mills and de Weck, 1988).

Definition of the Agent and Means of Exposure

Agent Definition and Biology

Purification of indoor allergens is dependent on the quality of source materials. Before the discovery of dust mites, several unsuccessful attempts were made to purify allergens from dust obtained from carpets and bedding. Even when dust mite cultures became available, the early attempts at purification used only a small amount of material, ~40 g. The first purification of the major allergen *Der p* I was by classical immunochemistry from 400 g of "spent" culture (Chapman and Platts-Mills, 1980). The spent culture was very rich in fecal material. Purification made it possible to measure the quantity of allergen in extracts, in dust samples, and airborne (Tovey et al., 1981a). In addition, the purified allergens have been used to study the immune response to allergens, as well as the properties of these proteins. Each of these approaches has relevance to the causes of lung symptoms.

Epidemiologic evidence about the relationship between sensitization to indoor allergens and asthma is based on skin tests or on serum assays of IgE antibodies. However, the immune response to inhalant allergens also includes IgG antibodies, IgA antibodies, and T cells. These T cells in allergic individuals are predominantly CD4+ cells of the TH2 type (Romagnani, 1992; Wierenga et al., 1990). There are two important questions about this response: (1) Can differences in the immune response predict which skin test-positive individuals will develop asthma? Most of the evidence suggests that skin test-positive asymptomatic individuals have very similar IgE antibodies and T cells (i.e., TH2) compared to symptomatic patients. (2) What, if anything, is the immune response to mite allergens in skin test-negative individuals? Here, the evidence is conflicting. Several groups are proposing that nonallergic individuals have made a TH1 response. This is supported by evidence that T cells in the cord blood of individuals who are going to become allergic produce less interferon (IFN-γ) in response to nonspecific stimuli, and also by evidence that it is possible to clone mite-specific TH1 T cells from the peripheral blood of nonallergic individuals (Miles et al., 1996; Prescott et al., 1998; Warner et al., 1997; Wierenga et al., 1990). On the other hand, most investigators have not found evidence of an immune response to mite proteins in nonallergic individuals. For example, most skin test-negative individuals have no detectable IgG antibodies to purified antigens, have poor or no in vitro T cell responses, and have no immediate or delayed skin responses. These results suggest (1) that the primary issue is who makes an IgE antibody response and (2) that most individuals who are nonallergic have not made any immune response to dust mites.

Almost all of the well-defined allergens are proteins or glycoproteins, and it is not surprising that many of them have amino acid sequence homology with known enzymes (Arruda et al., 1997; Stewart and Thompson, 1996). In some cases these proteins have enzymatic activity that could play a role in their immunogenicity. For example, *Der p* I is a potent protease that has been shown to cleave CD23 and CD25 on the surface of lymphocytes in vitro (Hewitt et al., 1995; Schulz et al., 1998). Furthermore, in animal experiments, it appears that the enzymatic activity of *Der p* I can influence immunogenicity (Comoy et al., 1998). However, enzymatic activity is not a prerequisite for a protein to be a significant allergen, and there are considerable doubts about whether these enzymes are active in vivo. Some of the allergens that have homology with enzymes have not been shown to have enzymatic activity. More significantly, it has not been established that any of these proteins have enzymatic activity under the conditions that occur in the human respiratory tract. In addition, many important allergens do not have homology with enzymes, including *Der p* II. It is certainly possible that enzymatic activity is relevant to the effects of some allergens; however, it is unlikely that this property plays an important role either in inducing IgE antibody responses or in causing symptoms.

Factors Influencing Exposure

The quantities of dust mite allergen that have been found in the air of houses range from <0.2 to ≥100 ng/m^3. Thus, accurate determination of the quantity and particle size of airborne allergen is dependent on immunoassays capable of accurately measuring quantities as small as 1 ng (Chapman et al., 1987; Luczynska et al., 1989; Sakaguchi et al., 1990b). After it became clear that *Der p* I was concentrated in mite fecal particles, experiments were designed to answer whether the allergen became airborne in this form. Using various approaches to air sampling, several conclusions became clear: (1) there is very little or no airborne mite allergen in an undisturbed room; (2) the allergen that becomes airborne during disturbance is predominantly on particles ≥10 µm; and (3) allergen falls rapidly after disturbance, in keeping with an aerodynamic size of ~10 µm (Tovey et al., 1981b). In addition, allergen-containing particles identified using a micro-immunodiffusion technique have been shown to be mite fecal particles (Tovey et al., 1981a). These properties are strikingly different from those of other indoor allergens (e.g., of cat or dog). However, the size of the particle is very similar to pollen grains. A small proportion (i.e., 5–15%) of the fecal particles that are breathed through the mouth would be expected to enter the lungs, and we assume that these particles are assumed to produce the inflammatory response (Bates et al., 1966; Svartengren et al., 1987). At present the question of how allergen particles enter the lungs is not resolved, and this issue is of considerable importance since it may well define the distribution of "inflammation" and airway obstruction.

Evidence Regarding Asthma
Exacerbation and Development

Association with Sensitization

Many studies have shown an association between dust mite

sensitization and asthma (Platts-Milis et al., 1997). These studies include case-control studies in clinics, emergency rooms, and hospitals (Gelber et al., 1993); population-based studies in schools (Peat et al., 1996; Squillace et al., 1997); and prospective studies (Sears et al., 1989; Sporik et al., 1990). The results consistently show odds ratios for asthma of 6 to >12 in individuals with dust mite sensitization. However, in all studies there are a significant number of individuals who are skin test positive but not symptomatic. Using multiple regression analysis, a prospective study in New Zealand demonstrated that mite sensitization was an independent risk factor for asthma (Sears et al., 1989). In the same analysis, pollen sensitization was not significantly associated with bronchial reactivity. This pattern has now been seen in many studies and in several different countries (Table 5-1). Furthermore, similar results have been found in areas where other indoor allergens are most important (Sporik et al., 1995). That is, the relationship between sensitization and asthma is not for allergens in general but specifically for those allergens to which patients are exposed perennially in the environment. Although most of the perennial allergens are found indoors, the airborne fungus *Alternaria* is also an important factor in areas of the world where exposure persists for many months of the year (Halonen et al., 1997).

In many of these studies the odds ratios for asthma for individuals who are allergic to dust mites are very high (i.e., ≥6) (Table 5-1). However there are several features of the prospective and case control studies that need to be emphasized. In all reported studies, there were a significant number of individuals who were skin test positive to dust mites (or other indoor allergens) but did not have symptoms. In many countries or areas, dust mites are the most important source of allergens. In the south, the southeast, and the west coast of the United States, as well as in New Zealand, Australia, Japan, and the UK, mites of the genus *Dermatophagoides* are dominant. By contrast in areas where dust mites do not thrive, other allergens are important (e.g., animal dander in the mountain states of the United States and in Scandinavia; or cockroach in the apartments of Chicago, Boston, New York, and Philadelphia). In addition the studies shown in Table 5-1 illustrate the fact that in most areas sensitization to outdoor pollens is not significantly associated with asthma.

TABLE 5-1 Sensitization as a Risk Factor for Asthma (symptomatic bronchial hyperreactivity)

Country	Study Type	Dominant Allergen	Odds Ratio	Pollen	Author
United States	Prospective	Mite (cat)	19.7[*]	n.s.	Sporik et al., 1990
New Zealand	Prospective	Mite (asp.)	6.6[*]	n.s.	Sears et al., 1989
Sweden	Population	Cat, dog	3.9[*]	Birch	Ronmark et al., 1998
Australia	School(s)	Mite	≥10.0[*]	n.s.	Peat et al., 1996
United States					
Virginia	School(s)	Mite (cat, cr.)	6.6[*]	n.s.	Squillace et al., 1997
Atlanta	Acute (ER)	Mite, cockroach	8.2[*]	n.s.	Call et al., 1992
Arizona	Prospective	Alternaria	[*]	n.s.	Halonen et al., 1997
New Mexico	School	Cat, dog	6.2[*]	n.s.	Sporik et al., 1995

NOTE: asp = Aspergillus; cr = cockroach; ER = emergency room; and n.s. = not studied.

[*] p < .001.

Sensitization to storage mites (*Lepidoglyphus destructor*, *Tyrophagus putrescentiae*, *Acarus siro*, and others) is an important risk factor in occupational asthma in agricultural settings (ATS, 1998). These mites may also play a role in nonoccupational asthma from exposures in indoor environments in rural and perhaps other areas, although research characterizing exposure levels and responses is lacking.

Association with Exposure

Sampling. The measurement of allergen entering the lungs or even the measurement of dust mite allergen inhaled has proved very difficult. The essential problem is that the allergen is carried on particles that behave aerodynamically as if they are 10–25 μm in diameter. This means that in still air, they remain airborne only for a few minutes. Thus, all measurements of airborne dust mite have required artificial disturbance, and it has not proved possible to standardize disturbance (Sakaguchi et al., 1990b; Tovey et al., 1981b). Because of this, other measurements of mite allergen have been used as an index of exposure. The accepted measurement is the concentration of Group 1 mite allergen, which is *Der p* I + *Der f* I in house dust expressed as micrograms per gram of dust. The immunoassays available use monoclonal antibodies in a two-site enzyme-linked immunosorbant assay (ELISA). The significance of this measurement has been endorsed by three international workshops (Platts-Mills et al., 1997). While there are many different ways of obtaining samples, the most widely used technique is to sample reservoir dust from bedding, bedroom floors, and the living room with a hand-held vacuum cleaner.

In areas of the world that are humid persistently or for at least eight months of the year, mites flourish inside houses, and it is not unusual to find 100 or even 500 mites per gram of dust. This translates to 2 or 10 μg of Group 1 allergen per gram of dust, respectively. In areas where most houses (i.e., at least 80%) have greater than 2 μg of mite allergen per gram of dust, sensitization to these allergens has consistently been found in a large proportion (45–85%) of children with asthma. In some studies the concentration in individual houses has been shown to be an important predictor of sensitization (Peat et al., 1996; Sporik et al., 1990).

In Germany, Kuehr and his colleagues demonstrated that the concentration necessary to convert "non-atopic" children to positive skin tests to dust mite was ~60 µg/g. In the same study, children who already had a positive skin test to another allergen were at risk of becoming sensitized with exposures greater than 2 µg/g (Kuehr et al., 1994). Taken together, the results strongly support a dose–response relationship between exposure and development of sensitization, with an approximate threshold of 2 µg/g. This threshold is a statistical concept representing the concentration at which exposure to mite allergen becomes a risk for sensitization. However, there is no sense in which the threshold for mite exposure is comparable to thresholds for airborne toxic gases: (1) high concentrations of mite allergen are not toxic to nonallergic individuals; (2) concentrations below this level may cause symptoms for highly allergic or highly reactive asthmatics; and (3) the measurement is an index of exposure, which may not reveal high levels of exposure elsewhere in the house and has only an indirect relationship to exposure of the lungs.

Experimental Evidence

The earliest experiments on bronchial provocation confirmed that allergens inhaled into the lungs could produce an asthmatic response (Blackley, 1873). With the development of the nebulizer it became much easier to challenge the lungs. The exposure represents ~10^8 droplets with 0.1–10 µg *Der p* I per milliliter in a 2-minute challenge giving approximately 0.5 mL inhaled. In allergic individuals this challenge consistently gives rise to an immediate fall in FEV_1 and often produces a late response as well (Cockcroft et al., 1979). Following a bronchial challenge, eosinophils are recruited into the lungs in keeping with the events that are thought to occur in asthma. These changes can also be produced by a segmental challenge, and in these experiments the eosinophils may persist in the lung for up to two weeks after challenge (Shaver et al., 1997).

The lungs of patients with asthma are characterized by inflammation and nonspecific BHR. In several different types of studies it has been shown that these changes can be reversed under conditions that include reduction of exposure to mite allergens. Mite allergic children moved from the Netherlands; Marseille, France; or Verona, Italy to sanatoriums in the Alps have consistently shown clinical improvement and decreased nonspecific BHR. These sanatoriums have many features that are different from the children's homes, including very low mite allergen levels (Platts-Mills and de Weck, 1988). Piacentini and his colleagues (1996) in Italy have also shown that the children who spend three months in the Dolomite Mountains have a decrease in eosinophils and eosinophil products in induced sputum in parallel with decreased BHR. With adults, similar data have been obtained by moving patients from their homes in London to mite-free hospital rooms. Again, there were many other changes associated with the move to mite-free rooms (e.g., no animals, very low spore counts, increased physical activity as patients felt better). However, it is clear that BHR is reversible in many mite-allergic patients.

Conclusions: Asthma Exacerbation and Development

In mite-sensitive asthmatics, continual exposure to mite allergens is a contributing cause of exacerbations and chronic bronchial hyperreactivity. The risk of disease, however, is not a function of either the magnitude of the positive response to skin tests or the level of IgE antibody measured in serum. Therefore, in summary:

• There is sufficient evidence of a causal relationship between dust mite allergen exposure and exacerbations of asthma individuals specifically sensitized to dust mites. Continual exposure to dust mite allergens is also a contributing cause of chronic bronchial hyperreactivity.

• There is sufficient evidence of a causal relationship between dust mite allergen exposure and the development of asthma in susceptible children.

It is very difficult to prove a causal relationship between an agent that is present in the environment every day and a chronic disease. Certainly, none of the studies reported here taken on its own can be considered to prove that dust mite antigens cause asthma. However, when all of these studies are taken together the case becomes compelling. In 1965, Bradford Hill presented the argument that many different types of study should be considered together to make a case for causality. He was addressing the specific case of a dust causing chronic lung disease (Hill, 1965). When these criteria are applied to the role of dust mites in asthma, the case becomes very strong (Box 5-1). It is the combination of association, biological plausibility, provocation experiments, and the results of avoidance that create the strength of the argument.

Evidence Regarding Exposure Mitigation and Prevention

Distribution of Mites in Houses

Most houses contain at least three of the four requirements for mite growth: (1) there are multiple sites that can provide a nest for mites (i.e., carpets, sofas, mattresses, pillows, bedding); (2) the presence of humans guarantees an abundant food source in the form of skin scales; (3) in the latter half of the twentieth century, most houses are kept close to the optimal temperature for mite growth. For these reasons, the fourth requirement—humidity—is the major and often the only factor that determines whether a house has high concentrations of mites and mite allergen. In areas where humidity is high for most of the year, mites will grow well and may be found in almost any fabric including drapes and clothing, as well as traditional nests. By contrast, in truly dry areas (e.g., the mountain states [≥5,000 feet elevation]), and in Chicago in winter, mites cannot grow and houses are generally devoid of mites (Rosenstreich et al., 1997; Sporik et al., 1995). It is in areas where humidity is marginal or raised for a prolonged season that the structural features of a house may make a large difference in the concentration of mites (see Box 5-2). The three biggest factors are the position of the living area within a building; the presence of carpeting on unventilated floors, which will both trap water leaks and lead to local condensation; and buildings whose ventilation rates are so low that humidity produced by the occupants will accumulate even though the outdoor humidity is not high. The scale of these effects is well illustrated by the fact that a sample of houses and duplexes in Boston was

BOX 5-1
Evidence That the Relationship Between Dust Mite
Exposure and Asthma Is Causal

1. The association between sensitization and asthma as judged by odds ratios (i.e., ≥6) is very strong in:

- population-based studies;
- case-control studies in emergency rooms or clinics; and
- prospective studies.

2. Consistent observations have been made in different populations:

- United Kingdom, Europe, United States, Australia, Japan, Hong Kong, etc.; and
- Caucasians, Chinese, African Americans, etc.

3. The relationship is specific, (i.e., mite sensitization is associated with asthma and not with any other lung disease).
4. There is a dose–response relationship between exposure to dust mite allergens and sensitization. However, the quantitative relationship between exposure and asthma is less clear.
5. Experimental evidence exists to support the relationship:

- Challenge studies—bronchial challenge can produce eosinophil recruitment, bronchospasm, and increased bronchial reactivity.
- Studies in which decreased exposure has been maintained for six months or more have consistently resulted in decreased symptoms and decreased BHR.

6. The mechanism proposed is biologically plausible. It is clearly plausible that persistent exposure of an allergic patient will give rise to chronic inflammation with consequent bronchial hyperreactivity.

BOX 5-2
Features of Houses That Can Produce Major Changes in Mite Growth in "Marginal" Climates

Structural Features
- Ground floor level
- Fitted carpets: especially those laid on unventilated or poorly ventilated flooring
- Poor ventilation (i.e., rates ≤ 0.3 air change per hour)

Modifiable Features
- Carpets, sofas, etc.
- Excess production of water in the house (e.g., unvented cooking, humidifiers, overcrowding)
- Water leakage
- Poor cleaning habits

NOTE: Less than 5% of the housing stock has a defined route for air entry.

---

between two and three times more likely to have "high" (in excess of 10 µg/g) dust mite allergen concentrations than apartments in the same city (Chew et al., 1998).

Strategies to Control Mite Growth According to Climate

Humid Climates (i.e., eight months of the year or more with outdoor air water content of ≥5 g/g). In these climates, controlling mite growth can be achieved only by air conditioning or reducing the nests for mite growth. Air conditioning to maintain indoor relative humidity below 50% requires tight housing and is expensive in terms of energy costs. Controlling or minimizing nests can be achieved by removing carpets; reducing furniture or using surfaces such as leather; and avoiding unnecessary fabrics such as drapes, soft toys, and excess unenclosed clothing. In these areas of the country, the normal practices for bedding are effective (e.g., covering mattresses and pillows; regular hot [>130°F or 55°C] washing of all bedding).

Areas of Moderate or Seasonal Humidity. Many strategies can be helpful in controlling mite growth in these areas (see Box 5-3). During dry seasons, simply opening windows for one hour per day will ensure removal of humidity from the house (Harving et al., 1994). In addition, moving to an upper-level apartment can dramatically decrease mite exposure. The simple physical strategies normally advised are effective. In particular, the recommendations for bedding, carpets, and reducing furnishings can all help. Chemical treatment of a carpet using benzyl benzoate can be effective in controlling mite growth for a period of months (Hayden et al., 1992).

Critical issues include the nature of the fabrics used to cover mattresses and pillows and whether fabrics that breathe but block passage of mites can prevent colonization of mattresses, pillows, or furniture (Vaughan et al., 1999).

Dry Areas. Here, control of mites can be achieved simply by ensuring ventilation on a daily basis. In these areas, only very tight housing will allow accumulation of humidity within the building. Thus, this is more likely to be a problem in cold areas where the temptation to control heat loss may prevent loss of water produced by the inhabitants of a house. In the Mountain States and the Southwest, mite growth in houses is unusual.

Conclusions: Exposure Mitigation and Prevention

• There is sufficient evidence of an association between the use of a combination of the physical measures described above and a reduction in dust mite allergen levels. As noted, the most appropriate measures vary according to the type and characteristics of the indoor environment and the prevailing climate.

• There is sufficient evidence of an association between the use of a combination of the physical measures described above and an improvement in symptoms or lung function in mite-sensitized asthmatics. These have been shown to be effective at reducing symptoms in controlled trials and should be part of normal management of asthma in mite-allergic individuals.

• There is inadequate or insufficient evidence to determine whether dust mite allergen mitigation strategies have an effect on asthma development. Because dust mites cannot survive in dry environments, living in such conditions does limit the opportunity for sensitization to mite allergens.

Summary

There are many reasons why dust mites of the genus *Dermatophagoides* have played such an important role in the evidence about the role of indoor allergens in asthma. These eight-legged arthropods are invisible to the eye and do not produce any odor that we can detect. Nonetheless, they grow extremely well in houses, requiring only humidity and a nest such as bedding, mattresses, or a carpet. As a result, a large proportion of the population is unknowingly exposed to high concentrations of these highly immunogenic proteins. In keeping with experimental responses to repeated low-dose antigen, the immune response is characterized by IgE antibodies. The factors that influence the development of asthma in allergic individuals are not fully understood. However, most of the allergens associated with asthma are perennial and are predominantly indoors. Indoor or perennial exposure characteristically produces symptoms that persist for much of the day and cannot be related to exposure. Indeed there is no characteristic history of symptoms that can be used to identify patients who are allergic to dust mites. It has been suggested that year-round exposure to dust mite allergens, which are inhaled as a "few" particles per day (i.e., ~100), may be the ideal way to establish chronic inflammation of the lungs and the associated bronchial hyperreactivity, without individuals' being aware that they are exposed. Given the evidence that decreasing exposure can help control the symptoms of an allergic patient, patients who are allergic should be identified and educated. However, it is equally important to try to identify the factors that have led to an increase in the number of individuals who have asthma associated with immediate hypersensitivity to dust mites. Although this may in part reflect an increased number of allergic individuals, it appears that much of the rise is due to an increase in asthma among allergic individuals.

Research Needs

Although more is known about dust mite allergen and its impact on asthma than most indoor exposures, research remains to be done. Particularly important is additional work on the effectiveness of specific environmental interventions in limiting asthma exacerbations and development (rather than simple measurement of allergen levels). Several studies now under way are evaluating whether aggressive allergen avoidance regimes have an effect on the subsequent development of asthma (Tovey and Marks, 1999). The results of such studies will inform the question of whether primary prevention of dust mite-induced asthma is possible, although the burdensome nature of such interventions suggests they may be difficult to implement in many circumstances. The development of methods to identify individuals, especially infants, at high risk would provide the information needed to focus primary prevention activities. A major issue in this regard is whether sensitization can occur before birth.

ENDOTOXINS

Endotoxins are components of some bacterial cell walls. They are released when the bacteria die or when the cell wall is damaged. Endotoxins originally came to the attention of physicians because of their potential to cause fevers; more recently it has been established that they can cause airway inflammation and airflow obstruction at higher exposure levels.

Research on endotoxin exposure and asthma is not as mature as that for many of the other exposures addressed in this chapter. Accordingly, this section focuses primarily on the background biologic information that underlies current research into endotoxin's role in the pathogenesis of asthma.

Definition of Agent and Means of Exposure

Agent Definition and Biology

Endotoxin is the substance responsible for certain characteristic toxic effects of gram-negative bacteria. The toxic compound, lipopolysaccharide (LPS), is a structural component of the outer membrane of these bacteria, the polysaccharide portion of which represents the antigenic surface (Sonesson et al., 1994). The lipid portion of the molecule (lipid A) is essential for its characteristic toxicity. The outer portion of the polysaccharide (O-specific antigen) varies among serotypes of a single bacterial species. The core polysaccharide and lipid A are conserved within species but vary in structure and composition between species and to a greater extent between genera. Variations in lipid A structure are associated with variations in toxic potency over a wide range, and there is some evidence for qualitative variations in toxicity. Gram-negative bacterial endotoxin should not be confused with *Bacillus thuringensis* delta endotoxin (Du et al., 1999; Potekhin et al., 1999), a protein from a gram-positive bacterium that has recently been genetically engineered into certain crops.

Factors Influencing Exposure

Because gram-negative bacteria are the natural surface flora for plants and are abundant in soil (Edmonds, 1979), endotoxin is ubiquitous in the outdoor environment, particularly during the growing season. High-level exposures to endotoxin occur when organic dust is generated in agriculture and related industries such as animal feed production and in cotton mills (Rylander and Morey, 1982). Recirculating water systems can also be sources of endotoxin, and high level exposures have been recorded in industries where machining fluids and recirculated wash water are used (Milton and Johnson, 1995; Walters et al., 1994). Humidification systems are also potentially abundant reservoirs for gram-negative bacteria and thus for endotoxin (Flaherty et al., 1984; Rylander et al., 1978). Therefore, even home humidifiers can generate high levels of endotoxin exposure (Tyndall et al., 1995).

Measurement of endotoxin exposure in homes is usually performed with a *Limulus* amebocyte lysate assay because this bioassay correlates well with endotoxin measured in the rabbit pyrogen assay. However, *Limulus*-based assays are prone to interference (Milton et al., 1997) and likely underestimate exposures in organic dusts, including house dust, compared to chemical assay (Saraf et al., 1999). Furthermore, endotoxin in organic dust contains a wide variety of lipid A structures, not all of which are equally reactive in the *Limulus*-based assay (Saraf et al., 1997). These compounds may not stimulate significant production of the inflammatory cytokine interleukin-1 (IL-1), but may be able to stimulate other effects such as reversing tolerance to certain antigens (Baker et al., 1990, 1992) that may be important in directing immune responses.

Endotoxin exposures in homes have not been extensively studied. A doctoral thesis described the association of airborne endotoxin levels with home characteristics recorded on questionnaires (Park, 1999). The strongest predictors of increased endotoxin in living room air were the presence of a dog, signs of mice, a concrete floor in the living room, and mold or mildew in the bedroom during the past year. Dehumidifiers were associated with reduced airborne endotoxin. However, airborne endotoxin levels in homes were similar to those in outdoor air. The particle size distribution of endotoxin in ambient and indoor air is not known. Inhaling typical home air, containing one endotoxin unit (EU) per cubic meter, for 24 hours (a total daily dose of approximately 1 EU in a small child and 10 EU in an adult) may be the source of the low levels of endotoxin present in bronchoalveolar lavage fluid (0.1 ng/ml) (Dubin et al., 1996).

In addition to endotoxin exposure from environmental sources, endogenous endotoxin exposure may arise from infection. It has been suggested that a strong association between periodontal disease and premature birth may be caused by endotoxin exposure from gram-negative bacteria present in subclinical infections (Damare et al., 1997). While periodontal disease is not likely a source of endotoxin exposure in children, other infectious agents common in children, such as *Hemophilus* sp., may be important sources of exposure, as may commensal coliforms.

Evidence Regarding Asthma Exacerbation and Development

Biologic Evidence

It has long been recognized that endotoxin is a potent stimulus for macrophage production of (TNFα), IL-1, and a variety of other cytokines; arachidonic and linoleic acid metabolites, and reactive oxygen species (Rietschel and Brade, 1992). The receptors responsible for binding endotoxin and triggering responses are still in the process of being described. The pathway for endotoxin binding and cell activation includes an opsonin, LPS binding protein (LBP), which presents LPS to CD14 (Wright et al., 1990). After binding LPS–LBP complexes, membrane-bound CD14, attached to myeloid cells via a glycosylphosphatidyl–inositol (GPI) anchor, activates myeloid cells and soluble CD14 activates nonmyeloid (endothelial or epithelial) cells (Pugin et al., 1993). The signal is apparently transduced via the human toll-like receptors TLR4 and TLR2 in myeloid and nonmyeloid cells, respectively (Chow et al., 1999; Kirschning et al., 1998; Ulevitch, 1999; Wright, 1999; Yang et al., 1998). There may also be a role in signaling for heterotrimeric G proteins associated with the GPI anchors for CD14 (Solomon et al., 1998).

CD14 and the toll-like receptors are pattern receptors and are also involved in the recognition of peptidoglycan and lipoteichoic acid, and other microbial products (Cleveland et al., 1996; Dziarski et al., 1998; Schwandner et al., 1999). However, opsonization by LBP appears to confer approximately a hundredfold greater sensitivity to LPS compared with peptidoglycan or lipoteichoic acid (Kirschning et al., 1998; Schwandner et al., 1999; Sugawara et al., 1999).

The monocyte lineage, including macrophages and dendritic cells, has abundant membrane-bound CD14 and exhibits sensitive and prolific responses to endotoxin. Cytokines produced include IL-1, 6, 8, 10, 12; TNFα, INF-β, TGF-β; MIP-1α; CSF-1; and GM-CSF. Expression of MHC-II and B7.1 are also upregulated (Medzhitov et al., 1997; Santiago-Schwarz et al., 1989; Weatherstone and Rich, 1989). LPS also is a B cell mitogen and promotes isotype switching from IgM to IgE in the presence of IL-4 (Snapper et al., 1991). Thus, endotoxin not only serves as a potent stimulus to innate immune responses but also serves as a stimulus and bridge to cognitive immunity.

The gene for CD14 maps to chromosome 5q31.1, a candidate region for loci regulating IgE expression. Baldini and colleagues (1999) observed that a C to T transposition in the promoter region of CD14 (bp-159) was associated, in TT homozygotes, with significantly higher sCD14 levels. Among white children with positive skin tests to local antigens, TT homozygotes had lower total IgE levels. Among those with any positive skin test, the TT homozygotes had significantly fewer positive tests.

Much experimentation suggests that the net effect of endotoxin exposure is to promote TH1-type immune responses such as those typically seen in bacterial infections (Baldini et al., 1999; Fearon and Locksley, 1996) (see also Chapter 4). However, endotoxin is noted for the peculiar patterns of response to repeated exposures known as the Schwartzman phenomenon and endotoxin tolerance, for which cellular mechanisms have been described (Berg et al., 1995; Bohuslav et al., 1998). It is clear from mechanistic studies of repeated exposure that endotoxin stimulation of production of IL-10, which suppressed the TH1 response, is at least as important a phenomenon as stimulation of IL-12, which enhances this response. Furthermore, much of the data indicating that endotoxin promotes TH1 responses, except Baldini et al. (1999), come from single-exposure in vitro studies. Human data on TH1/TH2 responses to endotoxin and studies of repeated experimental exposure in vivo are sparse. Two studies suggest that under at least certain circumstances, endotoxin promotes TH2 immune responses. A study of human volunteers injected with 4 ng/kg of endotoxin (Zimmer et al., 1996) demonstrated that IL-12 levels were unchanged following injection but IL-10 levels increased, suggesting that systemic endotoxin exposure promoted a TH2 immune response. A mouse model for periodontal disease, produced by repeatedly injecting LPS at 48-hour intervals, found that after the first several injections, gingival T cells were primarily of the TH1 subgroup, based on in situ hybridization. However, after 10 injections, T cell subgroups changed from a TH1 to a TH2 predominance (Iwasaki et al., 1998). Thus, chronic exposure may produce qualitatively different responses from a single acute exposure. The route of exposure as well as dose rate may be important factors, so that the net effect on development of immune responses following chronic low-level, airborne endotoxin exposure may be difficult to predict.

Epidemiologic Evidence

Several laboratory and cross-sectional epidemiologic studies suggest that asthmatics may be sensitive to the proinflammatory effects of endotoxin at lower levels of exposure than non-asthmatics. After endotoxin exposure, increased bronchial responsiveness to histamine challenge was observed among asthmatics, but not among nonasthmatic adults (Michel et al., 1989). In a cross-sectional study of asthmatic adults, the endotoxin content of house dust was associated with increased asthma severity (Michel et al., 1996).

High levels of endotoxin in agriculture and industry are associated with both acute and chronic effects on respiratory symptoms and lung function, regardless of preexisting lung disease. The increasing body of epidemiologic evidence for respiratory effects of occupational exposure to endotoxin at 10 or more times the ambient outdoor levels has been reviewed by Douwes and Heederik (1997) and Milton (1999). One occupational study suggested that the chronic airways disease associated with daily exposure to high levels of endotoxin is characterized by an increased peak flow amplitude and thus may represent a form of asthma (Milton et al., 1996). However, except for the use of contaminated humidifiers, most homes have airborne endotoxin levels similar to those found in ambient urban and suburban air. Thus, it is not clear what effect if any exposure at these levels can have on persons who do not already possess inflamed airways with increased LBP and sCD14.

If endotoxin influences the development of the asthmatic inflammatory response, it is unclear whether the influence will be as a risk factor or a protective factor; basic laboratory research (see above) suggests that either is possible. Both German and Swiss studies suggest associations between living on a farm and decreased risk of asthma (Braun-Fahrlander, 1999; von Mutius, 1994). The authors suggest that exposure to endotoxin early in life may be protective against asthma development and that there may be a gene-by-environment interaction in creating tolerance. At this time there are no data on home endotoxin exposure and the risk of asthma development in children.

Conclusions: Asthma Exacerbation and Development

Endotoxin is associated with occupational lung disease among workers exposed to high levels. Although endotoxin is suspected at lower levels to be a trigger for asthma, and may be either a disease promoter or a beneficial exposure depending on time course, dose, and route of exposure, there are too few data on low-level exposures to draw any firm conclusions at this time. Therefore, the committee concludes:

• There is inadequate or insufficient information to determine whether or not an association exists between low-level indoor endotoxin exposure and asthma exacerbation or development.

Evidence Regarding Exposure Mitigation and Prevention

There are few data on sources in the home environment and none on the effects of interventions aimed at altering domestic endotoxin exposure. However, experimental data do exist to demonstrate that cool mist (spinning disk and ultrasonic) humidifiers can emit very high levels of endotoxin aerosol, that filtration of the mist is not effective, and that warm mist or steam humidifiers do not emit aerosols of endotoxin or other pathogenic organisms (Tyndall et al., 1995). Attempts to prevent microbial contamination of cool mist humidifiers with antifouling agents have not been successful to date (Burge et al., 1980). Thus, prevention of high-level endotoxin exposure from humidifiers would appear to be best accomplished by elimination of cool mist units. However, the impact of these units on the severity and occurrence of asthma is not known. In summary:

• There is inadequate or insufficient evidence to determine whether or not an association exists between endotoxin interventions and reduction of endotoxin levels.

Conclusions: Exposure Mitigation and Prevention

No general conclusions about the means of altering exposure to low levels of endotoxin can be drawn at the present time. However, avoiding the use of cool mist humidifiers would appear to be a simple and effective means of eliminating the risk of high-level exposure to endotoxin at home as well as to organisms associated with hypersensitivity pneumonitis (Burke et al., 1977; Ganier et al., 1980; Seabury et al., 1976; Suda et al., 1995).

Research Needs

Given the significant body of data on the exquisite sensitivity of the innate immune system to small quantities of endotoxin, the hypotheses that domestic endotoxin exposure may influence the development of the immature immune system or affect the severity of asthma warrant further investigation.

This review suggests several avenues of research directed at understanding the role of endotoxin exposure and endotoxin susceptibility in the pathogenesis of asthma. These include studies of gene–environment interactions and the risk of developing atopy or asthma, preferably with prospective assessment of endotoxin exposure from birth, improved endotoxin exposure assessment across populations likely to have significant differences in exposure, and studies of endotoxin exposure and asthma severity.

Gene–environment interactions between the CD14 polymorphism and endotoxin exposure should take into account that CD14 is a pattern receptor and thus not specific for LPS–LBP complexes. Thus, future studies should include an assessment of exposure to other bacterial products that stimulate innate immunity via CD14 such as peptidoglycan. Prospective studies will be required to determine whether endotoxin exposure early in life plays a role in determining the direction of immune system development. Studies that can compare populations with possibly larger variations in exposure to endotoxin and other components of organic dusts than can be found within an urban or suburban area would likely have increased power to detect the effects of endotoxin exposure. Because the CD14 polymorphism is associated with atopy, a focus on specific and nonspecific IgE and TH phenotypes will likely be the most important variables for these studies.

Given that the *Limulus* bioassay has limitations and that "unusual" lipid A structures dominate the composition of house and other organic dusts, additional exposure assessment methods that can detect the range of environmental LPS should be employed along with *Limulus* assays in future studies. The possibility that endogenous sources of endotoxin exposure may be important in modulating the level of tolerance to environmental exposure (or vice versa) should also be examined.

FUNGI

Definition of Agent and Means of Exposure

Introduction

There are more than 1,000,000 species of fungi, 200 different types to which people are routinely exposed. Exposure occurs universally, both outdoors and indoors, and is impossible to avoid completely. Fungal exposure (even to one type of fungus) is complex with respect to disease agents and usually includes allergens, irritants, toxins, and sometimes potentially infectious units.

These factors have led to confusion, poorly constructed studies on the role of fungi especially in the area of allergic disease, inconclusive results, and avoidance of the field by the best investigators. However, clearly and unequivocally, fungal exposure does cause allergic, toxic, and infectious disease, and it remains only to document the extent of the problem, factors leading to disease, and approaches for control.

Fungi may play a role in asthma in several ways. The most obvious of these is via fungal allergen exposure that leads to sensitization, perhaps leads to the development of asthma, and exacerbates symptoms in sensitized people. Fungi also contain and release irritants that may enhance the potential for sensitization, potentiate allergen-induced symptoms, and (possibly) exacerbate asthma in nonsensitized people. Finally, fungal toxins could play a role in modulating the immune response and may cause direct lung damage leading to pulmonary diseases other than asthma.

Definition of Agents

Nature of Fungi Fungi are eukaryotic organisms characterized primarily by their filamentous morphology and saprobic lifestyle. Fungal cells are bounded by rigid cell walls that are usually composed of chitin fibrils embedded in a matrix of $(1\rightarrow3)$ β-D-glucans and/or mannans. Cell walls may be coated externally with waxes or hydrophilic extracellular polysaccharides that carry various levels of antigenic specificity. To digest food, fungi excrete enzymes into the environment initially as probes to evaluate food availability, then to digest complex carbon compounds. These enzymes are some of the major fungal allergens. While processing organic material, the fungi produce many ancillary metabolites, some of which are highly toxic (antibiotics, mycotoxins). These compounds may accumulate in the fungal body, in spores, and in the environment. The primary mode of fungal reproduction is by airborne spores, which form a major fraction of both the outdoor and the indoor large-particle aerosol. Fungi also colonize manmade environments, releasing both spores and metabolic materials.

Fungal Allergens Fungi produce an enormous array of compounds that are potentially allergenic. Each fungus produces many different allergens of a range of potency. Table 5-2 lists the major defined allergens isolated from fungi. Others have been identified, but they are generally "minor" (i.e., few patients react to them). Many others remain to be identified.

Sources and Variability Fungal allergen production varies by isolate (strain), species, and genera (Burge et al., 1989). Different allergen amounts and profiles are contained within spores, mycelium, and culture medium (Cruz et al., 1997; Fadel et al., 1992). In addition, substrate (growth) medium strongly influences the amount and patterns of allergen production. For example, the allergen content of *Alternaria* spores produced on ceiling tiles probably differs from that of spores produced on dead grass. Fungi release proteases during germination and growth, and fungal extracts contain sufficient protease to denature other allergens in mixtures. This has been demonstrated clearly for *Alternaria* extracts (Nelson HS et al., 1996).

Cross-Reactivity Patterns Patterns of cross-reactivity among

160

CLEARING THE AIR

TABLE 5-2 Major Defined Allergens Isolated from Fungi

Fungus	Major Allergen	Nature of Allergen(s)	Reference
Aspergillus fumigatus	Asp f I	18 kD; mitogillin	Arruda et al., 1990
	Asp f III	Peroxisomal membrane protein	Crameri, 1998
Aspergillus oryzae		Alkaline serine protease	Shen et al., 1998
Alternaria alternata	Alt a I		Yunginger et al., 1980
	Alt a II		Sanchez and Bush, 1994
Cladosporium herbarum	Cla h I	13-kD glycoprotein	Aukrust and Borch, 1979
Penicillium citrinum		33-kD protein	Shen et al., 1997
Penicillium chrysogenum		68-kD protein	Shen et al., 1995
Trichophyton tonsurans	Tri t I	30-kD protein	Deuell et al., 1991
Malassezia furfur	Mal f I	36-kD protein	Schmidt et al., 1997
Psilocybe cubensis	Psi c II	23-kD protein; cyclophilin	Horner et al., 1995

fungal allergens have been examined using in vitro methods in which inhibition of heterologous assays is assumed to indicate cross-reactivity (although nonspecific assay inhibition must be ruled out). RAST is the most often used assay (O'Neil et al., 1990), although allergens derived from many genera of fungi have also been shown to cross-react using IgE immunoblot techniques (Verma et al., 1995). In addition, cross-reactivity has also been inferred from patterns of skin reactivity to fungal extracts. Although this latter case could result from cross-reactivity, it could also happen with multiple exposures leading to (separate) multiple sensitizations (O'Neil et al., 1990). Unfortunately, data on fungal allergen cross-reactivity are inconsistent and appear to depend on the specific strains used and on methods of allergen extraction.

Other Fungal Agents Exposure to some kinds of fungal spores induces inflammatory changes in the lung independent of allergy or sensitization (Rao, 1999; Shahan et al., 1998). MIP-2, lactase dehydrogenase, and myeloperoxidase are released in response to fungal exposure, and blood cell patterns change. These effects are dependent on the type of fungus and on the concentration of spores administered. The role these processes might play in the development or exacerbation of asthma remains unknown.

Glucans Fungal cell walls are composed of acetylglucosamine polymer (chitin) fibrils embedded in a matrix of glucose polymers ((1→3) β-D-glucans). The glucans may be chemically bound to the chitin or may form a soluble matrix (Sietsma and Wessels, 1981). Potent T cell adjuvants, the (1→3) β-D-glucans have been investigated as antitumor agents (Kiho et al., 1991; Kitamura et al., 1994; Kraus and Franz, 1991). They increase resistance to gram-negative bacterial infection by stimulating macrophages and affecting the release of TNFα mediated by endotoxin (Adachi et al., 1994a, 1994b; Brattgjerd et al., 1994; Saito et al., 1992; Sakurai et al., 1994; Zhang and Petty, 1994). Soluble glucans have an effect in the lung similar to that of endotoxin (Fogelmark et al., 1994). Glucans may be involved in the development of fungal-induced hypersensitivity pneumonitis by affecting the inflammation-regulating capacity of airway macrophages. They also probably play a role in organic dust toxic syndrome in workers exposed to dust that includes high concentrations of fungal spores. The possible role of glucans in the development and/or exacerbation of asthma has not been studied.

Mycotoxins Most mycotoxins are cytotoxic and interfere with protein synthesis, causing cell lysis and death. Some mycotoxins are potent carcinogens, and a few affect cell division (cytochalasins) or are estrogenic (zearalenone) or vasoactive (ergot alkaloids). Some cross the blood–brain barrier and affect the central nervous system. Some mycotoxins selectively kill macrophages (Gerberick et al., 1984; Jakab et al., 1994; Nikulin et al., 1996, 1997; Richard and Thurston, 1975; Sorenson and Simpson, 1986; Sorenson et al., 1985, 1986). A toxin produced by *Aspergillus fumigatus* inhibits macrophage functioning and may play a role in allergic bronchopulmonary aspergillosis (ABPA) by facilitating colonization of the airways of asthmatics (Amitani et al., 1995; Murayama et al., 1996). Gliotoxin, another *A. fumigatus* toxin, causes fragmentation of DNA, especially in thymocytes, and may facilitate tissue invasion leading to aspergillosis in immunosuppressed patients (Sutton et al., 1996; Waring and Beaver, 1996; Waring et al., 1997). Some fungal components directly cause the release of mediators of inflammation, including cytokines, reactive oxygen metabolites, and chemotactic factors (Shahan et al., 1998). The role of mycotoxins in the development and exacerbation of asthma has not been studied.

Means of Exposure

Measurement of Fungal Exposure Visual observation is the most frequently used approach to estimate potential for fungal exposure, with observational data often obtained from occupant questionnaires (e.g., Brunekreef et al., 1989). Observational data are limited by the fact that fungi are microscopic and are not visible until growth is extensive. Culture of air or dust samples is also often used indoors (e.g., ACGIH, 1999; Burge and Solomon, 1987; Su et al., 1992; Verhoeff et al., 1992). Since the presence of allergens does not always depend on culturability, this type of measure is likely to underestimate actual allergen exposure, as well as (possibly) confounding results with high levels of nonallergenic types. Microscopic identification and counting of spores from air samples (e.g., Delfino et al., 1997) represent the usual approach for outdoor samples and constitute the only currently available method that assesses nonculturable spores. The method is limited by the relatively few species that can be identified and the time commitment required for analysis. Although a few assays are available, no epidemiological studies have emerged that use fungal allergen measures of exposure. Other methods have been proposed that evaluate total fungal biomass (ergosterol, glucan assays) or that identify the presence of specific fungi (PCR techniques). None of these has been used in studies of asthma.

Factors Influencing Exposure

Nature of the Particles Spore walls may be hydrophobic with a waxy outer coat, or hydrophilic with an outer surface of water-soluble polysaccharides. Most fungi release single spores that range in size from 2 to 10 μm. Some spores can exceed 100 μm in length, although aerodynamic diameters are usually much smaller. Some spores are released as chains or clumps. *Cladosporium* spores are frequently encountered as large branching chains, as well as individual spores, so that particle sizes of *Cladosporium* units can range from near 1 μm to many hundreds of micrometers in diameter. However, air sampling with particle size-separating cascade impactors indicates that most fungal spores are <5 μm in aerodynamic diameter. Fungal hyphae also become airborne with disturbances such as high winds. Studies are not available that document the natural presence (or absence) of fungal allergens on particles other than intact fungal spores.

Studies of Fungal Aerosol Release Fungal aerosols may be produced through intrinsic spore discharge mechanisms or mechanical agitation (Ingold, 1971). Most indoor release relies on active disturbance. Few studies document actual types and intensities of mechanical disturbance necessary to release spores from surface growth. At least for some types, strong agitation or even direct abrasion is necessary (e.g., *Stachybotrys chartarum*), whereas for others, air movement such as that produced by a fan may be sufficient (Madelin and Johnson, 1992).

Outdoor Exposure to Fungi Many studies document the almost continuous presence of fungal spores in outdoor air, and the factors affecting prevalence for different types (e.g., AAAAI, 1998; Cross, 1997; Li and Kendrick, 1995; Munuera et al., 1998; Takahashi, 1997). Fungal spores are always present in outdoor air, although levels can be very low during periods of snow cover (Cross, 1997). Patterns of prevalence depend on seasonal and climatic factors, geography, and to some extent, human activity, especially that associated with agriculture. Because fungi are continuously present outdoors, often in concentrations far higher than those indoors, it is difficult to rule out outdoor exposure as the primary determinant of sensitization and symptoms. It is also difficult to document the presence of indoor growth using air sampling.

How Does the Indoor Environment Affect or Influence Exposure?

Penetration of Outdoor Aerosols Fungal particles probably penetrate indoor environments following the same physical principles as other types of particles. Published studies that compare indoor–outdoor relationships for fungi have not clearly addressed the relative contribution of the outdoor aerosol. Although outdoor fungal spores readily enter through open windows, few penetrate into closed environments (e.g., buildings and automobiles) (Muilenberg et al., 1991; Solomon et al., 1980).

Indoor Sources Fungi are always present in dust and on surfaces. Fungal growth occurs only in the presence of moisture. Food materials and temperature affect the amount of water required, as does the strain of fungus. Several reports present data on laboratory conditions that lead to fungal growth on building materials (Chang et al., 1996; Grant et al., 1989). With relatively concentrated inocula, some xerophilic fungi increase at relatively low substrate water activities. However, extensive growth was reported only at humidities near 100%, and most fungi require very wet conditions (near saturation), lasting for many days, to extensively colonize an environment. Human habits may affect the numbers and types of active fungal sources in buildings. Poorly maintained water-based appliances (e.g., humidifiers, vaporizers) harbor and release fungi. Wood stored indoors contains many fungi, although exposure from this source has not been studied. Moldy food is an obvious source, and indoor recoveries from air may be dominated by these small sources if recently disturbed.

Evidence Regarding Asthma Exacerbation and Development

Sensitization to Fungal Allergens

Exposure to fungi clearly plays a role in asthma. Good-quality studies have been reported that document the sensitizing potential of fungal allergens and relate fungal sensitization to the existence of asthma. Challenge studies have documented asthmatic responses in sensitized patients, and several studies have begun to make the connection between natural exposure and symptom development.

Prevalence of Sensitization

It has been estimated that about 6–10% of the population and 15–50% of atopics are sensitized to fungal allergens (Table 5-3). In studies where *Alternaria* extracts were used alone (i.e., not in a mixture), overall rates for allergy patients range from 3 to 36% and for asthmatics, 7 to 39%.

Skin test surveys most often focus on broad panels of allergens, with fungi included either as a single representative extract (usually *Alternaria*), two or three extracts, or mixtures of several different fungi. However, Galant et al. (1998), in a survey of California allergy patients, revealed that nine different fungal extracts were necessary to detect 90% of mold allergy patients. Interestingly, most studies that focus specifically on fungal sensitivity also use only a few extracts, with *Alternaria* again being the dominant type. One study (Szantho et al., 1992) suggests that the prevalence of sensitivity to fungi is increasing and attributes the increase to an increase in concentrations of outdoor fungi due to growth on pollution-weakened plants. However, no monitoring data support this hypothesis. Fungal skin sensitivity rates increase with age (Erel et al., 1998). Production of IgG antibodies as a result of allergen exposure may block the skin test response even in patients clearly experiencing symptoms with exposure (Witteman et al., 1996). Fungal allergens can produce a strong IgG response, possibly making reported incidences of skin reactivity underestimates.

Fungal Sensitization and Asthma

In a population of adults in Sweden, sensitivity to *Cladosporium* or *Alternaria* (but not dust mites) was associated with a current asthma odds ratio (OR) of 3.4 (1.4–8.5) (Norbäck et al., 1999). In an Arizona population, responses to *A. alternata* were the most frequent among asthmatic children, and a positive *Alternaria* skin

TABLE 5-3 Immediate Skin Reactivity to Fungi

Geography	Population	N
Israel	Random adults	395
Isle of Wight	All children born 1989–1990	981
Saudi Arabia	Atopic Americans	1,159
California	Allergy patients	141
Spain	7- to 68-year-old allergy patients	171
Switzerland	<12-year-old allergy outpatients	1,207
Midwest	Allergic rhinitis referrals	100
Turkey	Allergy patients	614
Denmark	Allergy patients	292
Italy	Allergic children	253
New Orleans	Adult allergy patients	150
Italy	Italians with respiratory symptoms	2,942
United States	Asthmatic inner city children	500
Taiwan	Asthmatic children (5–15 years)	195
Hungary	Extrinsic asthmatic children	700
Japan	Asthmatic children (3–18 years)	94
Thailand	Asthmatic children	100

Fungi	Prevalence (%)	Author
Mold (*Aspergillus, Penicillium, Alternaria, Cladosporium*)	10.9	Katz et al., 1999
Alternaria, Cladosporium	6	Tariq et al., 1996
Alternaria	36	Suliaman et al., 1997
Molds	11–22	Galant et al., 1998
A. alternata	14.5	Sastre et al., 1996
Boletus, Coprinus, Pleurotus	1	Helbling et al., 1998
Mold (*Alternaria, Helminthosporium, Aspergillus, Candida, Curvularia*)	44	Corey et al., 1997
A. fumigatus	26	Guneser et al., 1994
A. iridis	3.1	Frost, 1988
C. herbarum	0.7	
Alternaria	9.48 only: 3.55	Angrisano et al., 1987
Basidiomycetes	32	Lehrer et al., 1986
Cladosporium,	6	
Alternaria,	13	
"Fungi"	58	
Alternaria	10.4–29.3	Corsico et al., 1998
Alternaria, Penicillium	39 (combined: 41)	Eggleston et al., 1998
Alternaria	24.4	Wang and Chen, 1992
Mold (*Alternaria, Phoma betae, Aspergillus, Cladosporium*)	1977: 10 1985: 30.4 1987–1988: 38.5	Szantho et al., 1992
Aspergillus (restrictus)	8.5	Sakamoto et al., 1990
A. fumigatus	8.5	
A. alternata	16.0	
Alternaria	7	Kongpanichkul et al., 1997
Cladosporium	7	
Penicillium	3	
Aspergillus	2	

test was the only independent association with asthma (Halonen et al., 1997). In a University of Virginia study, *Alternaria* sensitivity was also a significant independent risk factor for asthma in school children in Charlottesville, Virginia, and Los Alamos, New Mexico, but not in Albemarle County, Virginia (Perzanowski et al., 1998). Of the 1,218 children born on the Isle of Wight, 6% of the 918 4-year-old children tested had positive skin tests to *Alternaria* and/or *Cladosporium*. In these 61 children, a positive test to *Alternaria* was associated with a diagnosis of asthma (Tariq et al., 1996). In Costa Rica, where the prevalence of childhood asthma is 20–30%, dust mite, cat, *Alternaria*, and *Cladosporium*-specific IgEs predict asthma (Soto-Quiros et al., 1998). A case-control study in Denver (acute asthmatics 3–16 years old) versus matched controls used RAST to document specific IgE levels: 45% of asthmatics versus 4% of controls had high IgE to *Alternaria* (Nelson RP et al., 1996). In a large (6,394 children) questionnaire-based study, Peat et al. (1995) reveal that among asthmatic children, *Alternaria* sensitivity rates are higher inland in New South Wales, but in the damp coastal climate, sensitivity to dust mites is most prevalent. In 4,295 6- to 25-year-olds in the second National Health and Nutrition Examination Survey (NHANES II) cohort, asthma was associated with sensitivity to dust mite (OR = 2.9 [1.7–5]); and *Alternaria* (OR = 5.1 [2.9–8.9]) (Gergen and Turkeltaub, 1992). Retrospective investigation of asthma deaths in teenagers revealed that a large majority had positive skin tests to *Alternaria*, and the authors concluded that sensitivity to *Alternaria* is a risk for severe asthma and death from an attack (O'Hollaren et al., 1991). (The authors concluded that exposure to *Alternaria* is a risk factor, although they did not measure exposure.) Many studies focusing on asthma ignore fungi completely (e.g., Gottlieb et al., 1996).

Immunotherapy Data

Further evidence for the role of fungal sensitization in symptomatic asthma is provided in immunotherapy data. Bousquet and Michel (1994) have reviewed the overall literature on immunotherapy. The literature on mold immunotherapy has been reviewed by Dhillon (1991) and Bonifazi (1994). Immunotherapy with partially characterized and standardized *A. alternata* allergens in a small case-control study has been shown to decrease asthma symptoms overall and during challenge with relevant allergens (Cantani et al., 1988; Horst et al., 1990). In both of these studies, children who were sensitive only to *Alternaria* (by skin prick test) were treated. Dreborg et al. (1986) selected 30 children who were skin test, RAST, and provocation positive to *Cladosporium*. Most had other sensitivities as well. He treated 16 of these children; the remaining 14 served as controls. Although symptom scores remained similar for the two groups (probably because of multiple sensitivities), nasal and bronchial challenge responses and medication usage were significantly lower in the treated versus the control group.

Of the 16 children treated in the Dreborg (1986) study, 13 experienced general reactions during treatment. Kaad and Ostergaard (1982) report that 19% of 38 children treated with fungal extracts had to be withdrawn from immunotherapy because of apparent type III reactions. All had slightly elevated IgG (precipitating antibodies) to the relevant extracts before hyposensitization was begun and developed increased titers during therapy.

Overall, carefully controlled studies of mold immunotherapy show good effect. Negative studies are small, with too few patients and nonstandardized allergens. These data support the notion that there is a relationship between fungal allergen sensitization and symptoms of asthma. However, there are too few studies to confirm that mold immunotherapy is an effective public health intervention.

Fungal Exposure and Asthma

Challenge Experiments Challenge tests with *Stemphylium* in children with positive skin and RAST tests to *Stemphylium* extracts resulted in bronchial responses in 13 of 59 children (12 others had nasal responses) (Lelong et al., 1986). Malling (1986) used positive bronchial challenge in adult asthmatics with *Cladosporium* as a patient selection criterion for his studies on relationships between natural exposure and symptoms. Licorish et al. (1985) reproduced symptoms of asthma with *Penicillium* spore challenges.

Measured Natural Exposure

Outdoors Neas and colleagues (1996) associated pulmonary function changes in Pennsylvania school children with measured outdoor concentrations of several kinds of fungal spores. Peak flows, symptoms and time spent outdoors were recorded daily, and 24-hour average outdoor spore concentrations were measured using a Burkard spore trap. Changes in morning peak flow of –1 liter/min (CI –1.9 to 0.2) were associated with increments of 10,000 *Cladosporium* spores in unselected school children. Likewise, increases of 50 *Epicoccum* spores was associated with decrements in morning peak flows of 1.5 liter/min (CI –2.8 to –0.2). *Epicoccum* exposure was also associated with incidence of morning cough (OR = 1.8, CI = 1.0–3.2). Delfino et al. (1997) report associations between asthma severity (symptom scores and peak expiratory flow rates) and exposure to outdoor fungal spores measured with a spore trap. Symptoms were associated with total fungal spore increases of 4,000 spores per cubic meter, with a decrease in evening peak flow of 12 L per minute. Analysis of specific fungal genera increased the associations. The most important fungal correlates were *Alternaria*, basidiospores, and hyphal fragments. Malling (1986) related daily symptom scores to *Cladosporium* counts during the autumn mold season. He studied 24 adult asthmatics with positive provocation tests to *Cladosporium* extracts. Significant associations with spore counts were found for symptom scores and medication use. Although some patients were also sensitive to *Alternaria*, *Alternaria* counts did not relate to health outcomes in this population.

Indoors Su et al. (1992) describe relationships between indoor measured levels of culturable *Aspergillus* and asthma in Topeka, Kansas schoolchildren. They used culture plate impactors to collect short "grab" samples in homes of children recruited through the public schools. The relationship between *Aspergillus* and asthma symptoms was non-linear, possibly reflecting the different species of *Aspergillus* that are common in indoor environments. Garrett et al. (1998) report associations between measured concentrations of fungal spores (counted) and culturable fungi, observed indicators of dampness and molds, fungal sensitivity, and respiratory symptoms. Eighty households with 148 children (36% asthmatic) were sampled 6 times for airborne fungi. *Penicillium* exposure was associated with asthma, and *Aspergillus* exposure with atopy. *Cladosporium* and *Penicillium* exposure were associated with the presence of fungal allergy. No associations were seen for total fungal counts. Exposure measures to specific fungi showed a stronger association with symptoms than did dampness indicators.

Other studies of indoor fungi and asthma report comparisons between fungal levels in homes of asthmatics compared to homes of control subjects. Li and Hsu (1997) compared culturable fungi in homes of 46 asthmatic children, 20 atopic (presumably nonasthmatic) children, and 26 nonatopic controls. Although fungal concentrations were highest in the asthmatic and control homes, *Cladosporium* concentrations were higher in asthma homes than in control homes. In a study by Horak et al. (1996), *Bacillus, Aspergillus,* and total fungal concentrations were higher in homes of asthmatics than in control homes. These kinds of studies are compromised by the fact that families of asthmatic children often take measures to reduce exposure to allergens.

Visible Mold, Odors, and Dampness

The majority of studies that attempt to associate fungal exposure and asthma rely on reports of visible mold or other dampness indicators. Dampness is clearly associated with a variety of respiratory symptoms.

In addition, Williamson et al. (1997) report a correlation between mold growth indicators and asthma (196 age- and sex-matched subjects); $r = 0.23, p < .035$. These researchers also identified a relationship between severity of asthma and total dampness ($r = 0.3, p = .006$), and mold growth ($r = 0.23, p = .35$). A reduction in FEV_1 of 10.6% (CI 1.0–20.3) was associated with living in a damp home.

Dampness and even visible mold growth could be indicators for dust mite allergen exposure as well as fungal exposure. However, Garrett et al. (1998) report that homes with musty odor, water intrusion, high humidity, limited ventilation, and visible mold growth had higher fungal spore concentrations than dry homes.

TABLE 5-4 Dampness and Mold Indicators and Respiratory Symptoms

Reference	Population	Environmental Indicator	Symptom	OR (95% CI)
Jaakkola et al., 1993	2,568 children	Mold odor	Any respiratory symptom	2.54–8.67
		Any dampness or mold	Wheezing	2.62 (1.39–4.39)
			Phlegm	2.20 (1.27–3.82)
			Persistent cough	2.17 (1.39–3.39)
Slezak et al., 1998	1,085 head start children	Self-reported dampness or mold	Self-reported diagnosed asthma	1.94 (1.23–3.04)
Hu et al., 1997	2,041 young adults	Self-reported mold growth	Asthma prevalence	2.0 (1.2–3.2)
Brunekreef et al., 1989	4,625 children	Reported molds	Symptoms and	1.27–2.12
			pulmonary function	FEF 25-75 –1.6%
Strachan, 1988	873 children	Visible mold	Wheeze in past year	3.0 (1.72–5.25)
Nicolai et al., 1998	155 children with asthma	Dampness	Bronchial hyperreactivity in adolescence	16.14 (3.53–73.71)
Andriessen et al., 1998	1,614 children	Self-reported molds	PEF variability	1.92 (1.18–3.12) current home
Williamson et al., 1997	102 asthmatics,	Self-reported dampness	Asthma	2.11 (1.29–3.47), past home
	196 matched controls	Observed dampness		3.03 (1.65–5.57)

NOTE: CI = confidence interval; FEF = forced expiratory flow; OR = odds ratio; and PEF = peak expiratory flow.

Visible mold was associated with high concentrations of *Cladosporium* spores, but not total spores. Pasanen et al. (1992) report that airborne *Cladosporium* and yeast counts were significantly higher in damp than in dry residences. *Cladosporium*, in particular, appeared to be derived from indoor sources. Nicolai et al. (1998) report that the relationship between dampness and bronchial hyperreactivity in adolescents in Munich remains when controlled for dust mite allergen exposure, providing indirect evidence for a possible role of fungi.

Conclusions: Asthma Exacerbation and Development

The extent of fungal-related asthma remains unknown, and the exposure parameters leading to fungal-related asthma development or exacerbation are not clear. This is due (in part) to inadequate diagnostic and environmental testing procedures and lack of adequate study. Sensitization (i.e., positive skin test) to fungal allergens is associated with the presence of asthma in both children and adults. Although the evidence regarding asthma development is provocative, it is inclusive. In summary:

• There is sufficient evidence of an association between fungal exposure and symptom exacerbation in sensitized asthmatics. Exposure may also be related to nonspecific chest symptoms.

• There is inadequate or insufficient evidence to determine whether or not there is an association between fungal exposure and the development of asthma.

Evidence and Conclusions: Exposure Mitigation and Prevention

As with any relatively large-particle aerosol, fungal spores can be prevented from entering enclosed spaces. However, the health effects of such prevention have not been studied. Fungi are difficult to kill, and dead fungal material probably contains allergens, although this has not been thoroughly tested. Terleckyj and Axler (1987) discuss the effects of various disinfectants on fungi. Chlorine dioxide, glutaraldehyde, and ethyl alcohol affected most fungi after 15 minutes of contact time. However, *Aspergillus fumigatus* was highly resistant to most disinfectants, and a quaternary ammonium compound and an iodophor were the least effective for all fungi.

It is possible to physically remove active fungal growth from indoor environments. While such removal is undoubtedly advisable, its health impact has not been studied. Rautiala et al. (1998) studied spore levels during removal of fungal-contaminated material. They found that negatively pressurized containment is necessary to prevent the spread of spores to unaffected parts of the building and that, within the containment field, spore levels may remain high and workers should wear protection. Overall, mitigation and prevention of fungal growth in the indoor environment has not been studied, either from the viewpoint of effectiveness or with respect to health impact. In summary:

• There is limited or suggestive evidence of an association between the fungal removal measures described above and a reduction in the levels of fungi in the indoor environment. The paucity of studies with adequate exposure characterization and the possibility that cleaning may not eliminate the problematic component of fungal exposures prevent a more confident conclusion from being drawn.

• There is inadequate or insufficient evidence to determine whether or not an association exists between fungal control measures and improvement in symptoms or lung function in sensitized asthmatics.

Research Needs

Few fungal allergens have been identified, and patterns of cross-reactivity among fungal allergens have not been documented. Standardized methods for assessing exposure to fungal allergens are essential, preferably based on measurement of allergens rather than culturable or countable fungi. Acquisition of these data is a necessary step before adequate estimates of the role of fungal allergen in asthma can be documented.

Studies seeking to find environmental factors that either lead to the development of asthma or precipitate symptoms in existing asthmatics must include good measures of fungal exposure.

No studies have attempted to control exposure to fungal allergens either indoors or out. Intervention studies that seek to control indoor exposure to fungi are especially needed.

INFECTIOUS AGENTS

Infectious agents are unlike other indoor exposures addressed in this report because their source is the occupants themselves. They are nonetheless an appropriate exposure to address because they contribute to the overall risk of asthma disease outcomes from indoor exposures and may have some role in the observed increase in asthma incidence and mortality.

There are a number of infectious agents that have been associated with asthma (Johnston et al., 1995; Pattemore et al., 1992; Shaheen, 1995). This section focuses on four that have received particular attention from researchers: two viral agents—rhinovirus and respiratory syncytial virus—and two bacterial agents—chlamydia and mycoplasma.

Rhinovirus

Definition of the Agent and Means of Exposure

Rhinovirus is the medical term used to designate a large group of viruses responsible for a variety of respiratory infections including the common cold. There are more than 120 serotypes of the rhinovirus, of which few have been associated with lower-respiratory abnormality. Rhinovirus is transmitted through the respiratory route, with virus particles shed in nasal secretions and spread by coughing and sneezing. It can also be transmitted through direct contact with tissues or hands.

Evidence Regarding Asthma Exacerbation and Development

Widespread clinical experience and epidemiologic research indicate that rhinovirus is associated with wheezing and exacerbations of asthma in established asthmatics (Johnston, 1997; Johnston et al., 1995; Lemanske et al., 1989; Micillo et al., 1998; Rakes et al., 1999). The evidence of this is especially strong for children. The mechanism by which the virus traffics from the upper airway to the lower airway to cause asthma symptoms is not known. It is speculated that individuals with asthma may make more of a specific cytokine than those without asthma, leading to increased airway inflammation and the development of symptoms. Alternatively, individuals with asthma and those without asthma may make the same amount of cytokines in response to viral infections, but these cytokines may react differently in individuals with asthma and produce different effects in the airway. Individuals with asthma have a different cytokine profile from those without asthma, which may result in a different clinical phenotype.

Studies have not identified an association between rhinovirus infection and asthma development. Martinez (1995) observes that most infants who wheeze during viral infections become symptom free later in life and suggests that viral infections in infants are likely to play a minor role in the subsequent development of asthma. There is burgeoning research interest in this issue, however, with investigators suggesting both the possibility of a mechanistic association and the prospect that certain viruses may have a protective effect (Grunberg and Sterk, 1999; Martinez, 1994).

Respiratory Syncytial Virus

Definition of the Agent and Means of Exposure

Respiratory Syncytial Virus (RSV) belongs to the Paramyxoviridae family and to the genus *Pneumovirus*. Human paramyxoviruses are a common cause of respiratory disease in children, with RSV particularly important in infants. RSV has been found in every geographical area examined for evidence of infection (Glezen et al., 1981; Hall et al., 1979), and RSV infection appears to have similar characteristics in areas with widely differing climates. Outbreaks of infection occur yearly, typically in the spring and winter. Indeed, RSV is the only viral respiratory agent that can be relied on to produce a sizable crop of infection each year. Socioeconomic status and race or ethnicity appear to influence the risk of RSV bronchiolitis (Glezen et al., 1981).

Evidence Regarding Asthma Exacerbation and Development

The study of RSV and asthma exacerbations is complicated by the problematic diagnosis of asthma in infants and young children. However, viruses in general have been identified as responsible for most wheezing illnesses and asthma exacerbations occurring in childhood, with RSV among the predominant organisms recognized (Hegele, 1999; Pattemore et al., 1992). An association has also been noted for adults (Nicholson et al., 1993).

Significant attention has been paid to the possibility of an association between RSV bronchiolitis in infancy and later development of asthma. RSV bronchiolitis is associated with the development of high titers of virus-specific IgE in respiratory secretions during both the acute and the convalescent stages of illness in some patients (Robinson et al., 1993; Welliver and Duffy, 1993). An exaggerated RSV IgE response at the time of RSV bronchiolitis in infancy is associated with recurrent wheezing in later childhood (Scott et al., 1984). Various mechanisms may be proposed to explain this association. These include direct induction of release of inflammatory mediators from pulmonary mast cells, participation in a cascade of mediators of airway obstruction, reflection of general IgE hyperresponsiveness to multiple allergens, and induction of persistent T helper type 2 responses (Roman et al., 1997). Although an exaggerated RSV IgE response to RSV infection during infancy may correlate with recurrent wheezing during a child's early years, this may not represent asthma per se but rather repeated similar responses to multiple infections with RSV and other respiratory pathogens (Welliver et al., 1986). A study by Sigurs and associates (1995) tends to support the hypothesis that RSV infection may cause or promote asthma by inducing atopy in infected individuals. Since the rate of sensitization among controls with and without IgG antibodies to RSV is similar, an encounter with the virus without development of bronchiolitis does not seem sufficient to increase the risk of sensitization. Other investigators have reached different conclusions about the relationship between RSV infection in infancy and the development of atopy and asthma later in childhood. In a study by Pullan and Hey (1982), the frequency of positive skin tests for common allergens was similar for individuals with previous bronchiolitis and for controls. Murray and colleagues (1992) found no difference in skin test reactivity to dust mites between bronchiolitis patients and controls.

A precise mechanism by which RSV bronchiolitis might induce allergies and asthma has not been elucidated (Dezateux et al., 1997). However, one possible pathogenic mechanism could be through increasing the synthesis of IgE since it has been demonstrated that RSV bronchiolitis can induce an IgE antivirus response and that RSV-specific IgE responses in infancy are associated with later recurrence of wheezing. Welliver and colleagues (1981) found that IgE titers to RSV in nasal secretions were highest in patients with evidence of airway obstruction. Chang and colleagues (1990) prospectively followed 38 of these infants for 48 months after an initial episode of RSV bronchiolitis. Only 20% of infants with undetectable titers of RSV IgE had subsequent episodes of documented wheezing. In contrast, 70% of children with high RSV IgE antibody titers experienced wheezing. In a four-year prospective study of 13 infants of history-positive bilaterally allergic parents, Frick and colleagues (1979) noted that viral infections including RSV occurred one to two months before the onset of allergic sensitization in 10 of the 11 children who subsequently became atopic.

Despite the strong correlation between virus-specific IgE, mediator release, and the clinical pattern of acute and recurrent disease, it is premature to conclude that virus-specific IgE is causal in this process. Moreover, whether those infants who had persistent wheezing and/or airway obstruction were concurrently both sensitized and exposed to indoor allergens is not known.

Chlamydia

Definition of the Agent and Means of Exposure

There are two forms of the bacterium *Chlamydia* that have been examined in relation to asthma. *Chlamydia trachomatis* is a sexually transmitted infectious disease that may be spread from mother to child during birth and perhaps through close contact with secretions of an infected individual. *Chlamydia pneumoniae* is an important cause of adult respiratory disease including pneumonia, bronchitis, sinusitis, and pharyngitis and may be associated with atherosclerosis (Kalman et al., 1999). A 1991 editorial in the *Journal of the American Medical Association* suggested that *C. pneumoniae* infection might provide an explanation for the increased incidence of asthma (Bone, 1991).

Evidence Regarding Asthma Exacerbation and Development

An anecdotal report suggests that *C. trachomatis* produces asthma-like symptoms in infants that may be treated successfully with antichlamydial antibiotic therapy (Bavastrelli et al., 1992). Björnsson and colleagues (1996) report that serological signs of a previous *C. trachomatis* infection were found significantly more often in subjects who reported having had asthma at some time, asthma during the past year, wheezing during the past year, and bronchial hyperresponsiveness. It is unclear, however, whether such reports uniquely implicate *C. trachomatis* or are unrecognized *C. pneumoniae* infections since the two species are cross-reactive.

Chlamydia pneumoniae infection has been associated with asthma exacerbations in both adults and children. Björnsson and colleagues (1996) found a statistically significant relationship between current or recent *C. pneumoniae* infection and wheezing. Johnston (1997) found that immune responses to *C. pneumoniae* were positively associated with an increased frequency of asthma exacerbations in a group of 9- to 11-year-olds. Cunningham and colleagues (1998) report increased numbers of asthma exacerbations in children with higher local antibody responses to *C. pneumoniae*. Miyashita and colleagues (1998) found a significantly higher frequency and geometric mean titer of *C. pneumoniae* antibodies in adult patients with exacerbations of asthma. However, Cook and colleagues (1998) did not find an association between asthma exacerbations (termed "acute asthma" in the study) and seropositivity in a two-year study of 123 adult asthmatics and 1,518 controls.

Several groups of researchers have reported a serological association between *C. pneumoniae* and chronic adult asthma in studies that controlled for one or more known confounders. Hahn and colleagues (1991) found that patients with serologically confirmed *C. pneumoniae* infection were more likely than seronegative patients to develop bronchial asthma subsequent to their infection. A later study of 163 adolescents and adults by Hahn and McDonald (1998) had the same findings, leading the authors to conclude that acute *C. pneumoniae* respiratory tract infections in previously unexposed nonasthmatic individuals can result in chronic asthma. Cook and colleagues (1998) found a statistically significant association between severe chronic asthma and seropositivity in their two-year study of 123 adult asthmatics and 1,518 controls. Johnston (1999) reports a high prevalence of *C. pneumoniae* infection in a group of 9- to 11-year-old asthmatics. Von Hertzen and colleagues (1999) found asthma to be significantly associated with IgG antibody levels to *C. pneumoniae*, with the strongest association for nonatopic, long-standing asthma.

However, not all studies find this association. Kraft and colleagues (1998) did not find *C. pneumoniae* in the lungs or airways of 18 asthmatic subjects. Larsen and colleagues (1998) found that a group of 22 asthmatics did not differ from 55 controls with relation to two markers of *C. pneumoniae* infection. The authors note two weaknesses in many of the studies reporting a relationship: (1) a diagnosis of asthma was seldom well-established in these cases, and (2) no pathobiologic mechanism was proposed to account for initiation in adults (production of specific IgE was noted as a possible initiating mechanism in children).

At present (late 1999) there is a debate in the literature over the meaning of these studies. Although their results are consistent with an association between *C. pneumoniae* (and perhaps *C. trachomatis*) and asthma development, present data are insufficient to distinguish this premise from other reasonable hypotheses, notably whether asthma predisposes individuals to other chronic respiratory infections.

Mycoplasma

Definition of the Agent and Means of Exposure

Mycoplasma pneumoniae is a bacterial infection that is responsible for a number of respiratory diseases including tracheobronchitis, rhinitis, pharyngitis, otitis, and a form of pneumonia. Over the past several years there have been multiple anecdotal reports and epidemiologic studies suggesting an association between *M. pneumoniae* infection and exacerbations of asthma (e.g., Seggev et al., 1986). Among the more recent studies, Freymuth and colleagues (1999) found that *C. pneumoniae* and *M. pneumoniae* infections—alone or in combination—were present in nearly 82% of asthmatic exacerbations. However, Cunningham and colleagues (1998) and Johnston (1997, 1999) found no evidence of a role for *M. pneumoniae* in acute asthma exacerbations.

Evidence Regarding Asthma Exacerbation and Development

The literature regarding *M. pneumoniae* and asthma development is sparse. Yano and colleagues (1994) suggested in a case report that an adult male may have developed asthma as a result of *M. pneumoniae* infection. Kraft and colleagues (1998) found *M. pneumoniae* present in the lower airways of chronic stable asthmatics with significantly greater frequency (10 of 18) than in controls (1 of 11).

Conclusions: Asthma Exacerbation and Development

• There is sufficient evidence of an association between infection with rhinovirus and exacerbation of asthma.
• There is limited or suggestive evidence of an association between infection with RSV and exacerbation of asthma.
• There is limited or suggestive evidence of an association between infection with *Chlamydia pneumoniae* and the exacerbation of asthma.
• There is inadequate or insufficient evidence to determine whether or not there is an association between infection with *Chlamydia trachomatis* and exacerbation of asthma.
• There is limited or suggestive evidence of an association between *Mycoplasma pneumoniae* infection and exacerbation of asthma.
• There inadequate or insufficient information to determine whether or not there is an association between infection with rhinovirus and the development of asthma in infants. There is limited or suggestive evidence of __no__ association between infection with rhinovirus and the development of asthma in older children and adults.

• There is limited or suggestive evidence of an association between infection with RSV and the development of asthma.

• There is inadequate or insufficient information to determine whether or not there is an association between *Chlamydia trachomatis* or *Chlamydia pneumoniae* infections and the development of asthma.

• There is inadequate or insufficient information to determine whether or not there is an association between *Mycoplasma pneumoniae* infection and the development of asthma.

Evidence Regarding Exposure Mitigation and Prevention

A primary infection prevention strategy is the avoidance of transmission and exposure through proper personal hygiene practices. Vaccination has also been mentioned as a means of prevention. Although the data are conflicting, breast feeding offers a possible mechanism of passive immunization of infants that may play some role in protection against infection. Further studies are needed to evaluate this notion. A detailed discussion of these strategies and their effectiveness is outside the scope of this report.

Some data are available on the characteristics of RSV that may inform infection mitigation and prevention strategies for this agent. RSV withstands changes in temperature and pH relatively poorly. Only 10% of RSV remained after the virus was exposed to 55°C for 5 minutes. At 37°C, the virus was stable for 1 hour, and only 10% of the infectivity remained after 24 hours. RSV also withstands an acid medium poorly, and the optimal pH is 7.5. Ether, chloroform, and a variety of detergents such as sodium dodecyl sulfate quickly inactivate the virus. At room temperature, RSV in the secretions of patients may survive on nonporous surfaces, such as countertops for 3–30 hours, depending on the humidity. On porous surfaces such as cloth and paper, survival is shorter, usually lasting less than an hour. The infectivity of RSV on the hands is variable from person to person but usually lasts less than one hour. The survival of RSV in the environment appears to depend in part on the drying time as well as on the humidity. These data suggest that proper cleaning of objects and surfaces exposed to RSV may serve to limit exposure to the virus.

Aspects of the indoor environment may also increase the risk of respiratory infection. Building characteristics would have an indirect effect on asthma symptoms if these characteristics influenced the prevalence of the relevant respiratory illnesses among building occupants. Crowded living conditions—found more often among urban residents and lower-income individuals in the United States—may facilitate the spread of infectious disease. Certain indoor spaces such as day care facilities and schools may also present more favorable conditions for their spread. One study suggested higher rates of transmission of infectious pulmonary disease in more tightly constructed military barracks (Brundage et al., 1988). As reviewed by Fisk (1999), the results of ten studies suggest that the prevalence of respiratory illness can be influenced significantly by building characteristics such as ventilation rate, space sharing, or occupant density, with relative risks typically between 1.2 and 1.5. None of these studies have confirmed that the responsible agents were rhinoviruses, and only one older study involved children; therefore, this indirect linkage of building characteristics to asthma exacerbation remains theoretical. Nevertheless, the evidence of an indirect link is sufficient to warrant further investigation.

Conclusions: Exposure Mitigation and Prevention

• The committee declined to draw conclusions about the effectiveness of specific personal hygiene or medical practices in mitigating or preventing exposure to infectious agents because it believes this to be outside the scope of this report.
• There is inadequate or insufficient information to determine whether or not there is an association between building characteristics that reduce close contact between individuals and decreased spread of infectious diseases. There is also inadequate or insufficient evidence of an association with ventilation rates.

Research Needs

Numerous studies suggest an association between the infections discussed in this section and asthma exacerbations, although uncertain ascertainment of asthma and questions about the identity of the specific infections responsible limit the confidence with which some conclusions can be drawn. Advances in analysis techniques that allow more sensitive and confident identification of viruses, such as PCR and ELISA, will facilitate research on this topic. These advances will also aid studies of other viruses that may be associated with asthma such as adenovirus, coronavirus, cytomegalovirus, and parainfluenza.

Research on the possible association between infectious agents and asthma development is continuing and is encouraged. There are gaps in the knowledge concerning the mechanism(s) by which agents may promote asthma and whether particular interventions aimed at limiting infections result in decreased rates of asthma. Among the interesting questions are whether the lower respiratory tract acts as a potential reservoir for common respiratory viruses and whether maternal immunization has the potential to protect both the mother and the infant. Research on the impact of building characteristics on the transmission of infectious agents, which is in its infancy, may yield important public health benefits.

HOUSEPLANTS

Definition of the Agent and Means of Exposure

Houseplants have the potential to release pollen, sap, and other plant parts; arthropod pest allergens; and fungi.

Houseplant Pollen

The data on houseplant pollen allergy is restricted to the occupational literature and to case reports. Allergy to some flower pollens has been clearly documented in occupational situations and could occur in residential environments with cut flowers (de Jong et al., 1998). Co-sensitization with some houseplant pollen allergens and some outdoor allergens is extremely common (e.g., mugwort, daffodil) and may indicate that these plants share allergens. In cases of this sort, asthma related to outdoor pollen exposure could be exacerbated by houseplant pollen, although very little pollen is produced by most houseplants and close contact would be required.

Other Plant Parts and Fluids

Exposure to plant materials during the production of plant extracts as medicines has led to occupational asthma (Giavina-Bianchi et al., 1997). The expanding use of herbal medicines may lead to an increase in these types of exposures. Several papers report cases of houseplant (in particular, *Ficus*) latex causing contact sensitization and even inhalation allergy. The available data are restricted entirely to the case study literature, and only two reports were retrieved (Diez-Gomez et al., 1998; Schenkelberger et al., 1998).

Arthropods

Arthropods that inhabit plants (particularly spider mites) may release allergens. However, again, all data are in the occupational case study literature, and no information is available on the extent of the problem (Orta et al., 1998). The potential for cross-reactivity with dust mite allergens has not been evaluated.

Fungi

Several reports implicate houseplants as sources of fungi in indoor environments, with a focus on human infectious agents rather than allergens. These studies, in general, rely on source (soil) sampling and do not relate exposure to any allergic disease (Staib, 1992, 1996; Staib et al., 1978a, b, 1980; Summerbell et al., 1989). Burge and colleagues (1982) found no relationship between the presence of houseplants and either concentration or type of fungal spores or pollen in the air.

Conclusions: Asthma Exacerbation and Development

There is no evidence that exposures from houseplants lead to the development of asthma. Although patients occasionally complain of worsening symptoms related to houseplants or exposure to cut flowers, no studies have been conducted to document this connection. While some epidemiologic studies suggest that occupational exposure to plant extracts may lead to allergic diseases including asthma, the relevance of this literature to casual, non-occupational exposure is not clear. Mites or fungi that may be associated with houseplants could be involved in asthma development or exacerbation, but there is no evidence bearing on this hypothesis.

In summary:

• There is inadequate or insufficient evidence to determine whether or not an association exists between exposures from houseplants and the exacerbation or development of asthma.

Evidence and Conclusions: Exposure Mitigation and Prevention

The committee did not identify any literature regarding the effectiveness or impact on asthma outcomes of measures intended to limit houseplant exposure.

Research Needs

Further research is needed to determine whether or not houseplants release fungal spores into the air. This research will benefit both the allergy community and the infectious disease literature. Additionally, research should be conducted to determine what risks, if any, are associated with occupational exposure to plant materials.

POLLEN

Definition of the Agent and Means of Exposure

Introduction

Pollen exposure has long been recognized as a stimulant for symptoms of allergic disease. Allergic rhinitis (hay fever) has been considered the primary outcome, and outdoor environments are the major source of exposure. The role of pollen in the development and exacerbation of asthma has received relatively little direct attention, and very few studies are available that document indoor exposure to pollen allergens.

Nature of the Agent

Structure and Function Pollen is the male reproductive structure of flowering plants and gymnosperms. Pollen grains are usually unicellular, with a rigid and highly resistant cell wall formed of a complex polysaccharide-based substance called sporopollenin. Each pollen grain contains the systems required for recognition of genetically relevant female flowers and the production of a pollen tube that grows into contact with the egg cell to effect fertilization.

Pollen Allergens

The list of pollens from which extracts have been derived that produce positive skin tests in some fraction of the population is long. In one California study, 57 skin test allergens were necessary to detect 90% of the atopic patients. This set included 2 grasses, 16 weeds, and 27 trees (Galant et al., 1998).

The list of purified allergens from pollens is also growing. These include *Amb a* I (short ragweed, *Ambrosia artemisiifolia*), *Bet v* I (birch pollen, *Betula verrucosa*), and *Lol p* I (ryegrass, *Lolium perenne*). One study has estimated that a single pollen grain of *B. verrucosa* contains 0.006 ng *Bet v* I (Schappi et al., 1997).

Other Pollen-Associated Agents

Serine proteases that could directly affect respiratory function apart from allergic reactions have been detected in ragweed and mesquite pollen (Bagarozzi and Travis, 1998; Travis et al., 1996).

Characteristics of Pollen Exposure

Particle Size Considerations Intact pollen grains range from about 10 to 100 μm, with the most common types in the range of 15–30 μm. However, pollen allergens have been documented in air on much smaller particles. During light rain, 1.2 ng *Bet v* I/per cubic meter (200 birch pollen equivalents) were present on the particle fraction less than 7.5 μm in outdoor air (Schappi et al., 1997). Grass pollen grains have been shown to rupture during rainfall, and allergen (*Lol p* V) has been recovered in particle fractions less than 5 μm both as free molecules and attached to other particles such as starch grains and combustion (diesel exhaust) particles (Suphioglu, 1998). Pollen allergens have been visualized on the surface of diesel particles under laboratory conditions (Knox et al., 1997). Ragweed allergens have also been recovered from small particles (Habenicht et al., 1984; Solomon et al., 1983).

Patterns of Pollen Prevalence

Outdoors An enormous body of literature documents prevalence patterns for pollens throughout the world. In the United States, the American Academy of Allergy, Asthma, and Immunology has been collecting pollen prevalence data for more than 30 years and currently publishes an annual report containing data from about 100 stations across the nation (e.g., AAAAI, 1998). In the United States, the pollen types most often considered important are mountain cedar, birch, and oak (trees); grasses (which cross-react broadly across genera); and ragweed (a late-summer weed).

Pollen is produced seasonally. In general, tree pollens are released early in the year, grasses during late spring and early summer, and weed pollens in the late summer and fall. Major exceptions occur. For example, some grass pollen is produced throughout the year in some areas.

Meteorological variables directly affect pollen concentrations in air (in addition to their effects on pollen production). Most pollen types are released during the morning hours, but dispersal may occur later in the day with increase in wind and other disturbances. Wind is a well-known dispersal agent for pollens, which may travel long distances (hundreds of miles) in traveling air masses (Levetin and Buck, 1986). Rain generally washes pollen from the air.

Indoors Outdoor infiltrate is the primary source of pollen in indoor environments. Birch pollen allergen (*Bet v* I) has been found associated with suspended particles indoors (Holmquist and Vesterberg, 1999; Ormstad et al., 1998).

Evidence Regarding Asthma Exacerbation and Development

Pollen Exposure and Asthma

Sensitization to Pollen Allergens In 3,371 Canadian allergy patients, skin tests indicated that 52% were sensitive to grass allergens and 45% to ragweed allergens (Boulet et al., 1997). In the Swiss population (random, more than 8,000 individuals studied) 32% were atopic, with sensitivity to grass allergen the most prevalent (12%), followed by dust mite (8.9%) and birch pollen (7.9%) (Wuthrich et al., 1995). In 1,159 random participants in Hamburg and Ehrfurt, respectively, grass and birch allergen sensitivities were 24 versus 19% and 19 versus 8% (Nowak et al., 1996). Of inner city asthmatics in Chicago, 45% were sensitized to ragweed pollen (76% were sensitive to indoor allergens) (Kang et al., 1993). In 7,079 retrospectively observed patients with asthma and/or rhinitis, 44% had one or more positive skin tests, with grass, cat, and birch allergens eliciting the greatest number of positive tests. Of the 35% of sensitized patients who were monosensitized, 7.4% reacted only to mite allergens and 7.0% to grass allergen (Eriksson and Holmen, 1996). On the other hand, Subiza and colleagues (1994) report that 92% of Madrid allergy patients are sensitive to grass allergens, 63% to olive allergens, and 56% to sycamore (*Platanus*) allergens.

Relationships to Asthma

Pollen Sensitization Related to Asthma In general, pollen sensitization has been associated primarily with hay fever rather than asthma, and this opinion is supported in several studies. In a study reported by Eriksson and Holmen (1996), patients with grass allergen sensitivity were more likely to have rhinitis than asthma. In a Tucson population, sensitivity to ryegrass and mulberry pollen independently predicted rhinitis but not asthma (Halonen et al., 1997). In a case-control study of 343 random children, 35% were atopic, and 90% of those with five or more episodes of wheezing had one or more positive skin tests (Henderson et al., 1995). Pollen sensitivity was not associated with wheezing in this population.

On the other hand, in an Australian population of 745 young adults, sensitivity to ryegrass pollen was among the four top allergen sensitivities that posed a risk for the existence of current asthma (Abramson et al., 1996). Winter respiratory symptoms (including asthma) have been related to positive prick, RAST, or endonasal challenge tests with alder and hazel pollen allergens (Laurent et al., 1994). Grass allergen elicited the most positive skin tests among black asthmatics in Johannesburg (Luyt et al., 1995).

Pollen Exposure and Asthma Ordaz and colleagues (1998) followed symptoms in 104 asthmatic patients and related exacerbation events to airborne pollen counts. When infectious disease-related events were excluded, there was a good correlation between event days and pollen counts (r = 0.7, $p < .01$). In a prospective study of mild to moderate asthmatics ($N = 139$), Epton and colleagues (1997) were unable to distinguish a relationship between measured pollen concentrations and measured peak flow.

Baraldi and colleagues (1999) documented a twofold increase in exhaled NO in grass-sensitive asthmatic children during the grass pollen season. Natural allergen exposure was shown to induce T cell, mast cell, and eosinophil inflammatory response in grass-sensitive patients undergoing seasonal exacerbation ($N = 17$) (Djukanovic et al., 1996). Rosas and colleagues (1998) report an association between grass pollen counts and asthma admissions in Mexico City in both dry and wet seasons. Celenza and colleagues (1996) related changes in grass pollen concentrations and decreases in temperature to epidemic asthma. Their patients, who were involved in a thunderstorm-associated epidemic in England, had high serum IgE levels specific for grass pollen allergens (Venables et al., 1997). Newson and colleagues (1997) counted asthma admissions in two age groups (0–14 and >14 years) and related these events to lightning episodes and grass pollen concentrations. Although there were excess densities of lightning and grass pollen counts in epidemics, many (in fact, most) epidemics were not preceded by thunderstorms. In a follow-up study, they report that thunderstorms following periods of high pollen counts are more likely to lead to asthma epidemics (Newson et al., 1998).

In an analysis of 59,624 asthma hospitalizations in Finland, a peak was observed in May that was attributed to the exposure of sensitized people to birch pollen allergens (Harju et al., 1997). As part of the increase in doctor-diagnosed asthma and episodes of breathlessness in children in Norway, symptoms with exposure to birch pollen increased from 3.7 to 6.1% of children between 1981 and 1993 (Skjonsberg et al., 1995).

Woodcock and Custovic (1998) suggest that exposure to ragweed allergen reduces corticosteroid binding capacity in sensitized asthmatics, contributing to poor asthma control.

In 20 subjects, airway responsiveness to histamine challenge was highest during the privet season, and symptom scores and bronchodilator use increased. However, challenge with privet pollen allergens induced no immediate asthmatic responses, and late responses occurred in only 6 of 17 tested patients (Richards et al., 1995).

Pollen–Air Pollutant Interactions

Laboratory Studies Strand and colleagues (1997) report that in a group of 18 pollen-sensitive asthmatics, short exposure in the laboratory to concentrations of NO_2 representative of those that occur in outdoor air enhances pollen allergen-induced late asthmatic responses. Repeated short exposures to NO_2 enhanced effects of otherwise nonsymptomatic doses of pollen allergen (Strand et al., 1998). However, in similar studies, Hanania and colleagues (1998) found no change in allergen response following ozone challenge.

Field Studies Anderson and colleagues (1998) report a synergistic effect for pollen and SO_2 with respect to hospital admissions of children (but not adults) for asthma. No other pollen–pollutant interactions were observed. In a comparison between randomly selected participants in Hamburg and Ehrfurt, Germany, the prevalence of atopy and pollen sensitization was highest in Hamburg, while air pollutant exposures were highest in Ehrfurt (Nowak et al., 1996).

Conclusions: Asthma Exacerbation and Development

Evidence from studies of outdoor exposure indicates that pollen exacerbates existing asthma in sensitized individuals. Information does not permit a conclusion concerning whether or not there is an overall role for pollen allergens in the development of asthma. Although pollen allergens have been documented in both dust and indoor air, data is lacking to draw a informed conclusion. Thus:

There is inadequate or insufficient evidence to determine whether or not an association exists between pollen exposure in the indoor environment and the exacerbation or development of asthma.

The committee concludes:

• There is inadequate or insufficient evidence to determine whether or not an association exists between interventions to lower pollen concentrations in indoor environments and improvement of symptoms or lung function in pollen-allergic asthmatics. Indoors is typically a protective environment for such individuals.

Evidence and Conclusions:
Exposure Mitigation and Prevention

There is relatively little information on the impact of ventilation and air-cleaning measures on indoor pollen levels, although it is axiomatic that shutting windows and other measures that generally limit outdoor infiltrate can be effective. Chapter 10 discusses the studies identified by the committee and draws conclusions about the effectiveness of ventilation and air-cleaning measures in reducing indoor concentrations in general.

Research Needs

Studies should be conducted to evaluate the ambiguous relationship between pollen exposure, sensitivity, and asthma. Additional research is also needed to discover the extent of indoor pollen allergen exposure and the interactions between pollen sensitivity and air pollutants.

REFERENCES

AAAAI. 1998. Pollen and Spore Report. American Academy of Allergy, Asthma, and Immunology, Milwaukee, WI.

Åberg N. 1989. Asthma and allergic rhinitis in Swedish conscripts. Clinical and Experimental Allergy 19(1):59–63.

Abramson M, Kutin JJ, Raven J, Lanigan A, Czarny D, Walters EH. 1996. Risk factors for asthma among young adults in Melbourne, Australia. Respirology 1(4):291–297.

ACGIH. 1999. Bioaerosols: Assessment and Control. American Conference of Governmental Industrial Hygienists, Cincinnati.

Adachi Y, Ohno N, Yadomae T. 1994a. Preparation and antigen specificity of an anti-(1→3)-beta-D-glucan antibody. Biological and Pharmaceutical Bulletin 17(11):1508–1512.

Adachi Y, Okazaki M, Ohno N, Yadomae T. 1994b. Enhancement of cytokine production by macrophages stimulated with (1→3)-beta-D-glucan, grifolan (GRN) isolated from *Grifola frondosa*. Biological and Pharmaceutical Bulletin 17(12):1554–1560.

Amitani R, Taylor G, Elezis EN, Llewellyn-Jones C, Mitchell J, Kuze F, Cole PJ, Wilson R. 1995. Purification and characterization of factors produced by *Aspergillus fumigatus* which affect human ciliated respiratory epithelium. Infection and Immunity 63(9):3266–3271.

Anderson HR, Ponce de Leon A, Bland JM, Bower JS, Emberlin J, Strachan DP. 1998. Air pollution, pollens, and daily admissions for asthma in London 1987–92. Thorax 53(10):842–848.

Andriessen JW, Brunekreef B, Roemer W. 1998. Home dampness and respiratory health status in European children. Clinical and Experimental Allergy 28(10):1191–1200.

Angrisano A, Di Berardino L, Montrasio G, Compostella G. 1987. Allergy caused by *Alternaria* in children (Italian). Pediatria Medica e Chirurgica 9(2):159–160.

Arlian LG, Bernstein IL, Gallagher JS. 1982. The prevalence of house dust mites *Dermatophagoides* spp, and associated environmental conditions in homes in Ohio. Journal of Allergy and Clinical Immunology 69(6):527–532.

Arruda LK, Platts-Mills TAE, Fox JW, Chapman MD. 1990. *Aspergillus fumigatus*. Allergen I, a major IgE-binding protein, is a member of the mitogillin family of cytotoxins. Journal of Experimental Medicine 172(5):1529–1532.

194   CLEARING THE AIR

Arruda LK, Vailes LD, Platts-Mills TA, Hayden ML, Chapman MD. 1997. Induction of IgE antibody responses by glutathione-S-transferase from the German cockroach (*Blattella germanica*). Journal of Biological Chemistry 272(33):20907–20912.

ATS (American Thoracic Society). 1998. Respiratory health hazards in agriculture (Official conference report). American Journal of Respiratory and Critical Care Medicine 158(5):S1–S76.

Aukrust L, Borch SM. 1979. Partial purification and characterization of two *Cladosporium herbarum* allergens. International Archives of Allergy and Applied Immunology 60(1):68–79.

Avner DB, Perzanowski MS, Platts-Mills TA, Woodfolk JA. 1997. Evaluation of different techniques for washing cats: quantitation of allergen removed from the cat and the effect on airborne *Fel d* I [see comments]. Journal of Allergy and Clinical Immunology 100(3):307–312.

Bagarozzi DA Jr, Travis J. 1998. Ragweed pollen proteolytic enzymes: possible roles in allergies and asthma. Phytochemistry 47(4):593–598.

Baker PJ, Taylor CE, Stashak PW, Fauntleroy MB, Haslov K, Qureshi N, Takayama K. 1990. Inactivation of suppressor T cell activity by the nontoxic lipopolysaccharide of *Rhodopseudomonas sphaeroides*. Infection and Immunity 58(9):2862–2868.

Baker PJ, Hraba T, Taylor CE, Myers KR, Takayama K, Qureshi N, Stuetz P, Kusumoto S, Hasegawa A. 1992. Structural features that influence the ability of lipid A and its analogs to abolish expression of suppressor T cell activity. Infection and Immunity 60(7):2694–2701.

Baldini M, Carla Lohman I, Halonen M, Erickson RP, Holt PG, Martinez FD. 1999. A Polymorphism* in the 5′ flanking region of the CD14 gene is associated with circulating soluble CD14 levels and with total serum immunoglobulin E. American Journal of Respiratory Cell and Molecular Biology 20(5):976–983.

Baraldi E, Carra S, Dario C, Azzolin N, Ongaro R, Marcer G, Zacchello F. 1999. Effect of natural grass pollen exposure on exhaled nitric oxide in asthmatic children. American Journal of Respiratory and Critical Care Medicine 159(1):262–266.

Bates DV, Fish BR, Hatch TF, Mercer TT, Morrow PE. 1966. Deposition and retention models for internal dosimetry of the human respiratory tract. Task group on lung dynamics. Health Physics 12(2):173–207.

Bavastrelli M, Midulla M, Rossi D, Salzano M. 1992. *Chlamydia trachomatis* infection in children with wheezing simulating asthma. Lancet 339(8802):1174.

Becher R, Hongslo JK, Jantunen MJ, Dybing E. 1996. Environmental chemicals relevant for respiratory hypersensitivity: the indoor environment. Toxicology Letters 86(2–3):155–162.

Beeson MF, Dewdney JM, Edwards RG, Lee D, Orr RG. 1983. Prevalence and diagnosis of laboratory animal allergy. Clinical Allergy 13(5):433–442.

Berg DJ, Kuhn R, Rajewsky K, Muller W, Menon S, Davidson N, Grunig G, Rennick D. 1995. Interleukin-10 is a central regulator of the response to LPS in murine models of endotoxic shock and the Shwartzman reaction but not endotoxin tolerance. Journal of Clinical Investigation 96(5):2339–2347.

Berge M, Munir AK, Dreborg S. 1998. Concentrations of cat (*Fel d* I), dog (*Can f* I) and mite (*Der f* I and *Der p* I) allergens in the clothing and school environment of Swedish school children with and without pets at home. Pediatric Allergy and Immunology 9(1):25–30.

Bernton HS, Brown H. 1964. Insect allergy: preliminary studies of the cockroach. Journal of Allergy 35:506–513.

Bernton HS, Brown H. 1967. Cockroach allergy II: the relation of infestation to sensitization. Southern Medical Journal 60(8):852–855.

Bernton HS, McMahon TF, Brown H. 1972. Cockroach asthma. British Journal of Diseases of the Chest 66(1):61–66.

Björnsson E, Hjelm E, Janson C, Fridell E, Boman G. 1996. Serology of chlamydia in relation to asthma and bronchial hyperresponsiveness. Scandinavian Journal of Infectious Disease 28(1):63–69.

Blackley CH. 1873. Experimental research in the causes and nature of catarrhus aestivus (hay fever or hay asthma). London, England: Balliere, Tindall and Cox.

Bohuslav J., Kravchenko VV, Parry GC, Erlich JH, Gerondakis S, Mackman N, Ulevitch RJ. 1998. Regulation of an essential innate immune response by the p50 subunit of NF-kappaB. Journal of Clinical Investigation 102(9):1645–1652.

Bollinger ME, Eggleston PA, Flanagan E, Wood RA. 1996. Cat antigen in homes with and without cats may induce allergic symptoms. Journal of Allergy and Clinical Immunology 97(4):907–914.

Bone RC 1991. *Chlamydial pneumonia* and asthma: a potentially important relationship. Journal of the American Medical Association 266(2):265.

Bonifazi F. 1994. Immunotherapy in pollen and mould asthma [Review]. Monaldi Archives for Chest Disease 49(2):150–153.

Boulet LP, Turcotte H, Laprise C, Lavertu C, Bedard PM, Lavoie A, Hebert J. 1997. Comparative degree and type of sensitization to common indoor and outdoor allergens in subjects with allergic rhinitis and/or asthma. Clinical and Experimental Allergy 27(1):52–59.

Bousquet J, Michel FB. 1994. Specific immunotherapy in asthma [Review]. Allergy Proceedings 15(6):329–33.

Brattgjerd S, Evensen O, Lauve A. 1994. Effect of injected yeast glucan on the activity of macrophages in Atlantic salmon, *Salmo salar L.*, as evaluated by *in vitro* hydrogen peroxide production and phagocytic capacity. Immunology 83(2):288–294.

Braun-Fahrlander C, Gassner M, Grize L, Neu U, Sennhauser FH, Varonier HS, Vuille JC, Wuthrich B. 1999. Prevalence of hay fever and allergic sensitization in farmer's children and their peers living in the same rural community. SCARPOL team. Swiss Study on Childhood Allergy and Respiratory Symptoms with Respect to Air Pollution. Clinical and Experimental Allergy 29(1):28–34.

Brundage JF, Scott RM, Lednar WM, Smith DW, Miller RN. 1988. Building-associated risk of febrile acute respiratory diseases in Army trainees. Journal of the American Medical Association 259(14):2108–2112.

Brunekreef B, Dockery DW, Speizer FE, Ware JH, Spengler JD, Ferris BG. 1989. Home dampness and respiratory morbidity in children. American Review of Respiratory Disease 140(5):1363–1367.

Burge HA, Solomon WR, Boise JR. 1980. Microbial prevalence in domestic humidifiers. Applied and Environmental Microbiology 39(4):840–844.

Burge HA, Solomon WR, Muilenberg ML. 1982. Evaluation of indoor plantings as allergen exposure sources. Journal of Allergy and Clinical Immunology 70(2):101–108.

Burge HA, Solomon WR. 1987. Sampling and analysis of biological aerosols. Atmospheric Science 21(2):451.

Burge HA, Hoyer ME, Solomon WR, Simmons EG, Gallup J. 1989. Quality control factors for *Alternaria* allergens. Mycotaxon 34(1):55–63.

Burke GW, Carrington CB, Strauss R, Fink JN, Gaensler EA. 1977. Allergic alveolitis caused by home humidifiers. Unusual clinical features and electron microscopic findings. Journal of the American Medical Association 238(25): 2705–2708.

Call RS, Smith TF, Morris E, Chapman MD, Platts-Mills TA. 1992. Risk factors for asthma in inner city children. Journal of Pediatrics 121(6):862–866. [Comment in: J Pediatr 1993. 123(1):171.]

Campbell EM, Kunkel SL, Strieter RM, Lukacs NW. 1998. Temporal role of chemokines in a murine model of cockroach allergen-induced airway hyperreactivity and eosinophilia. Journal of Immunology 161(12):7047–7053.

Cantani A, Businco E, Maglio A. 1988. *Alternaria* allergy: a three-year controlled study in children treated with immunotherapy. Allergologia et Immunopathologia 16(1):1–4.

Celenza A, Fothergill J, Kupek E, Shaw RJ. 1996. Thunderstorm associated asthma: a detailed analysis of environmental factors. British Medical Journal 312(7031):604–607. [Comment in BMJ 1990. 312(7031):590–591.]

Chang JCS, Foarde KK, Van Osdell DW. 1996. Assessment of fungal (*Penicillium chrysogenum*) growth on three HVAC duct materials. Environment International 22(4):425–431.

Chang TL, Shea CM, Urioste S, Thompson RC, Boom WH, Abbas AK. 1990. Heterogeneity of helper/inducer T lymphocytes. III. Responses of IL-2- and IL-4- producing (TH1 and TH2) clones to antigens presented by different accessory cells. Journal of Immunology 145(9):2803–2808.

Chapman, MD. 1993. Dissecting cockroach allergens. Clinical and Experimental Allergy 23(6):459–461.

Chapman MD, Platts-Mills TA. 1980. Purification and characterization of the major allergen from *Dermatophagoides pteronyssinus*-antigen P1. Journal of Immunology 125(2):587–592.

Chapman MD, Heymann PW, Wilkins SR, Brown MJ, Platts-Mills TA. 1987. Monoclonal immunoassays for the major dust mite (*Dermatophagoides*) allergens, Der p I and Der f I, and quantitative analysis of the allergen content of mite and house dust extracts. Journal of Allergy and Clinical Immunology 80(2):184–194.

Chew GL, Burge HA, Dockery DW, Muilenberg ML, Weiss ST, Gold DR. 1998. Limitations of a home characteristics questionnaire as a predictor of indoor allergen levels. American Journal of Respiratory and Critical Care Medicine 157(5 Pt 1):1536–1541.

Chow JC, Young DW, Golenbock DT, Christ WJ, Gusovsky F. 1999. Toll-like receptor-4 mediates lipopolysaccharide-induced signal transduction. Journal of Biological Chemistry 274(16):10689–10692.

Christensen LT, Schmidt CD, Robbins L. 1975. Pigeon breeders' disease—a prevalence study and review. Clinical Allergy 5(4):417–430.

Clarke CW, Aldons PM. 1979. The nature of asthma in Brisbane. Clinical Allergy 9(2):147–152.

Cleveland MG, Gorham JD, Murphy TL, Tuomanen E, Murphy KM. 1996. Lipoteichoic acid preparations of gram-positive bacteria induce interleukin-12 through a CD14-dependent pathway. Infection and Immunity 64(6):1906–1912.

Cochran DG. 1995. Toxic effects of boric acid on the German cockroach. Experimentia 51(6):561–563.

Cochran DG. 1996. Relevance of resistance ratios to operational control in the German cockroach (*Dictyoptera, Blattellidae*). Journal of Economic Entomology 89(2):318–321.

Cockcroft DW, Ruffin RE, Frith PA, Cartier A, Juniper EF, Dolovich J, Hargreave FE. 1979. Determinants of allergen-induced asthma: dose of allergen, circulating IgE antibody concentration, and bronchial responsiveness to inhaled histamine. American Review of Respiratory Disease 120(5):1053–1058.

Cockcroft A, Edwards J, McCarthy P, Andersson N. 1981. Allergy in laboratory animal workers. Lancet 1(8224):827–830.

Comoy EE, Pestel J, Duez C, Stewart GA, Vendeville C, Fournier C, Finkelman F, Capron A, Thyphrontis G. 1998. The house dust mite allergen, *Dermatophagoides pteronyssinus*, promotes type 2 responses by modulating the balance between IL-4 and IFN-gamma. Journal of Immunology 160(5):2456–2462.

Cook PJ, Davies P, Tunnicliffe W, Ayres JG, Honeybourne D, Wise R. 1998. *Chlamydia pneumoniae* and asthma. Thorax 53(4):254–259. [Comment in: Thorax 1998 December 53(12):1095–1096.]

Corey JP, Kaiseruddin S, Gungor A. 1997. Prevalence of mold-specific immunoglobulins in a midwestern allergy practice. Otolaryngology—Head and Neck Surgery 117(5):516–520.

Corsico R, Cinti B, Feliziani V, Gallesio MT, Liccardi G, Loreti A, Lugo G, Marcucci F, Marcer G, Meriggi A, Minelli M, Gherson G, Nardi G, Negrini AC, Piu G, Passaleva A, Pozzan M, D'Ambrosio FP, Venuti A, Zanon P, Zerboni R. 1998. Prevalence of sensitization to *Alternaria* in allergic patients in Italy. Annals of Allergy, Asthma, and Immunology 80(1):71–76.

Crameri R. 1998. Recombinant *Aspergillus fumigatus* allergens: from nucleotide sequences to clinical applications. International Archives of Allergy and Immunology 115(2):99–114.

Crane J, Kemp T, Siebers R, Rains N, Fishwick D, Fitzharris P. 1997. Increased house dust mite allergen in synthetic pillows may explain increased wheezing [letter]. British Medical Journal 314(7096):1763–1764.

Cross S. 1997. Mould spores: the unusual suspects in hay fever. Community Nurse 3(4):25–26.

Cruz A, Saenz de Santamaria M, Martinez J, Martinez A, Guisantes J, Palacios R. 1997. Fungal allergens from important allergenic fungi imperfecti [Review]. Allergologia et Immunopathologia 25(3):153–158.

Cuijpers CE, Swaen GM, Wesseling G, Sturmans F, Wouters EF. 1995. Diverse effects of the indoor environment on respiratory health in primary school children. Environmental Research 68(1):11–23.

Cullinan P, Lowson D, Nieuwenhuijsen MJ, Gordon S, Tee RD, Venables KM, McDonald JC, Newman Taylor AJ. 1994. Work related symptoms, sensitisation, and estimated exposure in workers not previously exposed to laboratory rats. Occupational and Environmental Medicine 51(9):589–592.

Cunningham AF, Johnston SL, Julious SA, Lampe FC, Ward ME. 1998. Chronic *Chlamydia pneumoniae* infection and asthma exacerbations in children. European Respiratory Journal 11(2):345–349.

Custovic A, Green R, Fletcher A, Smith A, Pickering CA, Chapman MD, Woodcock A. 1997. Aerodynamic properties of the major dog allergen *Can f* I: distribution in homes, concentration, and particle size of allergen in the air. American Journal of Respiratory and Critical Care Medicine 155(1):94–98.

Custovic A, Fletcher A, Pickering CA, Francis HC, Green R, Smith A, Chapman M, Woodcock A. 1998a. Domestic allergens in public places. III: house dust mite, cat, dog and cockroach allergens in British hospitals. Clinical and Experimental Allergy 28(1):53–59.

Custovic A, Simpson A, Pahdi H, Green RM, Chapman MD, Woodcock A. 1998b. Distribution, aerodynamic characteristics, and removal of the major cat allergen *Fel d* I in British homes. Thorax 53(1):33–38.

Damare SM, Wells S, Offenbacher S. 1997. Eicosanoids in periodontal diseases: potential for systemic involvement. Advances in Experimental Medicine and Biology 433:23–35.

Davies GE, McArdle LA. 1981. Allergy to laboratory animals: a survey by questionnaire. International Archives of Allergy and Applied Immunology 64:302–307.

De Andrade AD, Birnbaum J, Magalon C, Magnol JP, Lanteaume A, Charpin D, Vervloet D. 1996. *Fel d* I levels in cat anal glands. Clinical and Experimental Allergy 26(2):178–180.

de Blay F, Chapman MD, Platts-Mills TA. 1991a. Airborne cat allergen (*Fel d* I). Environmental control with the cat in situ [see comments]. American Review of Respiratory Disease 143(6):1334–1339.

de Blay F, Pauli G, Bessot JC. 1991b. Cross-reactions between respiratory and food allergens. Allergy Proceedings 12(5):313–317.

de Groot H, van Swieten P, van Leeuwen J, Lind P, Aalberse RC. 1988. Monoclonal antibodies to the major feline allergen *Fel d* I. I. Serologic and biologic activity of affinity-purified *Fel d* I and of *Fel d* I-depleted extract. Journal of Allergy and Clinical Immunology 82(5 Pt 1):778–786.

de Jong NW, Vermeulen AM, Gerth van Wijk R, de Groot H. 1998. Occupational allergy caused by flowers. Allergy 53(2):204 209.

Delfino RJ, Zeiger RS, Seltzer JM, Street DH, Matteucci RM, Anderson PR, Koutrakis P. 1997. The effect of outdoor fungal spore concentrations on daily asthma severity. Environmental Health Perspectives 105(6):622–635.

Deuell B, Arruda LK, Hayden ML, Chapman MD, Platts-Mills TAE. 1991. *Trichophyton tonsurans* allergen I. Characterization of a protein that causes immediate but not delayed hypersensitivity. Journal of Immunology 147(1): 96–101.

Dezateux C, Fletcher ME, Dundas I, Stocks J. 1997. Infant respiratory function after RSV-proven bronchiolitis. American Journal of Respiratory and Critical Care 155(4):1349–1355. [Published erratum appears in Am J Respir Crit Care Med 1997. 156(2 Pt 1):675.]

Dhillon M. 1991. Current status of mold immunotherapy [Review]. Annals of Allergy 66(5):385–392.

Diez-Gomez ML, Quirce S, Aragoneses E, Cuevas M. 1998. Asthma caused by *Ficus benjamina* latex: evidence of cross-reactivity with fig fruit and papain. Annals of Allergy, Asthma, and Immunology 80(1):24–30.

Djukanovic R, Feather I, Gratziou C, Walls A, Peroni D, Bradding P, Judd M, Howarth PH, Holgate ST. 1996. Effect of natural allergen exposure during the grass pollen season on airways inflammatory cells and asthma symptoms. Thorax 51(6):575–581.

Douwes J, Heederik D. 1997. Epidemiologic investigations of endotoxins. International Journal of Occupational Environmental Health 3(1):S26–S31.

Dreborg S, Agrell B, Foucard T, Kjellman NI, Koivikko A, Nilsson S. 1986. A double-blind, multicenter immunotherapy trial in children, using a purified and standardized *Cladosporium herbarum* preparation. I. Clinical results. Allergy 41(2):131–140.

Du J, Knowles BH, Li J, Ellar DJ. 1999. Biochemical characterization of *Bacillus thuringiensis* cytolytic toxins in association with a phospholipid bilayer. Biochemical Journal 338(Pt 1):185–193.

Dubin W, Martin TR, Swoveland P, Leturcq DJ, Moriarty AM, Tobias PS, Bleecker ER, Goldblum SE, Hasday JD. 1996. Asthma and endotoxin: lipopoly-saccharide-binding protein and soluble CD14 in bronchoalveolar compartment. American Journal of Physiology 270(5 Pt 1):L736–744.

Duffort O, Carreira J, Lombarder M. 1987. Characterization of the main IgE binding components of cat dander. International Archives of Allergy and Applied Immunology 84:339–344.

Dybendal T, Hetland T, Vik H, Apold J, Elsayed S. 1989a. Dust from carpeted and smooth floors. I. Comparative measurements of antigenic and allergenic proteins in dust vacuumed from carpeted and non-carpeted classrooms in Norwegian schools. Clinical and Experimental Allergy 19(2):217–224.

Dybendal T, Vik H, Elsayed S. 1989b. Dust from carpeted and smooth floors. II. Antigenic and allergenic content of dust vacuumed from carpeted and smooth floors in schools under routine cleaning schedules. Allergy 44(6):401–411.

Dybendal T, Wedberg WC, Elsayed S. 1991. Dust from carpeted and smooth floors. IV. Solid material, proteins and allergens collected in the different filter stages of vacuum cleaners after ten days of use in schools. Allergy 46(6):427–435.

Dybendal T, Elsayed S. 1994. Dust from carpeted and smooth floors. VI. Allergens in homes compared with those in schools in Norway. Allergy 49(4):210–216.

Dziarski R, Tapping RI, Tobias PS. 1998. Binding of bacterial peptidoglycan to CD14. Journal of Biological Chemistry 273(15): 8680–8690.

Edmonds RL, Ed. 1979. Aerobiology, The Ecological Systems Approach.. Stroudsburg: Dowden, Hutchinson & Ross.

Eggleston PA, Rosenstreich D, Lynn H, Gergen P, Baker D, Kattan M, Mortimer KM, Mitchell H, Ownby D, Slavin R, Malveaux F. 1998. Relationship of indoor allergen exposure to skin test sensitivity in inner-city children with asthma. Journal of Allergy and Clinical Immunology 102(4 Pt 1):563–570.

Eggleston PA, Wood RA, Rand C, Nixon WJ, Chen PH, Lukk P. 1999. Removal of cockroach allergen from inner-city homes. Journal of Allergy and Clinical Immunology 104(4 Pt 1):842-846.

Epton MJ, Martin IR, Graham P, Healy PE, Smith H, Balasubramaniam R, Harvey IC, Fountain DW, Hedley J, Town GI. 1997. Climate and aeroallergen levels in asthma: a 12 month prospective study. Thorax 52(6):528–534.

Erel F, Karaayvaz M, Caliskaner Z, Ozanguc N. 1998. The allergen spectrum in Turkey and the relationships between allergens and age, sex, birth month, birthplace, blood groups and family history of atopy. Journal of Investigational Allergology and Clinical Immunology 8(4):226–233.

Eriksson NE, Holmen A. 1996. Skin prick tests with standardized extracts of inhalant allergens in 7,099 adult patients with asthma or rhinitis: cross-sensitizations and relationships to age, sex, month of birth and year of testing. Journal of Investigational Allergology and Clinical Immunology 6(1):36–46.

Fadel R, David B, Paris S, Guesdon JL. 1992. Alternaria spore and mycelium sensitivity in allergic patients: in vivo and in vitro studies. Annals of Allergy 69(4):329–335.

Fearon DT, Locksley RM. 1996. The instructive role of innate immunity in the acquired immune response. Science 272(5258):50–53.

Fisk WJ. 1999. Estimates of potential nationwide productivity and health benefits from better indoor environments: an update. In: Indoor Air Quality Handbook. Spengler J, Samet JM, McCarthy JF, Eds. New York: McGraw Hill.

Flaherty DK., Deck FH, Cooper J, Bishop K, Winzenburger PA, Smith LR, Bynum L, Witmer WB. 1984. Bacterial endotoxin isolated from a water spray air humidification system as a putative agent of occupation-related lung disease. Infection and Immunity 43(1):206–212.

Fogelmark B, Sjostrand M, Rylander R. 1994. Pulmonary inflammation induced by repeated inhalations of beta(1,3)-D-glucan and endotoxin. International Journal of Experimental Pathology 75(2):85–90.

Freymuth F, Vabret A, Brouard J, Toutain F, Verdon R, Petitjean J, Gouarin S, Duhamel JF, Guillois B. 1999. Detection of viral, *Chlamydia pneumoniae* and *Mycoplasma pneumoniae* infections in exacerbations of asthma in children. Journal of Clinical Virology 13(3):131–139.

Frick OL, German DF, Mills J. 1979. Development of allergy in children. I. Association with virus infections. Journal of Allergy and Clinical Immunology 63(4):228–241.

Frost A. 1988. Frequency of allergy to *Alternaria* and *Cladosporium* in a specialist clinic. Allergy 43(7):504–507.

Fukuda T, Mochida S, Fuckushima Y, Makino S. 1995. Detection of allergen-induced genes in peripheral blood mononuclear cells of patients with allergic asthma using subtractive hybridization. Journal of Allergy and Clinical Immunology 96(6 Pt 2):1076–1082.

Galant S, Berger W, Gillman S, Goldsobel A, Incaudo G, Kanter L, Machtinger S, McLean A, Prenner B, Sokol W, Spector S, Welch M, Ziering W. 1998. Prevalence of sensitization to aeroallergens in California patients with respiratory allergy. Allergy Skin Test Project Team. Annals of Allergy, Asthma, and Immunology 81(3):203–210.

Ganier M, Lieberman P, Fink J, Lockwood DG. 1980. Humidifier lung. An outbreak in office workers. Chest 77(2):183–187.

Garcia DP, Corbett ML, Sublett JL, Pollard SJ, Meiners JF, Karibo JM, Pence HL, Petrosko JM. 1994. Cockroach allergy in Kentucky: a comparison of inner city, suburban, and rural small town populations. Annals of Allergy 72(3):203–208.

Garrett MH, Rayment PR, Hooper MA, Abramson MJ, Hooper BM. 1998. Indoor airborne fungal spores, house dampness and associations with environmental factors and respiratory health in children. Clinical and Experimental Allergy 28(4):459–467.

Gelber LE, Seltzer LH, Bouzoukis JK, Pollart SM, Chapman MD, Platts-Mills TA. 1993. Sensitization and exposure to indoor allergens as risk factors for asthma among patients presenting to hospital. American Review of Respiratory Disease 147(3):573–578.

Gerberick GF, Sorenson WG, Lewis DM. 1984. The effects of T-2 toxin on alveolar macrophage function in vitro. Environmental Research 33(1):246–260.

Gergen PJ, Turkeltaub PC. 1992. The association of individual allergen reactivity with respiratory disease in a national sample: data from the second National Health and Nutrition Examination Survey, 1976–80 (NHANES II). Journal of Allergy and Clinical Immunology 90(4 Pt 1):579–588.

Gergen PJ, Mortimer KM, Eggleston PA, Rosenstreich D, Mitchell H, Ownby D, Kattan M, Baker D, Wright EC, Slavin R, Malveaux F. 1999. Results of the National Cooperative Inner-City Asthma Study (NCICAS) environmental intervention to reduce cockroach allergen exposure in inner-city homes. Journal of Allergy and Clinical Immunology 103(3 Pt 1):501–506.

Giavina-Bianchi PF Jr, Castro FF, Machado ML, Duarte AJ. 1997. Occupational respiratory allergic disease induced by *Passiflora alata* and *Rhamnus purshiana*. Annals of Allergy, Asthma, and Immunology 79(5):449–454.

Glezen WP, Paredes A, Allison JE, Taber LH, Frank AL. 1981. Risk of respiratory syncytial virus infection for infants from low-income families in relationship to age, sex, ethnic group, and maternal antibody level. Journal of Pediatrics 98(5):708–715.

Gold DR, Burge HA, Carey V, Milton DK, Platts-Mills T, Weiss ST. 1999. Predictors of repeated wheeze in the first year of life. The relative roles of cockroach, birth weight, acute lower respiratory illness, and maternal smoking. American Journal of Respiratory and Critical Care Medicine 160(1):227–236.

Gottlieb DJ, Sparrow D, O'Connor GT, Weiss ST. 1996. Skin test reactivity to common aeroallergens and decline of lung function. The Normative Aging Study. American Journal of Respiratory and Critical Care Medicine 153(2):561–566.

Grant C, Hunter CA, Flannigan B, Bravery AF. 1989. The moisture requirements of moulds isolated from domestic dwellings. International Biodeterioration and Biodegradation 25:259–284.

Griffith IJ, Craig S, Pollock J, Yu XB, Morgenstern JP, Rogers BL. 1992. Expression and genomic structure of the genes encoding FdI, the major allergen from the domestic cat. Gene 113(2):263–268.

Gross NJ. 1980. Allergy to laboratory animals: epidemiologic, clinical, and physiologic aspects, and a trial of cromolyn in its management. Journal of Allergy and Clinical Immunology 66(2):158–165.

Grunberg K, Sterk PJ. 1999. Rhinovirus infections: induction and modulation of airways inflammation in asthma. Clinical and Experimental Allergy. 29(Supplement 2):65–73.

Guneser S, Atici A, Koksal F, Yaman A. 1994. Mold allergy in Adana, Turkey. Allergologia et Immunopathologia 22(2):52–54.

Habenicht HA, Burge HA, Muilenberg ML, Solomon WR. 1984. Allergen carriage by atmospheric aerosol. II. Ragweed-pollen determinants in submicronic atmospheric fractions. Journal of Allergy and Clinical Immunology 74(1):64–67.

Hahn DL, Dodge RW, Golubjatnikov R. 1991. Association of *Chlamydia pneumoniae* (strain TWAR) infection with wheezing, asthmatic bronchitis, and adult-onset asthma. Journal of the American Medical Association 266(2):225–230. [Comment in: JAMA 1991. 266(2):265.]

Hahn DL, McDonald R. 1998. Can acute *Chlamydia pneumoniae* respiratory tract infection initiate chronic asthma? Annals of Allergy, Asthma, and Immunology 81(4):339–344.

Hall CB, Kopelman AE, Douglas RG Jr, Geiman JM, Meagher MP. 1979. Neonatal respiratory syncytial virus infection. New England Journal of Medicine 300(8):393–396.

Halonen M, Stern DA, Wright AL, Taussig LM, Martinez FD. 1997. *Alternaria* as a major allergen for asthma in children raised in a desert environment. American Journal of Respiratory and Critical Care Medicine 155(4):1356–1361.

Hanania NA, Tarlo SM, Silverman F, Urch B, Senathirajah N, Zamel N, Corey P. 1998. Effect of exposure to low levels of ozone on the response to inhaled allergen in allergic asthmatic patients. Chest 114(3):752–756.

Harju T, Keistinen T, Tuuponen T, Kivela SL. 1997. Seasonal variation in childhood asthma hospitalisations in Finland, 1972–1992. European Journal of Pediatrics 156(6):436–439.

Harving H, Korsgaard J, Dahl R. 1994. House-dust mite exposure reduction in specially designed mechanically ventilated "healthy" homes. Allergy 49(9): 713–718.

Hayden ML, Rose G, Diduch KB, Domson P, Chapman MD, Heymann PW, Platts-Mills TA. 1992. Benzyl benzoate moist powder: investigation of acaricidal activity in cultures and reduction of dust mite allergens in carpets. Journal of Allergy and Clinical Immunology 89(2):536–545.

Hayden ML, Perzanowski M, Matheson L, Scott P, Call RS, Platts-Mills TA. 1997. Dust mite allergen avoidance in the treatment of hospitalized children with asthma. Annals of Allergy, Asthma, and Immunology 79(5):437–442.

Hedlin G, Heilborn H, Lilja G, Norrlind K, Pegelow KO, Schou C, Lowenstein H. 1995. Long-term follow-up of patients treated with a three-year course of cat or dog immunotherapy. Journal of Allergy and Clinical Immunology 96(6 Pt 1):879–885.

Hegele RG. 1999. Infection by asthma-associated viruses: clinical implications. Medscape Respiratory Care 3(1). URL:http://www.medscape.com/Medscape/RespiratoryCare/journal/1999/v03.n01/mrc4638.volc/mrc5212.hege/mrc5212.hege-01.html. Accessed August 25, 1999.

Helbling A, Gayer F, Pichler WJ, Brander KA. 1998. Mushroom (*Basidiomycete*) allergy: diagnosis established by skin test and nasal challenge. Journal of Allergy and Clinical Immunology 102(5):853–858.

Henderson FW, Henry MM, Ivins SS, Morris R, Neebe EC, Leu SY, Stewart PW. 1995. Correlates of recurrent wheezing in school-age children. The Physicians of Raleigh Pediatric Associates. American Journal of Respiratory and Critical Care Medicine 151(6):1786–1793.

Hendrick DJ, Faux JA, Marshall R. 1978. Budgerigar-fancier's lung: the commonest variety of allergic alveolitis in Britain. British Medical Journal 2(6130):81–84.

Hesselmar B, Åberg N, Åberg B, Eriksson B, Björkstén B. 1999. Does early exposure to cat or dog protect against later allergy development? Clinical and Experimental Allergy 29(5):611–617.

Hewitt CR, Brown AP, Hart BJ, Pritchard DI. 1995. A major house dust mite allergen disrupts the immunoglobulin E network by selectively cleaving CD23: innate protection by antiproteases. Journal of Experimental Medicine 182(5):1537–1544.

Hide DW, Matthews S, Matthews L, Stevens M, Ridout S, Twiselton R, Gant C, Arshad SH. 1994. Effect of allergen avoidance in infancy on allergic manifestations at age two years. Journal of Allergy and Clinical Immunology 93(5):842–846.

Hill AB. 1965. The environment and disease: association or causation. Proceedings of the Royal Society of Medicine 58:295–300.

Hizawa N, Freidhoff LR, Ehrlich E, Chiu YF, Duffy DL, Schou CP, Dunston GM, Beaty TH, Marsh DG, Barnes KC, Huang SK. 1998. Genetic influences of chromosomes 5q31-q33 and 11q13 on specific IgE responsiveness to common inhaled allergens among African American families. Collaborative Study on the Genetics of Asthma (CSGA). Journal of Allergy and Clinical Immunology 102(3):449–453.

Hodson T, Custovic A, Simpson A, Chapman M, Woodcock A, Green R. 1999. Washing the dog reduces dog allergen levels, but the dog needs to be washed twice a week. Journal of Allergy and Clinical Immunology 103(4):581–585.

Hollander A, Doekes G, Heederik D. 1996. Cat and dog allergy and total IgE as risk factors of laboratory animal allergy. Journal of Allergy and Clinical Immunology 98(3):545–554.

Holmquist L, Vesterberg O. 1999. Quantification of birch and grass pollen allergens in indoor air. Indoor Air 9(2):85–91.

Horak B, Dutkiewicz J, Solarz K. 1996. Microflora and acarofauna of bed dust from homes in Upper Silesia, Poland. Annals of Allergy, Asthma, and Immunology 76(1):41–50.

Horner WE, Reese G, Lehrer SB. 1995. Identification of the allergen *Psi c* 2 from the basidiomycete *Psilocybe cubensis* as a fungal cyclophilin. International Archives of Allergy and Immunology 107(1–3):298–300.

Horst M, Hejjaoui A, Horst V, Michel FB, Bousquet J. 1990. Double-blind, placebo-controlled rush immunotherapy with a standardized *Alternaria* extract. Journal of Allergy and Clinical Immunology 85(2):460–472.

Hu FB, Persky V, Flay BR, Richardson J. 1997. An epidemiological study of asthma prevalence and related factors among young adults. Journal of Asthma 34(1):67–76.

Ingold CT. 1971. Fungal spores: their liberation and dispersal. Oxford: Clarenden Press.

Ingram JM, Sporik R, Rose G, Honsinger R, Chapman MD, Platts-Mills TA. 1995. Quantitative assessment of exposure to dog (*Can f* I) and cat (*Fel d* I) allergens: relation to sensitization and asthma among children living in Los Alamos, New Mexico. Journal of Allergy and Clinical Immunology 96(4):449–456.

IOM (Institute of Medicine). 1993. Indoor Allergens. Pope AM, Patterson R, Burge H, eds. Washington, DC. National Academy Press.

Iversen M, Pedersen B. 1990. The prevalence of allergy in Danish farmers. Allergy 45(5):347–353.

Iwasaki Y, Hara Y, Koji T, Shibata Y, Nakane PK, Kato I. 1998. Differential expression of IFN-gamma, IL-4, IL-10, and IL-1beta mRNAs in decalcified tissue sections of mouse lipopolysaccharide-induced periodontitis mandibles assessed by in situ hybridization. Histochemistry and Cell Biology 109(4):339–347.

Jaakkola JJ, Jaakkola N, Ruotsalainen R. 1993. Home dampness and molds as determinants of respiratory symptoms and asthma in pre-school children. Journal of Exposure Analysis and Environmental Epidemiology 3(Suppl 1):129–142.

Jakab GJ, Hmieleski RR, Zarba A, Hemenway DR, Groopman JD. 1994. Respiratory aflatoxicosis—suppression of pulmonary and systemic host defenses in rats and mice. Toxicology and Applied Pharmacology 125(2): 198–205.

Johnston SL. 1997. Influence of viral and bacterial respiratory infections on exacerbations and symptom severity in childhood asthma. Pediatric Pulmonology Supplement 16:88–89.

Johnston SL. 1999. The role of viral and atypical bacterial pathogens in asthma pathogenesis. Pediatric Pulmonology Supplement 18:141–143.

Johnston SL, Pattemore PK, Sanderson G, Smith S, Lampe F, Josephs L, Symington P, O'Toole S, Myint SH, Tyrrell DA, Holgate ST. 1995. Community study of role of viral infections in exacerbations of asthma in 9–11 year old children. British Medical Journal 310(6989):1225–1229. [Comment in BMJ 1995. 311(7005):629–630.]

Kaad PH, Ostergaard PA. 1982. The hazard of mould hyposensitization in children with asthma. Clinical Allergy 12(3):317–320.

Kalman S, Mitchell W, Marathe R, Lammel C, Fan J, Hyman RW, Olinger L, Grimwood J, Davis RW, Stephens RS. 1999. Comparative genomes of *Chlamydia pneumoniae* and *C. trachomatis*. Nature Genetics 21(4):385–389.

Kang B. 1976. Study on cockroach antigen as a probable causative agent in bronchial asthma. Journal of Allergy and Clinical Immunology 58(3):357–365.

Kang BC, Johnson J, Veres-Thorner C. 1993. Atopic profile of inner–city asthma with a comparative analysis on the cockroach-sensitive and ragweed-sensitive subgroups. Journal of Allergy and Clinical Immunology 92(6):802–811.

Kang BC, Zhou K, Lai YL, Hong CB. 1996. Experimental asthma developed by room air contamination with cockroach allergen. International Archives of Allergy and Immunology 111(3):299–306.

Karn RC. 1994. The mouse salivary androgen-binding protein (ABP) alpha subunit closely resembles chain 1 of the cat allergen *Fel d* I. Biochemical Genetics 32(7–8):271–277.

Kattan M, Mitchell H, Eggleston P, Gergen P, Crain E, Redline S, Weiss K, Evans R III, Kaslow R, Kercsmar C, Leickly F, Malveaux F, Wedner HJ. 1997. Characteristics of inner-city children with asthma: the National Cooperative Inner-City Asthma Study. Pediatric Pulmonology 24(4):253–262.

Katz Y, Verleger H, Barr J, Rachmiel M, Kiviti S, Kuttin ES. 1999. Indoor survey of moulds and prevalence of mould atopy in Israel. Clinical and Experimental Allergy 29(2):186–192.

Kemp TJ, Siebers RW, Fishwick D, O'Grady GB, Fitzharris P, Crane J. 1996. House dust mite allergen in pillows [see comments]. British Medical Journal 313(7062):916.

Kerrebijn KF. 1970. Endogenous factors in childhood CNSLD: methodological aspects in population studies. In: Bronchitis III. Orie NGM, van der Lende R, Eds. The Netherlands: Royal Vangorcum Assen, pp. 38–48.

Kiho T, Sakushima M, Wang SR, Nagai K, Ukai S. 1991. Polysaccharides in fungi. XXVI. Two branched (1→3)-beta-D-glucans from hot water extract of Yu er. Chemical and Pharmaceutical Bulletin 39(3):798–800.

Kilpeläinen M, Terho EO, Koskenvuo M. 1997. Asthma and atopic diseases among Finish university students. European Respiratory Journal 10(Supplement):143s.

Kirschning CJ, Wesche H, Merrill Ayres T, Rothe M. 1998. Human toll-like receptor 2 confers responsiveness to bacterial lipopolysaccharide. Journal of Experimental Medicine 188(11):2091–2097.

Kitamura S, Hori T, Kurita K, Takeo K, Hara C, Itoh W, Tabata K, Elgasaeter A, Stokke BT. 1994. An antitumor, branched (1→3)-beta-D-glucan from a water extract of fruiting bodies of *Cryptoporus volvatus*. Carbohydrate Research 263(1):111–121.

Klucka CV, Ownby DR, Green J, Zoratti E. 1995. Cat shedding of *Fel d* I is not reduced by washings, Allerpet-C spray, or acepromazine. Journal of Allergy and Clinical Immunology 95(6):1164–1171.

Knox RB, Suphioglu C, Taylor P, Desai R, Watson HC, Peng JL, Bursill LA. 1997. Major grass pollen allergen *Lol p* I binds to diesel exhaust particles: implications for asthma and air pollution. Clinical and Experimental Allergy 27(3):246–251.

Koehler PG, Patterson RS, Brenner RJ. 1987. German cockroach (*Orthoptera, Blattellidae*) infestations in low-income apartments. Journal of Economic Entomology 80(2):446–450.

Kohler F, Kohler C, Patris A, Grillat JP. 1983. Fréquence de l'allergie pollinique chez les agriculteurs par rapport aux autres catégories socio-professionnelles. Revue Francaise D Allergologie 23(119–124).

Kongpanichkul A, Vichyanond P, Tuchinda M. 1997. Allergen skin test reactivities among asthmatic Thai children. Journal of the Medical Association of Thailand 80(2):69–75.

Kraft M, Cassell GH, Henson JE, Watson H, Williamson J, Marmion BP, Gaydos CA, Martin RJ. 1998. Detection of *Mycoplasma pneumoniae* in the airways of adults with chronic asthma. American Journal of Respiratory and Critical Care Medicine 158(3):998–1001. [Published erratum appears in Am J Respir Crit Care Med 1998. 158(5 Pt 1):1692.]

Kraus J, Franz G. 1991. β(1-3)Glucans: anti-tumor activity and immuno-stimulation. In: Fungal cell wall and immune response. Latge JP, Boucias D, Eds. Berlin, Heidelberg: SpringerVerlag. NATO ASI series, H53:431–444.

Kuehr J, Frischer T, Meinert R, Barth R, Forster J, Schraub S, Urbanek R, Karmaus W. 1994. Mite allergen exposure is a risk factor for the incidence of specific sensitization. Journal of Allergy and Clinical Immunology 94(1):44–52.

Larsen FO, Norn S, Mordhorst CH, Skov PS, Milman N, Clementsen P. 1998. *Chlamydia pneumoniae* and possible relationship to asthma. Serum immuno-globulins and histamine release in patients and controls. APMIS 106(10):928–934.

Laurent J, Decoux L, Ickovic MR, Le Gall C, Gacouin JC, Sauvaget J, Lafay M. 1994. Winter pollinosis in Paris. Allergy 49(9):696–701.

Lehrer SB, Lopez M, Butcher BT, Olson J, Reed M, Salvaggio JE. 1986. Basidiomycete mycelia and spore-allergen extracts: skin test reactivity in adults with symptoms of respiratory allergy. Journal of Allergy and Clinical Immunology 78(3 Pt 1):478–485.

Lelong M, Henard J, Wattre P, Duprey J, Thelliez P, Miersman R. 1986. Does immediate-type respiratory allergy occur regarding *Stemphylium*? Evaluation of 39 challenge tests [French]. Allergie et Immunologie 18(8):21, 23, 25–6.

Lemanske RF Jr, Dick EC, Swenson CA, Vrtis RF, Busse WW. 1989. Rhinovirus upper respiratory infection increases airway hyperreactivity and late asthmatic reactions. Journal of Clinical Investigation. 83(1):1–10.

Levetin E, Buck P. 1986. Evidence of mountain cedar pollen in Tulsa. Annals of Allergy 56(4):295–299.

Li CS, Hsu LY. 1997. Airborne fungus allergen in association with residential characteristics in atopic and control children in a subtropical region. Archives of Environmental Health 52(1):72–79.

Li DW, Kendrick B. 1995. A year-round study on functional relationships of airborne fungi with meteorological factors. International Journal of Biometeorology 39(2):74–80.

Licorish K, Novey HS, Kozak P, Fairshter RD, Wilson AF. 1985. Role of *Alternaria* and *Penicillium* spores in the pathogenesis of asthma. Journal of Allergy and Clinical Immunology 76(6):819–825.

Lind P, Norman PS, Newton M, Lowenstein H, Schwartz B. 1987. The prevalence of indoor allergens in the Baltimore area: house dust-mite and animal-dander antigens measured by immunochemical techniques. Journal of Allergy and Clinical Immunology 80(4):541–547.

Litonjua AA, Sparrow D, Weiss ST, O'Connor GT, Long AA, Ohman JL Jr. 1997. Sensitization to cat allergen is associated with asthma in older men and predicts new-onset airway hyperresponsiveness. The Normative Aging Study. American Journal of Respiratory and Critical Care Medicine 156(1):23–27.

Longbottom JL, Austwick PK. 1987. Allergy to rats: quantitative immuno-electrophoretic studies of rat dust as a source of inhalant allergen. Journal of Allergy and Clinical Immunology 80(3 Pt 1):243–251.

Luczynska CM, Arruda LK, Platts-Mills TA, Miller JD, Lopez M, Chapman MD. 1989. A two-site monoclonal antibody ELISA for the quantitation of the major *Dermatophagoides* spp. allergens, Der p I and Der f I. Journal of Immunological Methods 118(2):227–235.

Luyt DK, Davis G, Dance M, Simmank K, Patel D. 1995. Clinical characteristics of black asthmatic children. South African Medical Journal 85(10):999–1001.

Madelin TM, Johnson HE. 1992. Fungal and actinomycete spore aerosols measured at different humidities with an aerodynamic particle sizer. Journal of Applied Bacteriology 72:400–409.

Malling HJ. 1986. Diagnosis and immunotherapy of mould allergy. IV. Relation between asthma symptoms, spore counts and diagnostic tests. Allergy 41(5):342–350.

Martinez, FD. 1994. Role of viral infections in the inception of asthma and allergies during childhood: Could they be protective? Thorax 49(12):1189–1191.

Martinez FD. 1995. Viral infections and the development of asthma. American Journal of Respiratory and Critical Care Medicine 151(5):1644–1648.

Medzhitov R., Preston-Hurlburt P, Janeway CA Jr. 1997. A human homologue of the *Drosophila* Toll protein signals activation of adaptive immunity. Nature 388(6640):394–397.

Michel O, Duchateau J, Sergysels R. 1989. Effect of inhaled endotoxin on bronchial reactivity in asthmatic and normal subjects. Journal of Applied Physiology 66(3):1059–1064.

Michel O, Kips J, Duchateau J, Vertongen F, Robert L, Collet H, Pauwels R, Sergysels R. 1996. Severity of asthma is related to endotoxin in house dust. American Journal of Respiratory and Critical Care Medicine 154:1641–1646.

Micillo E, Marcatili P, Palmieri S, Mazzarella G. 1998. Viruses and asthmatic syndromes. Monaldi Archives for Chest Disease 53(1):88–91.

Miles EA, Warner JA, Jones AC, Colwell BM, Bryant TM, Warner JO. 1996. Peripheral blood mononuclear cell proliferative responses in the first year of life in babies born to allergic parents. Clinical and Experimental Allergy 26(7):780–788.

Milton DK. 1999. Endotoxin and other bacterial cell-wall components. In: Bioaerosols: Assessment and Control. Macher J, Milton DK, Burge HA, Morey P, Eds. Cincinnati, OH: American Conference of Governmental Industrial Hygienists.

Milton DK, Johnson DK. 1995. Endotoxin exposure assessment in machining operations. In: The Industrial Metalworking Environment: Assessment and Control. Dearborn, MI: American Automobile Manufacturers Association.

Milton DK, Wypij D, Kriebel D, Walters M, Hammond SK, Evans JS. 1996. Endotoxin exposure–response in a fiberglass manufacturing plant. American Journal of Industrial Medicine 29(1):3–13.

Milton DK, Johnson DK, Park JH. 1997. Environmental endotoxin measurement: interference and sources of variation in the *Limulus* assay of house dust. American Industrial Hygiene Association Journal 58:861–867.

Miyamoto T, Oshima S, Ishizaka T, Sato SH. 1968. Allergenic identity between the common floor mite (*Dermatophagoides farinae*, Hughes 1961) and house dust as a causative antigen in bronchial asthma. Journal of Allergy and Clinical Immunology 42(1):14–28.

Miyashita N, Kubota Y, Nakajima M, Niki Y, Kawane H, Matsushima T. 1998. *Chlamydia pneumoniae* and exacerbations of asthma in adults. Annals of Allergy, Asthma, and Immunology 80(5):405–409.

Morgenstern JP, Griffith IJ, Brauer AW, Rogers BL, Bond JF, Chapman MD, Kuo MC. 1991. Amino acid sequence of *Fel d* I, the major allergen of the domestic cat: protein sequence analysis and cDNA cloning. Proceedings of the National Academy of Sciences of the United States of America 88(21):9690–9694.

Muilenberg ML, Skellenger WS, Burge HA, Solomon WR. 1991. Particle penetration into the automotive interior. I. Influence of vehicle speed and ventilatory mode. Journal of Allergy and Clinical Immunology 87(2):581–585.

Munir AK, Andersson R, Einersson R, Schou C, Dreborg S. 1992. The amount of dog allergen in dust from Swedish schools is much higher than that of cat. Allergy 47:223.

Munir AK, Bjorksten B, Einarsson R, Schou C, Ekstrand-Tobin A, Warner A, Kjellman NI. 1994. Cat (*Fel d* I), dog (*Can f* I), and cockroach allergens in homes of asthmatic children from three climatic zones in Sweden. Allergy 49(7):508–516.

Munuera Giner M, Carrion Garcia JS, Garcia Selles J. 1998. Incidence of *Alternaria* spores in the atmosphere of Murcia (SE Spain). Seasonal, monthly and intradiurnal variations. Journal of Investigational Allergology and Clinical Immunology 8(5):304–308.

Murayama T, Amitani R, Ikegami Y, Nawada R, Lee WJ, Kuze F. 1996. Suppressive effects of *Aspergillus fumigatus* culture filtrates on human alveolar macrophages and polymorphonuclear leucocytes. European Respiratory Journal 9(2):293–300.

Murray AB, Ferguson AC. 1983. Dust-free bedrooms in the treatment of asthmatic children with house dust or house dust mite allergy: a controlled trial. Pediatrics 71(3):418–422.

Murray M, Webb MS, O'Callaghan C, Swarbrick AS, Milner AD. 1992. Respiratory status and allergy after bronchiolitis. Archives of Disease in Childhood 67(4):482–487.

Neas LM, Dockery DW, Burge H, Koutrakis P, Speizer FE. 1996. Fungus spores, air pollutants, and other determinants of peak expiratory flow rate in children. American Journal of Epidemiology 143(8):797–807.

Nelson HS, Ikle D, Buchmeier A. 1996. Studies of allergen extract stability: the effects of dilution and mixing. Journal of Allergy and Clinical Immunology 98(2):382–388.

Nelson RP Jr, DiNicolo R, Fernandez-Caldas E, Seleznick MJ, Lockey RF, Good RA. 1996. Allergen-specific IgE levels and mite allergen exposure in children with acute asthma first seen in an emergency department and in nonasthmatic control subjects. Journal of Allergy and Clinical Immunology 98(2):258–263.

Newman-Taylor AJ. 1982. Laboratory animal allergy. European Journal of Respiratory Diseases 63:60–64.

Newson R, Strachan D, Archibald E, Emberlin J, Hardaker P, Collier C. 1997. Effect of thunderstorms and airborne grass pollen on the incidence of acute asthma in England, 1990–94. Thorax 52(8):680–685. [Comment in Thorax 1997. 52(8):669–670.]

Newson R, Strachan D, Archibald E, Emberlin J, Hardaker P, Collier C. 1998. Acute asthma epidemics, weather and pollen in England, 1987–1994. European Respiratory Journal 11(3):694–701.

Nicholson KG, Kent J, Ireland DC. 1993. Respiratory viruses and exacerbations of asthma in adults. British Medical Journal 307(6910):982–986.

Nicolai T, Illi S, von Mutius E. 1998. Effect of dampness at home in childhood on bronchial hyperreactivity in adolescence. Thorax 53(12):1035–1040.

Nikulin M, Reijula K, Jarvis BB, Hintikka EL. 1996. Experimental lung mycotoxicosis in mice induced by *Stachybotrys atra*. International Journal of Experimental Pathology 77(5):213–218.

Nikulin M, Reijula K, Jarvis BB, Veijalainen P, Hintikka EL. 1997. Effects of intranasal exposure to spores of *Stachybotrys atra* in mice. Fundamental and Applied Toxicology 35(2):182–188.

Norbäck D, Björnsson E, Janson C, Palmgren U, Boman G. 1999. Current asthma and biochemical signs of inflammation in relation to building dampness in dwellings. International Journal of Tuberculosis and Lung Disease 3(5):368–376.

Norman PS, Ohman JL Jr, Long AA, Creticos PS, Gefter MA, Shaked Z, Wood RA, Eggleston PA, Hafner KB, Rao P, Lichtenstein LM, Jones NH, Nicodemus CF. 1996. Treatment of cat allergy with T-cell reactive peptides. American Journal of Respiratory and Critical Care Medicine 154(6 Pt 1):1623–1628.

Nowak D, Heinrich J, Jorres R, Wassmer G, Berger J, Beck E, Boczor S, Claussen M, Wichmann HE, Magnussen H. 1996. Prevalence of respiratory symptoms, bronchial hyperresponsiveness and atopy among adults: West and East Germany. European Respiratory Journal 9(12):2541–2552.

O'Hollaren MT, Yunginger JW, Offord KP, Somers MJ, O'Connell EJ, Ballard DJ, Sachs MI. 1991. Exposure to an aeroallergen as a possible precipitating factor in respiratory arrest in young patients with asthma. New England Journal of Medicine 324(6):359–363.

O'Neil CE, Horner WE, Reed MA, Lopez M., Lehrer SB. 1990. Evaluation of basidiomycete and deuteromycete *(Fungi imperfecti)* extracts for shared allergenic determinants. Clinical and Experimental Allergy 20(5):533–538.

Ordaz VA, Castaneda CB, Campos CL, Rodriguez VM, Saenz JG, Rios PC. 1998. Asthmatic exacerbations and environmental pollen concentration in La Comarca Lagunera (Mexico). Revista Alergia Mexico 45(4):106–111.

Ormstad H, Johansen BV, Gaarder PI. 1998. Airborne house dust particles and diesel exhaust particles as allergen carriers. Clinical and Experimental Allergy 28(6):702–708.

Orta JC, Navarro AM, Bartolome B, Delgado J, Martinez J, Sanchez MC, Martinez A, Valverdu A, Conde J, Palacios R. 1998. Comparative allergenic study of *Tetranychus urticae* from different sources. Journal of Investigational Allergology and Clinical Immunology 8(3):149–154.

Park JH. 1999. Endotoxin in the home: exposure assessment and health effects. Environmental Health. Harvard University School of Public Health, Boston, MA.

Pasanen AL, Niininen M, Kalliokoski P, Nevalainen A, Jantunen MJ. 1992. Airborne *Cladosporidium* and other fungi in damp versus reference residences. Atmospheric Environment 26B(1):121–124.

Pattemore PK, Johnston SL, Bardin PG. 1992. Viruses as precipitants of asthma symptoms. I. Epidemiology. Clinical and Experimental Allergy 22(3):325–336.

Peat JK, Toelle BG, Gray EJ, Haby MM, Belousova E, Mellis CM, Woolcock AJ. 1995. Prevalence and severity of childhood asthma and allergic sensitisation in seven climatic regions of New South Wales. Medical Journal of Australia 163(1):22–26 [see comments].

Peat JK, Tovey E, Toelle BG, Haby MM, Gray EJ, Mahmic A, Woolcock AJ. 1996. House dust mite allergens. A major risk factor for childhood asthma in Australia. American Journal of Respiratory and Critical Care Medicine 153(1):141–146.

Perzanowski MS, Sporik R, Squillace SP, Gelber LE, Call R, Carter M, Platts-Mills TA. 1998. Association of sensitization to *Alternaria* allergens with asthma among school-age children. Journal of Allergy and Clinical Immunology 101(5):626–632.

Piacentini GL, Martinati L, Mingoni S, Boner AL. 1996. Influence of allergen avoidance on the eosinophil phase of airway inflammation in children with allergic asthma. Journal of Allergy and Clinical Immunology 97(5):1079–1084.

Platts-Mills TA, Tovey ER, Mitchell EB, Moszoro H, Nock P, Wilkins SR. 1982. Reduction of bronchial hyperreactivity during prolonged allergen avoidance. Lancet 2(8300):675–678.

Platts-Mills TA, Longbottom J, Edwards J, Cockroft A, Wilkins S. 1987. Occupational asthma and rhinitis related to laboratory rats: serum IgG and IgE antibodies to the rat urinary allergen. Journal of Allergy and Clinical Immunology 79(3):505–515.

Platts-Mills TA, de Weck AL. 1988. Dust mite allergens and asthma: a worldwide problem. Bulletin of the World Health Organization 66(6):769–780.

Platts-Mills TA, Vervloet D, Thomas WR, Aalberse RC, Chapman MD. 1997. Indoor allergens and asthma: report of the Third International Workshop, Cuenca, Spain. Journal of Allergy and Clinical Immunology 100(6 Pt 1):S2–S24.

Pollart SM, Mullins DE, Vailes LD, Hayden ML, Platts-Mills TA, Sutherland WM, Chapman MD. 1991a. Identification, quantitation, and purification of cockroach allergens using monoclonal antibodies. Journal of Allergy and Clinical Immunology 87(2):511–521.

Pollart SM, Smith TF, Morris EC, Gerber LE, Platts-Mills TA. 1991b. Environmental exposure to cockroach allergens: analysis with monoclonal antibody-based enzyme immunoassays. Journal of Allergy and Clinical Immunology 87(2):505–510.

Potekhin SA, Loseva OI, Tiktopulo EI, Dobritsa AP. 1999. Transition state of the rate-limiting step of heat denaturation of Cry3A delta-endotoxin. Biochemistry 38(13):4121–4127.

Prahl P, Roed-petersen J. 1979. Type I allergy from cows in veterinary surgeons. Contact Dermatitis 5(1):33–38.

Prescott SL, Macaubas C, Smallacombe T, Holt BJ, Sly PD, Loh R, Holt PG. 1998. Reciprocal age-related patterns of allergen-specific T-cell immunity in normal vs. atopic infants. Clinical and Experimental Allergy 5(Supplement):39–44; discussion 50–51.

Pugin J, Schurer-Maly CC, Leturcq D, Moriarty A, Ulevitch RJ, Tobias PS. 1993. Lipopolysaccharide activation of human endothelial and epithelial cells is mediated by lipopolysaccharide-binding protein and soluble CD14. Proceedings of the National Academy of Sciences of the United States of America 90(7):2744–2748.

Pullan CR, Hey EN. 1982. Wheezing, asthma, and pulmonary dysfunction 10 years after infection with respiratory syncytial virus in infancy. British Medical Journal (Clinical Research Edition) 284(6330):1665–1669.

Rains N, Siebers R, Crane J, Fitzharris P. 1999. House dust mite allergen (*Der p* I) accumulation on new synthetic and feather pillows [see comments]. Clinical and Experimental Allergy 29(2):182–185.

Rakes GP, Arruda E, Ingram JM, Hoover GE, Zambrano JC, Hayden FG, Platts-Mills TAE, Heymann PW. 1999. Rhinovirus and respiratory syncytial virus in wheezing children requiring emergency care. American Journal of Respiratory and Critical Care Medicine 159(3):785–790.

Rao CY. 1999. A quantitative risk assessment for *Stachybotrys chartarum* exposure. D. Sc. Thesis, Department of Environmental Health, Harvard School of Public Health, Boston. 152 pages.

Rautalahi M, Terho EO, Vohlonen I, Husman K. 1987. Atopic sensitization of dairy farmers to work-related and common allergens. European Journal of Respiratory Diseases 71(Supplement 152):155–164.

Rautiala S, Reponen T, Nevalainen A, Husman T, Kalliokoski P. 1998. Control of exposure to airborne viable microorganisms during remediation of moldy buildings; report of three case studies. American Industrial Hygiene Association Journal 59(7):455–460.

Richard JL, Thurston JR. 1975. Effect of aflatoxin on phagocytosis of *Aspergillus fumigatus* spores by rabbit alveolar macrophages. Applied Microbiology 30(1):44–47.

Richards G, Kolbe J, Fenwick J, Rea H. 1995. The effects of privet exposure on asthma morbidity. New Zealand Medical Journal 108(996):96–99.

Rietschel ET, Brade H. 1992. Bacterial endotoxins. Scientific American 267(2): 55–61.

Robinson D, Hamid Q, Bentley A, Ying S, Kay AB, Durham SR. 1993. Activation of CD4+ T cells, increased TH2-type cytokine mRNA expression, and eosinophil recruitment in bronchoalveolar lavage after allergen inhalation challenge in patients with atopic asthma. Journal of Allergy and Clinical Immunology 92(2):313–324.

Romagnani S. 1992. Human TH1 and TH2 subsets: regulation of differentiation and role in protection and immunopathology. International Archives of Allergy Immunology 98(4):279–285.

Roman M, Calhoun WJ, Hinton KL, Avendano LF, Simon V, Escobar AM, Gaggero A, Diaz PV. 1997. Respiratory syncytial virus infection in infants is associated with predominant Th-2-like response. American Journal of Respiratory and Critical Care Medicine 156(1):190–195.

Ronmark E, Lundback B, Jonsson E, Platts-Mills T. 1998. Asthma, type-1 allergy and related conditions in 7-and 8-year-old children in northern Sweden: prevalence rates and risk factor pattern. Respiratory Medicine 92(2):316–324.

Rosas I, McCartney HA, Payne RW, Calderon C, Lacey J, Chapela R, Ruiz-Velazco S. 1998. Analysis of the relationships between environmental factors (aeroallergens, air pollution, and weather) and asthma emergency admissions to a hospital in Mexico City. Allergy 53(4):394–401.

Rosenstreich DL, Eggleston P, Kattan M, Baker D, Slavin RG, Gergen P, Mitchell H, McNiff-Mortimer K, Lynn H, Ownby D, Malveaux F. 1997. The role of cockroach allergy and exposure to cockroach allergen in causing morbidity among inner-city children with asthma. New England Journal of Medicine 336(19):1356–1363. [Comment in N Engl J Med 1997. 336(19):1382–1384 and 337(11):791–792.]

Rylander R, Haglind P, Lundholm M, Mattsby I, Stenqvist K. 1978. Humidifier fever and endotoxin exposure. Clinical Allergy 8(5):511–516.

Rylander R, Morey P. 1982. Airborne endotoxin in industries processing vegetable fibers. American Industrial Hygiene Association Journal 43(11):811–812.

Saito K, Nishijima M, Ohno N, Nagi N, Yadomae T, Miyazaki T. 1992. Activation of complement and *Limulus* coagulation system by an alkali-soluble glucan isolated from *Omphalia lapidescens* and its less branched derivatives. Chemical and Pharmaceutical Bulletin 40:1227–1230.

Sakaguchi M, Inouye S, Miyazawa H, Kamimura H, Kimura M, Yamazaki S. 1990a. Evaluation of countermeasures for reduction of mouse airborne allergens. Laboratory Animal Science 40(6):613–615.

Sakaguchi M, Inouye S, Yasueda H, Irie T, Yoshizawa S, Shida T. 1990b. Measurement of allergens associated with dust mite allergy. II. Concentrations of airborne mite allergens (Der I and Der II) in the house. International Archives of Allergy and Applied Immunology 90(2):190–193.

Sakamoto T, Ito K, Yamada M, Iguchi H, Ueda M, Matsuda Y, Torii S. 1990. Allergenicity of the osmophilic fungus *Aspergillus restrictus* evaluated by skin prick test and radioallergosorbent test [Japanese]. Arerugi—Japanese Journal of Allergology 39(11):1492–1498.

Sakurai T, Ohno N, Yadomae T. 1994. Changes in immune mediators in mouse lung produced by administration of soluble (1→3)-beta-D-glucan. Biological and Pharmaceutical Bulletin 17(5):617–622.

Sanchez H, Bush RK. 1994. Complete sequence of a cDNA encoding an *Alternaria* allergen. Journal of Allergy and Clinical Immunology 93:208 [abstract].

Santiago-Schwarz F, McHugh DM, Fleit HB. 1989. Functional analysis of monocyte-macrophages derived from nonadherent cord blood progenitor cells: correlation with the ontogeny of cell surface proteins. Journal of Leukocyte Biology 46(3):230–238.

Saraf A, Larsson L, Burge H, Milton D. 1997. Quantification of ergosterol and 3-hydroxy fatty acids in settled house dust by gas chromatography mass spectrometry—comparison with fungal culture and determination of endotoxin by a *Limulus* amebocyte lysate assay. Applied and Environmental Microbiology 63:2554–2559.

Saraf A, Park JH, Milton DK, Larsson L.1999. Use of quadrupole GC-MS and iontrap GC-MSMS for determining 3-hydroxy fatty acids in settled house dust: relation to endotoxin activity. Journal of Environmental Monitoring 2:163–168.

Sarpong SB, Hamilton RG, Eggleston PA, Adkinson NF Jr. 1996a. Socioeconomic status and race as risk factors for cockroach allergen exposure and sensitization in children with asthma. Journal of Allergy and Clinical Immunology 97(6):1393–1401.

Sarpong SB, Wood RA, Eggleston PA. 1996b. Short-term effect of extermination and cleaning on cockroach allergen *Bla g* II in settled dust. Annals of Allergy, Asthma, and Immunology 76(3):257–260.

Sarpong SB, Wood RA, Karrison T, Eggleston PA. 1997. Cockroach allergen (*Bla g* 1) in school dust. Journal of Allergy and Clinical Immunology 99(4):486–492.

Sarpong SB, Karrison T. 1998a. Season of birth and cockroach allergen sensitization in children with asthma. Journal of Allergy and Clinical Immunology 101(4 Pt 1):566–568.

Sarpong SB, Karrison T. 1998b. Skin test reactivity to indoor allergens as a marker of asthma severity in children with asthma. Annals of Allergy, Asthma, and Immunology 80(4):303–308.

Sarpong SB, Han Y. 1999. A threshold of cockroach allergen (*Bla g* II) exposure and sensitization. American Journal of Respiratory and Critical Care Medicine 156(3):A128.

Sastre J, Ibanez MD, Lombardero M, Laso MT, Lehrer S. 1996. Allergy to cockroaches in patients with asthma and rhinitis in an urban area (Madrid). Allergy 51(8):582–586.

Schappi GF, Suphioglu C, Taylor PE, Knox RB. 1997. Concentrations of the major birch tree allergen *Bet v* I in pollen and respirable fine particles in the atmosphere. Journal of Allergy and Clinical Immunology 100(5):656–661.

Schenkelberger V, Freitag M, Altmeyer P. 1998. *Ficus benjamina*—the hidden allergen in the house [German]. Hautarzt 49(1):2–5.

Schmidt M, Zargari A, Holt P, Lindbom L, Hellman U, Whitley P, Van Der ploey I, Harfast B, Scheynius A. 1997. The complete cDNA sequence and expression of the first major allergenic protein of *Malasseziu furfur*, Mal f I. European Journal of Biochemistry 246(1):181–185.

Schou C. 1993. Defining allergens of mammalian origin. Clinical and Experimental Allergy 23(1):7–14.

Schou C, Lind P, Fernandez-Caldas E, Lockey RF, Lowenstein H. 1990. Identification and purification of an important cross-reactive allergen from American (*Periplaneta americana*) and German (*Blattella germanica*) cockroach. Journal of Allergy and Clinical Immunology 86(6 Pt 1):935–946.

Schulz O, Sewell HF, Shakib F. 1998. Proteolytic cleavage of CD25, the alpha subunit of the human T cell interleukin 2 receptor, by Der p I, a major mite allergen with cysteine protease activity. Journal of Experimental Medicine 187(2):271–275.

Schumacher MJ, Tait BD, Holmes MC. 1981. Allergy to murine antigens in a biological research institute. Journal of Allergy and Clinical Immunology 68(4):310–318.

Schwandner R, Dziarski R, Wesche H, Rothe M, Kirschning CJ. 1999. Peptidoglycan- and lipoteichoic acid-induced cell activation is mediated by toll-like receptor 2. Journal of Biological Chemistry 274(25):17406–17409.

Scott R, Pullan CR, Scott M, McQuillin J. 1984. Cell-mediated immunity in respiratory syncytial virus disease. Journal of Medical Virology 13(1):105–114.

Seabury J, Becker B, Salvaggio J. 1976. Home humidifier thermophilic actinomycete isolates. Journal of Allergy and Clinical Immunology 57(2):174–176.

Sears MR, Herbison GP, Holdaway MD, Hewitt CJ, Flannery EM, Silva PA. 1989. The relative risks of sensitivity to grass pollen, house dust mite, and cat dander in the development of childhood asthma. Clinical and Experimental Allergy 19(4):419–424.

Seggev JS, Lis I, Siman-Tov R, Gutman R, Abu-Samara H, Schey G, Naot Y. 1986. *Mycoplasma pneumoniae* is a frequent cause of exacerbation of bronchial asthma in adults. Annals of Allergy 57(4):263–265.

Shahan TA, Sorenson WG, Paulauskis JD, Morey R, Lewis DM. 1998. Concentration- and time-dependent upregulation and release of the cytokines MIP-2, KC, TNF, and MIP-1alpha in rat alveolar macrophages by fungal spores implicated in airway inflammation. American Journal of Respiratory Cell and Molecular Biology 18(3):435–440.

Shaheen SO. 1995. Changing patterns of childhood infection and the rise in allergic disease. Clinical and Experimental Allergy 25(11):1034–1037.

Shaver JR, Zangrilli JG, Cho SK, Cirelli RA, Pollice M, Hastie AT, Fish JE, Peters SP. 1997. Kinetics of the development and recovery of the lung from IgE-mediated inflammation: dissociation of pulmonary eosinophilia, lung injury, and eosinophil-active cytokines. American Journal of Respiratory and Critical Care Medicine 155(2):442–448.

Shen HD, Liaw SF, Lin WL, Ro LH, Yang HL, Han SH. 1995. Molecular cloning of cDNA encoding for the 68 kDA allergen of *Penicillium notatum* using MoAbs. Clinical and Experimental Allergy 25(4):350–356.

Shen HD, Au LC, Lin WL, Liaw SF, Tsai JJ, Han SH. 1997. Molecular cloning and expression of *Penicillium citrinum* allergen with sequence homology and antigenic crossreactivity to a hsp 70 human heat shock protein. Clinical and Experimental Allergy 27(6):682–690.

Shen HD, Lin WL, Tam MF, Wang SR, Tsai JJ, Chou H, Han SH. 1998. Alkaline serine proteinase: a major allergen of *Aspergillus oryzae* and its cross-reactivity with *Penicillium citrinum*. International Archives of Allergy and Immunology 116(1):29–35.

Shulaner FA. 1970. Sensitivity to the cockroach in three groups of allergic children. Pediatrics 45(3):465–466.

Sicherer SH, Wood RA, Eggleston PA. 1997. Determinants of airway responses to cat allergen: comparison of environmental challenge to quantitative nasal and bronchial allergen challenge. Journal of Allergy and Clinical Immunology 99(6 Pt 1):798–805.

Sietsma JH, Wessels JG. 1981. Solubility of (1 leads to 3)-beta-D/(1 leads to 6)-beta-D-glucan in fungal walls: importance of presumed linkage between glucan and chitin. Journal of General Microbiology 125(Pt 1):209–212.

Sigsgaard T, Hjort C, Omland O, Miller MR, Pedersen OF. 1996. Skin prick test to house dust mite, asthma and hyperreactivity in a cohort of young rurals. European Respiratory Journal 9(Supplement):379s.

Sigurs N, Bjarnason R, Sigurbergsson F, Kjellman B, Bjorksten B. 1995. Asthma and immunoglobulin E antibodies after respiratory syncytial virus bronchiolitis: a prospective cohort study with matched controls. Pediatrics 95(4):500–505.

Skjonsberg OH, Clench-Aas J, Leegaard J, Skarpaas IJ, Giaever P, Bartonova A, Moseng J. 1995. Prevalence of bronchial asthma in schoolchildren in Oslo, Norway. Comparison of data obtained in 1993 and 1981. Allergy 50(10): 806–810.

Slezak JA, Persky VW, Kviz FJ, Ramakrishnan V, Byers C. 1998. Asthma prevalence and risk factors in selected Head Start sites in Chicago. Journal of Asthma 35(2):203–212.

Smith JM, Disney ME, Williams JD, Goels ZA. 1969. Clinical significance of skin reactions to mite extracts in children with asthma. British Medical Journal 1(659):723–726.

Snapper CM, Pecanha LM, Levine AD, and Mond JJ. 1991. IgE class switching is critically dependent upon the nature of the B cell activator, in addition to the presence of IL-4. Journal of Immunology 147(4):1163–1170.

Solomon KR., Kurt-Jones EA, Saladino RA, Stack AM, Dunn IF, Ferretti M, Golenbock D, Fleisher GR, Finberg RW. 1998. Heterotrimeric G proteins physically associated with the lipopolysaccharide receptor CD14 modulate both *in vivo* and *in vitro* responses to lipopolysaccharide. Journal of Clinical Investigations 102(11):2019–2027.

Solomon WR, Burge HA, Boise JR. 1980. Exclusion of particulate allergens by window air conditioners. Journal of Allergy and Clinical Immunology 65(4):305–308.

Solomon WR, Burge HA, Muilenberg ML. 1983. Allergen carriage by atmospheric aerosol. I. Ragweed pollen determinants in smaller micronic fractions. Journal of Allergy and Clinical Immunology 72(5 Pt 1):443–447.

Sonesson HRA, Zähringer U, Grimmecke HD, Westphal O, Rietschel ET. 1994. Bacterial Endotoxin: chemical structure and biological activity. In: Endotoxin and the Lungs. Brigham K., Ed. New York: Marcel Dekker, Inc.

Sorenson WG, Simpson J, Castranova V. 1985. Toxicity of the mycotoxin patulin for rat alveolar macrophages. Environmental Research 38(2):407–416.

Sorenson WG, Simpson J. 1986. Toxicity of penicillic acid for rat alveolar macrophages in vitro. Environmental Research 41(2):505–513.

Sorenson WG, Gerberick GF, Lewis DM, Castranova V. 1986. Toxicity of mycotoxins for the rat pulmonary macrophage in vitro. Environmental Health Perspectives 66:45–53.

Soto-Quiros M, Gutierrez I, Calvo N, Araya C, Karlberg J, Hanson LA, Belin L. 1998. Allergen sensitization of asthmatic and nonasthmatic schoolchildren in Costa Rica. Allergy 53(12):1141–1147.

Sporik R, Holgate ST, Platts-Mills TA, Cogswell JJ. 1990. Exposure to house-dust mite allergen (*Der p* I) and the development of asthma in childhood. A prospective study. New England Journal of Medicine 323(8):502–507. [Comment in: N Engl J Med 1991. 324(5):337–338.]

Sporik R, Ingram JM, Price W, Sussman JH, Honsinger RW, Platts-Mills TA. 1995. Association of asthma with serum IgE and skin test reactivity to allergens among children living at high altitude. Tickling the dragon's breath. American Journal of Respiratory and Critical Care Medicine 151(5):1388–1392. [Comment in: Am J Respir Crit Care Med 1997. 155(2):769–770.]

Squillace SP, Sporik RB, Rakes G, Couture N, Lawrence A, Merriam S, Zhang J, Platts-Mills TA. 1997. Sensitization to dust mites as a dominant risk factor for asthma among adolescents living in central Virginia. Multiple regression analysis of a population-based study. American Journal of Respiratory and Critical Care Medicine 156(6):1760–1764.

Staib F. 1992. Pathogenic fungi in human dwellings. Mycoses 35(11–12):289–292.

Staib F. 1996. Fungi in the home and hospital environment [German]. Mycoses 39(Suppl 1):26–29.

Staib F, Folkens U, Tompak B, Abel T, Thiel D. 1978a. A comparative study of antigens of *Aspergillus fumigatus* isolates from patients and soil of ornamental plants in the immunodiffusion test. Zentralblatt für Bakteriologie, Parasitenkunde, Infektionskrankheiten und Hygiene—Erste Abteilung Originale—Reihe A: Medizinische Mikrobiologie und Parasitologie 242(1): 93–99.

Staib F, Tompak B, Thiel D, Blisse A. 1978b. *Aspergillus fumigatus* and *Aspergillus niger* in two potted ornamental plants, cactus (*Epiphyllum truncatum*) and clivia (*Clivia miniata*). Biological and epidemiological aspects. Mycopathologia 66(1–2):27–30.

Staib F, Mishra SK, Blisse A. 1980. Interaction between *Aspergilli* and *Streptomycetes* in the soil of potted indoor plants: a preliminary report (contribution to the epidemiology of human aspergillosis). Mycopathologia 70(1):9–12.

Stewart GA, Thompson PJ. 1996. The biochemistry of common aeroallergens. Clinical and Experimental Allergy 26(9):1020–1044. [Comment in: Clin Exp Allergy 1997. 27(6):714–715.]

Strachan DP. 1988. Damp housing and childhood asthma: validation of reporting of symptoms. British Medical Journal 297(6658):1223–1226.

Strachan DP, Carey IM. 1995. Home environment and severe asthma in adolescence: a population based case-control study. British Medical Journal 311(7012):1053–1056.

Strand V, Rak S, Svartengren M, Bylin G. 1997. Nitrogen dioxide exposure enhances asthmatic reaction to inhaled allergen in subjects with asthma. American Journal of Respiratory and Critical Care Medicine 155(3):881–887.

Strand V, Svartengren M, Rak S, Barck C, Bylin G. 1998. Repeated exposure to an ambient level of NO$_2$ enhances asthmatic response to a nonsymptomatic allergen dose. European Respiratory Journal 12(1):6–12.

Su HJ, Rotnitzky A, Burge HA, Spengler JD. 1992. Examination of fungi in domestic interiors by using factor analysis: correlations and associations with home factors. Applied and Environmental Microbiology 58:181–186.

Subiza J, Cabrera M, Valdivieso R, Subiza JL, Jerez M, Jimenez JA, Narganes MJ, Subiza E. 1994. Seasonal asthma caused by airborne *Platanus* pollen. Clinical and Experimental Allergy 24(12):1123–1129.

Suda T, Sato A, Ida M, Gemma H, Hayakawa H, Chida K. 1995. Hypersensitivity pneumonitis associated with home ultrasonic humidifiers. Chest 107(3): 711–717.

Sue MA, Gordon EH, Freund LH. 1993. Utility of additional skin testing in "nonallergic" asthma. Annals of Allergy 68(5):395–397.

Sugawara S, Arakaki R, Rikiishi H, Takada H. 1999. Lipoteichoic acid acts as an antagonist and an agonist of lipopolysaccharide on human gingival fibroblasts and monocytes in a CD14-dependent manner. Infection and Immunology 67(4):1623–1632.

Suliaman FA, Holmes WF, Kwick S, Khouri F, Ratard R. 1997. Pattern of immediate type hypersensitivity reactions in the Eastern Province, Saudi Arabia. Annals of Allergy 78(4):415–418.

Summerbell RC, Krajden S, Kane J. 1989. Potted plants in hospitals as reservoirs of pathogenic fungi. Mycopathologia 106(1):13–22.

Suoniemi I, Bjorksten F, Haahtela T. 1981. Dependence of immediate hypersensitivity in the adolescent period on factors encountered in infancy. Allergy 36(4):263–268.

Suphioglu C. 1998. Thunderstorm asthma due to grass pollen. International Archives of Allergy and Immunology 116(4):253–260. [Published erratum appears in Int Arch Allergy Immunol 1999. 119(2):37.]

Sutton P, Waring O, Mullbacher A. 1996. Exacerbation of invasive aspergillosis by the immunosuppressive fungal metabolite, gliotoxin. Immunology and Cell Biology 74(4):318–322.

Svanes C, Jarvis D, Chinn S, Burney P. 1999. Childhood environment and adult atopy: results from the European Community Respiratory Health Survey. Journal of Allergy and Clinical Immunology 103(3 Pt 1):415–420.

Svartengren M, Falk R, Linnman L, Philipson K, Camner P. 1987. Deposition of large particles in human lung. Experimental Lung Research 12(1):75–88.

Swanson MC, Agarwal MK, Reed CE. 1985. An immunochemical approach to indoor aeroallergen quantitation with a new volumetric air sampler: studies with mite, roach, cat, mouse, and guinea pig antigens. Journal of Allergy and Clinical Immunology 76(5):724–729.

Szantho A, Osvath P, Horvath Z, Novak EK, Kujalek E. 1992. Study of mold allergy in asthmatic children in Hungary. Journal of Investigational Allergology and Clinical Immunology 2(2):84–90.

Takahashi T. 1997. Airborne fungal colony-forming units in outdoor and indoor environments in Yokohama, Japan. Mycopathologia 139(1):23–33.

Tariq SM, Matthews SM, Stevens M, Hakim EA. 1996. Sensitization to *Alternaria* and *Cladosporium* by the age of 4 years. Clinical and Experimental Allergy 26(7):794–798.

Tauer-Reich I, Fruhmann G, Czuppon AB, Baur X. 1994. Allergens causing bird fancier's asthma. Allergy 49(6):448–453.

Terleckyj B, Axler DA. 1987. Quantitative neutralization assay of fungicidal activity of disinfectants. Antimicrobial Agents and Chemotherapy 31(5):794–798.

Tovey ER, Chapman MD, Platts-Mills TA. 1981a. Mite faeces are a major source of house dust allergens. Nature 289(5798):592–593.

Tovey ER, Chapman MD, Wells CW, Platts-Mills TA. 1981b. The distribution of dust mite allergen in the houses of patients with asthma. American Review of Respiratory Disease 124(5):630–635.

Tovey E, Marks G. 1999. Methods and effectiveness of environmental control. Journal of Allergy and Clinical Immunology 103(2 Pt 1):179–191.

Travis J, Whitworth T, Matheson N, Bagarozzi D Jr. 1996. Proteinases from pollen and pests. Acta Biochimica Polonica 43(3):411–417.

Twiggs JT, Agarwal MK, Dahlberg MJ, Yunginger JW. 1982. Immunochemical measurement of airborne mouse allergens in a laboratory animal facility. Journal of Allergy and Clinical Immunology 69(6):522–526.

Tyndall RL, Lehman E, Bowman EK, Milton DK, Barbaree J. 1995. Home humidifiers as a potential source of exposure to microbial pathogens, endotoxins and allergens. Indoor Air 5:171–178.

Ulevitch RJ. 1999. Endotoxin opens the tollgates to innate immunity. Nature Medicine 5(2):144–145.

Valovirta E, Koivikko A, Vanto T, Viander M, Ingeman L. 1984. Immunotherapy in allergy to dog: a double-blind clinical study. Annals of Allergy 53(1):85–88.

van der Heide S, van Aalderen WM, Kauffman HF, Dubois AE, de Monchy JG. 1999. Clinical effects of air cleaners in homes of asthmatic children sensitized to pet allergens. Journal of Allergy and Clinical Immunology 104(2 Pt 1): 447–451.

Vanto T, Viander M, Koivikko A. 1980. Skin prick test in the diagnosis of dog dander allergy: a comparison of different extracts with clinical history, provocation tests and RAST. Clinical Allergy 10(2):121–132.

Vanto T, Koivikko A. 1983. Dog hypersensitivity in asthmatic children. A clinical study with special reference to the relationship between the exposure to dogs and the occurrence of hypersensitivity symptoms. Acta Paediatrica Scandinavica 72(4):571–575.

Vaughan JW, McLaughlin TE, Perzanowski MS, Platts-Mills TA. 1999. Evaluation of materials used for bedding encasement: effect of pore size in blocking cat and dust mite allergens. Journal of Allergy and Clinical Immunology 103(2 Pt 1):227–231.

Venables KM, Tee RD, Hawkins ER, Gordon DJ, Wale CJ, Farrer NM, Lam TH, Baxter PJ, Newman Taylor AJ. 1988. Laboratory animal allergy in a pharmaceutical company. British Journal of Industrial Medicine 45(10): 660–666.

Venables KM, Allitt U, Collier CG, Emberlin J, Greig JB, Hardaker PJ, Highham JH, Laing-Morton T, Maynard RL, Murray V, Strachan D, Tee RD. 1997. Thunderstorm-related asthma—the epidemic of 24/25 June 1994. Clinical and Experimental Allergy 27(7):725–736.

Verhoeff AP, van Wijnen JH, Brunekreef B, Fischer P, van Reenen-Hoekstra ES, Samson RA. 1992. Presence of viable mould propagules in indoor air in relation to house damp and outdoor air. Allergy 47(2 Pt 1):83–91.

Verma J, Sridhara S, Singh BP, Gangal SV. 1995. Studies on shared antigenic/ allergenic components among fungi. Allergy 50(10):811–816.

von Hertzen L, Töyrylä M, Gimishanov A, Bloigu A, Leinonen M, Saikku P, Haahtela T. 1999. Asthma, atopy and Chlamydia pneumoniae antibodies in adults. Clinical and Experimental Allergy 29(4):522–528.

von Mutius E, Martinez FD, Fritzsch C, Nicolai T, Reitmeir P, Thiemann HH. 1994. Skin test reactivity and number of siblings [see comments]. British Medical Journal 308(6930):692–695.

Voorhorst R, Spieksma FThM, Varekamp N. 1969. House dust mite atopy and the house dust mite Dermatophagoides pteronyssinus (Troussart, 1897). Leiden: Stafleu's Scientific Publishing Co.

Walls AF, Longbottom JL. 1985. Comparison of rat fur, urine, saliva, and other rat allergen extracts by skin testing, RAST, and RAST inhibition. Journal of Allergy and Clinical Immunology 75(2):242–251.

Walters M, Milton DK, Larsson L, Ford T. 1994. Airborne environmental endotoxin: a cross-validation of sampling and analysis techniques. Applied and Environmental Microbiology 60(3):996–1005.

Wang JY, Chen WY. 1992. Inhalant allergens in asthmatic children in Taiwan: comparison evaluation of skin testing, radioallergosorbent test and multiple allergosorbent chemiluminescent assay for specific IgE. Journal of the Formosan Medical Association 91(12):1127–1132.

Waring P, Beaver J. 1996. Gliotoxin and related epipolythiodioxopiperazines. General Pharmacology 27(8):1311–1316.

Waring P, Khan T, Sjaarda A. 1997. Apoptosis induced by gliotoxin is preceded by phosphorylation of histone H3 and enhanced sensitivity of chromatin to nuclease digestion. Journal of Biological Chemistry 272(29):17929–17936.

Warner JA. 1992. Environmental allergen exposure in homes and schools [editorial]. Clinical and Experimental Allergy 22(12):1044–1045.

Warner JA, Little SA, Pollock I, Longbottom JL, Warner JO. 1991. The influence of exposure to house dust mite on sensitization in asthma. Pediatric Allergy and Immunology 1:79–86.

Warner JA, Jones AC, Miles EA, Colwell BM, Warner JO. 1997. Prenatal origins of asthma and allergy. Ciba Foundation Symposium 206:220–208; discussion 228–232.

Weatherstone KB, Rich EA. 1989. Tumor necrosis factor/cachectin and interleukin-1 secretion by cord blood monocytes from premature and term neonates. Pediatric Research 25(4):342–346.

Welliver RC, Wong DT, Sun M, Middleton E Jr, Vaughan RS, Ogra PL. 1981. The development of respiratory syncytial virus-specific IgE and the release of histamine in nasopharyngeal secretions after infection. New England Journal of Medicine 305(15):841–846.

Welliver RC, Sun M, Rinaldo D, Ogra PL. 1986. Predictive value of respiratory syncytial virus–specific IgE responses for recurrent wheezing following bronchiolitis. Journal of Pediatrics 109(5):776–780.

Welliver RC, Duffy L. 1993. The relationship of RSV-specific immunoglobulin E antibody responses in infancy, recurrent wheezing, and pulmonary function at age 7–8 years. Pediatric Pulmonology 15(1):19–27.

Wierenga EA, Snoek M, de Groot C, Chretien I, Bos JD, Jansen HM, Kapsenberg ML. 1990. Evidence for compartmentalization of functional subsets of CD2+ T lymphocytes in atopic patients. Journal of Immunology 144(12):4651–4656.

Williams LW, Reinfried PA, Brenner RJ. 1999. Cockroach extermination does not rapidly reduce allergen in settled dust. Journal of Allergy and Clinical Immunology 104(3 Pt 1):702–703.

Williamson IJ, Martin CJ, McGill G, Monie RD, Fennerty AG. 1997. Damp housing and asthma: a case-control study. Thorax 52(3):229–234.

Wilson NW, Robinson NP, Hogan MB. 1999. Cockroach and other inhalant allergies in infantile asthma. Annals of Allergy, Asthma, and Immunology 83(1):27–30.

Witteman AM, Stapel SO, Sjamsoedin DH, Jansen HM, Aalberse RC, van der Zee JS. 1996. *Fel d* I-specific IgG antibodies induced by natural exposure have blocking activity in skin tests. International Archives of Allergy and Immunology 109(4):369–375.

Wood RA, Chapman MD, Adkinson NF Jr, Eggleston PA. 1989. The effect of cat removal on allergen content in household-dust samples. Journal of Allergy and Clinical Immunology 83(4):730–734.

Wood RA, Johnson EF, Van Natta ML, Chen PH, Eggleston PA. 1998. A placebo-controlled trial of a HEPA air cleaner in the treatment of cat allergy. American Journal of Respiratory and Critical Care Medicine 158(1):115–120.

Woodcock A, Custovic A. 1998. Role of the indoor environment in determining the severity of asthma. Thorax 53 (Suppl 2):S47–51.

Wright SD. 1999. Toll, a new piece in the puzzle of innate immunity. Journal of Experimental Medicine 189(4):605–609.

Wright SD, Ramos RA, Tobias PS, Ulevitch RJ, Mathison JC. 1990. CD14, a receptor for complexes of lipopolysaccharide (LPS) and LPS binding protein. Science 249(4975):1431–1433.

Wuthrich B, Schindler C, Leuenberger P, Ackermann-Liebrich U. 1995. Prevalence of atopy and pollinosis in the adult population of Switzerland (SAPALDIA study). Swiss Study on Air Pollution and Lung Diseases in Adults. International Archives of Allergy and Immunology 106(2):149–156.

Yang RB, Mark MR, Gray A, Huang A, Xie MH, Zhang M, Goddard A, Wood WI, Gurney AL, Godowski PJ. 1998. Toll-like receptor–2 mediates lipopolysaccharide-induced cellular signaling. Nature 395(6699):284–288.

Yano T, Ichikawa Y, Komatu S, Arai S, Oizumi K. 1994. Association of _Mycoplasma pneumoniae_ antigen with initial onset of bronchial asthma. American Journal of Respiratory and Critical Care Medicine 149(5):1348–1353.

Yman L, Brandt R, Ponterius G. 1973. Serum albumin—an important allergen in dog epithelia extract. International Archives of Allergy 44:358–368.

Young RP, Hart BJ, Merrett TG, Read AF, Hopkins JM. 1992. House dust mite sensitivity: interaction of genetics and allergen dosage. Clinical and Experimental Allergy 22:205–211.

Young RP, Dekker JW, Wordsworth BP, Schou C, Pile KD, Matthiesen F, Rosenberg WM, Bell JI, Hopkin JM, Cookson WO. 1993. HLA and HLA-DP genotypes and immunoglobulin E responses to common major allergens. Clinical and Experimental Allergy 24:431–439.

Yunginger JW, Jones RT, Nesheim ME, Geller M. 1980. Studies on _Alternaria_ allergens. III. Isolation of a major allergenic fraction (ALT-I). Journal of Allergy and Clinical Immunology 66(2):138–147.

Zhang K, Petty HR. 1994. Influence of polysaccharides on neutrophil function: Specific antagonists suggest a model for cooperative saccharide-associated inhibition of immune complex-triggered superoxide production. Journal of Cellular Biochemistry 56(2):225–235.

Zielonka TM, Charpin D, Berbis P, Luciani P, Casanova D, Vervloet D. 1994. Effects of castration and testosterone on _Fel d_ I production by sebaceous glands of male cats: I—Immunological assessment [see comments]. Clinical and Experimental Allergy 24(12):1169–1173.

Zimmer S, Pollard V, Marshall GD, Garofalo RP, Traber D, Prough D, Herndon DN. 1996. The 1996 Moyer Award. Effects of endotoxin on the TH1/TH2 response in humans. Journal of Burn Care and Rehabilitation 17(6 Pt 1):491–496.

6
INDOOR CHEMICAL Exposures

Concern in recent years regarding the potential health effects of indoor air exposures, as well as the marked increase in the prevalence of asthma in industrialized countries, has prompted a burgeoning of scientific research on exposure to airborne agents and asthma.

The committee was charged with the task of evaluating the strength of the scientific evidence concerning the possible association between these agents and asthma prevalence and severity. The committee was also tasked with examining possible means of mitigating or preventing exposure to these agents. In this chapter the committee evaluates indoor exposure to chemical agents, addressing the following to the extent permitted by available research:

1. which factors influence exposures to the agent;
2. whether a relationship exists between the agent and asthma prevalence or severity, taking into account the strength of the scientific evidence and the appropriateness of the methods used to detect the relationship;
3. what type of relationship exists between the agent and asthma;
4. whether there are special considerations regarding the

agent (for example, subpopulations at risk and interactions with other exposures);

5. which strategies effectively mitigate or prevent exposure to the agent;

6. whether these strategies only reduce exposures, or decrease the occurrence or exacerbation of asthma; and

7. whether these strategies are reasonable for use by the target populations.

Each section begins by providing a definition of the agent and a summary of the factors that influence exposure. The evidence concerning the possible association between the agent and asthma is discussed, followed by the committee's conclusions regarding the health impacts. Where information is available, evidence regarding possible means of mitigating or preventing exposure to the agent is addressed. Each section concludes with any committee recommendations for general or specific areas in which additional research is needed with respect to the agent. Because there are great differences in the amount and type of information available on specific agents, the sections vary in their depth and focus.

NITROGEN DIOXIDE

Definition of Agent and Means of Exposure

Nitrogen dioxide (NO_2) is a common indoor and outdoor pollutant that is produced, along with other oxides of nitrogen, whenever high-temperature combustion occurs. NO_2 is one of six "criteria" air pollutants for which National Ambient Air Quality Standards are set by the U.S. Environmental Protection Agency (EPA). The current standard is 50 parts per billion (ppb) averaged over one year. Much higher indoor concentrations are sometimes observed when indoor sources are present due to the limited dilution often observed in confined spaces. Indoor sources include gas stoves and space heaters, kerosene space heaters, and poorly vented furnaces and fireplaces. In homes with indoor combustion sources, personal NO_2 exposures are usually driven by indoor concentration in the home, even in urban areas with elevated outdoor levels.

In addition to nitrogen oxides, indoor combustion appliances may also emit CO, SO_2, formaldehyde, volatile organic compounds (VOCs), and submicron particulate matter (PM). Some of these pollutants, including SO_2 and PM, are known respiratory irritants. Most epidemiologic studies reviewed in this section have assessed NO_2 exposure based on the presence or absence of gas appliances in the home, rather than on the basis of NO_2 measurements. Few if any studies have simultaneously measured NO_2 and associated co-pollutants such as PM. As a result, it is usually not possible to attribute health effects associated with gas appliance use to NO_2 exposures per se.

Factors Influencing Exposure

Indoor exposure to NO_2 resulting from the use of gas appliances is common. On average, about half of U.S. homes have gas stoves or ovens, and much higher percentages of gas appliances exist in some urban areas (Samet et al., 1987).

Considerable data exist on indoor NO_2 exposures in U.S. homes and the factors that influence them (Goldstein et al., 1988; Leaderer et al., 1986; Quackenboss et al., 1986; Ryan et al., 1988; Spengler et al., 1983, 1994, 1996). Indoor NO_2 concentrations depend on the presence and emission strength of indoor sources, the ventilation rate of the home, and the penetration of outdoor NO_2 (Drye et al., 1989; Spengler et al., 1996). Important indoor sources include unvented cooking and heating appliances that burn gas or kerosene.

The use of a gas range results in an increase of about 25 ppb in the background NO_2 concentration in a home, with peaks as high as 200 to 400 ppb in the kitchen during cooking (Samet et al., 1987). In a large study carried out in Albuquerque, New Mexico, the highest two-week average indoor NO_2 concentrations were observed in homes with gas stoves that had continuous pilot lights and in homes where the stove or unvented gas or kerosene heaters were used for supplemental heating (Spengler et al., 1996). In the same study, higher indoor NO_2 concentrations were observed in gas cooking than in electric cooking homes, but this difference was much less pronounced in the summer when home ventilation rates were higher. Regardless of stove type, indoor concentrations are generally higher in locations with high outdoor concentrations, such as large urban areas, due to the infiltration of outdoor air; however, indoor sources still explain most of the variance in personal exposures in urban areas. There is some evidence to suggest that homes with gas stoves in underprivileged, inner city communities may have uniquely high NO_2 levels, possibly due to higher frequency and longer duration of cooking, small home volume, and use of stoves for supplemental winter heating (Goldstein et al., 1988). Further research into this question is warranted.

Data collected in multiple locations in homes with gas stoves show that strong NO_2 concentration gradients exist, with kitchen levels higher than elsewhere (Matti et al., 1999). Kitchen concentrations during cooking may exceed long-term average concentrations by an order of magnitude or more. These spatial and temporal characteristics imply differential exposures for different residents depending on time spent in the kitchen while cooking is taking place.

Evidence Regarding Asthma Exacerbation and Development

Numerous epidemiologic studies have examined whether respiratory health effects are associated with exposures to typical indoor NO_2 concentrations. Most such studies have addressed respiratory symptoms and/or lung function variables as the primary outcomes, both of which include measures (i.e., wheeze or decline in FEF_{25-75}) that are usually associated with the clinical diagnosis of asthma. However, few studies have focused on asthma as an outcome or on respiratory effects among asthmatic subjects specifically. This limits the utility of the existing literature for assessing the impact of NO_2 on the development or exacerbation of asthma.

Epidemiologic study designs have included cross-sectional surveys, case-control studies, longitudinal cohort studies, and time-series panel studies assessing acute impacts. Exposures in most studies have been assigned based on answers to questions regarding the presence of gas stoves or other indoor combustion appliances. Some studies have included limited indoor or occasionally personal NO_2 measurements using passive diffusion samplers. As a group, epidemiologic studies have the advantage of studying realistic levels and patterns of NO_2 exposures. However, results can be difficult to interpret due to possible confounding.

There have also been several clinical experiments involving brief controlled exposures of humans in environmental chambers. Such studies enable careful control of experimental conditions including exposure level and duration. They also make it more feasible to examine asthma-related respiratory outcomes other than symptoms and lung function, such as airway hyperresponsiveness (AHR), pulmonary cells and cytokines obtained by bronchoscopy, and effects on allergen responsiveness. A limitation of the experimental chamber studies is that they usually employ NO_2 concentrations much higher than those typically observed in the indoor environment. Also, the study populations have typically been small and, to some extent, unrepresentative of the more sensitive members of the general population.

Very few epidemiologic studies have evaluated asthma diagnosis as an outcome in relation to NO_2 exposure, and results from these few studies have been mixed. A well-conducted case-control study found no association between the presence of a gas cooking appliance in the home and incident asthma (odds ratio [OR]=1.33; 95% confidence interval [CI] 0.68–2.58) among children 3–4 years old in Montreal (Infante-Rivard, 1993). The main study analyzed 457 cases (children with a first diagnosis of asthma made by a pediatrician) and 457 controls (matched on age and census tract), and found significant associations between incident asthma and several factors other than gas appliances, including heavy maternal smoking, childhood atopy, and others. A subset of 140 subjects provided a 24-hour personal NO_2 sample. In an unmatched analysis of this subset, there was a significant association with asthma for subjects in the highest category of NO_2 exposure (>15 ppb); however, the small size of the exposed group and the post hoc nature of the analysis preclude any meaningful inferences regarding the causality of this association. Strachan and Carey (1995) analyzed environmental risk factors for severe wheezing among school children aged 11–16 years in Sheffield, England; 486 cases and 475 age- and school-matched controls were analyzed, where cases were children whose parents reported 12 or more wheezing episodes in the past 12 months or an attack of wheezing that limited speech. Controls had no history of asthma or wheezing at any age. Non-feather bedding and ownership of furry pets were both significantly associated with case status; however, the use of gas for cooking was not.

In contrast, two large questionnaire-based cross-sectional surveys, one in Canada and the other in Australia, reported significant associations between asthma and gas stove use. The first study analyzed data from a nationwide survey of Canadian parents. It found a significant association between current, doctor-diagnosed asthma and gas cooking in the homes of children between 5 and 8 years of age, controlling for age, race, sex, parental education, environmental tobacco smoke (ETS), and other factors (Dekker et al., 1991). The adjusted odds ratio for gas cooking was 1.95 (95% CI 1.41–2.68) when 634 subjects with asthma were compared to 9,207 with neither asthma, chest symptoms, nor other respiratory diseases. However, the authors cautioned against over-interpretation given the small number of asthma cases exposed to gas cooking (N = 60 out of 634 total asthma cases). Associations were also reported with ETS, living in a damp home, and use of a humidifier. These factors, but not gas cooking, were also associated with reports of wheezing. A second, similar study in South Australia analyzed data from 14,124 families with a child aged 4.25 to 5 years of age (Volkmer et al., 1995). Gas versus electric stove use was associated with slightly increased prevalence of asthma (OR = 1.24, 95% CI 1.07–1.42) and wheezing in the preceding 12 months (OR = 1.16, 95% CI 1.01–1.32). The cross-sectional associations found in these two surveys suggest the possibility of small impacts of NO_2 on asthma risk; however, prospective cohort studies would be needed to rigorously test this hypothesis.

Respiratory symptoms that are associated with asthma, such as coughing, wheezing, and shortness of breath, have been studied in relation to gas stove usage and/or NO_2 measurements in a large number of cross-sectional surveys, longitudinal panel studies, and a more limited number of prospective cohort studies. Symptom outcomes have also been analyzed in one chamber exposure study.

Cross-sectional surveys assess association between symptoms and measures of exposure collected simultaneously by questionnaires in general population samples, usually controlling for covariates with known or suspected effects on symptoms, such as ETS, home dampness, allergies, and so forth. Although results have been mixed, the accumulating evidence supports the existence of small associations between respiratory symptoms and gas appliance use. At least ten survey studies have reported significant associations between symptoms and gas stove exposure in adults and/or children (Dodge, 1982; Garrett et al., 1998; Jarvis et al., 1996, 1998; Jedrychowski et al., 1995; Koo et al., 1990; Melia et al., 1977; Viegi et al., 1991, 1992; Volkmer et al., 1995). For example, in a survey of 1,159 men and women in England, Jarvis et al. (1996) detected significant associations for women only between symptoms in the past 12 months and the use of gas appliances, adjusting for covariates. The adjusted odds ratios were 2.07 (95% CI 1.41–3.05) for wheezing, 2.32 (95% CI 1.25–4.34) for waking with shortness of breath, and 2.60 (95% CI 1.20–5.65) for asthma attacks. Other cross-sectional survey studies have reported no associations between respiratory symptoms and gas appliances (Braun-Fahrlander et al., 1992; Dekker et al., 1991; Dijkstra et al., 1990; Hosein et al., 1989). There are no obvious differences between the two groups of studies that would explain the differences in results.

Several of the cross-sectional studies mentioned above have assessed exposure using actual NO_2 measurements, as well as the more common questionnaire-based assessments of gas appliance usage (Brunekreef et al., 1990; Garrett et al., 1998). Interestingly, such studies have generally found no analytical advantage to the actual NO_2 measurements. This counterintuitive finding may have several explanations, including non-representativeness of NO_2 measurement times or locations, and/or that NO_2 per se is not the causal agent responsible for gas stove associations with symptoms.

In a prospective cohort study, Neas et al. (1991) followed respiratory symptoms over a 12-month period in 1,567 white children aged 7–11 living in six U.S. cities. Indoor home NO_2 measurements were collected over two weeks in the winter and in the summer. Incident symptoms were analyzed in relation to mean

NO_2 levels. A 15 ppb increase in average household NO_2 concentrations was associated with increased cumulative incidence of any of several lower respiratory symptoms (OR = 1.4, 95% CI 1.1–1.7). The effect was larger for girls than boys. Two 1999 studies of asthmatic children, available only in abstract form when this report was completed, appear to generally support these findings (Kattan, 1999; Smith et al., 1999).

Longitudinal panel studies and experimental chamber studies assess the acute relationship between brief (episodic) NO_2 exposures and respiratory symptoms. These studies have not provided strong evidence for acute effects on symptoms at relevant indoor concentrations. Chamber studies involving one- to three-hour exposures to 50–1,500 ppb NO_2 did not detect effects on respiratory symptoms among normal or asthmatic subjects (Salome et al., 1996; Utell et al., 1991). In a panel study involving daily recording of symptoms and gas stove use among 164 asthmatic adults over the winter months in Denver, Colorado, Ostro et al. (1994) reported statistically significant associations between a variety of respiratory symptoms and stove use. However, interpretation of these results as a causal effect of gas stove emissions (e.g., NO_2) is hampered by questions of biologic plausibility as well as concerns about potential reporting bias. One 1999 abstract suggests acute effects of NO_2 on the severity of symptoms associated with respiratory syncytial virus (RSV) infections (Chauhan et al., 1999), and an environmental chamber study concluded that NO_2 exposure increases the susceptibility of airway epithelial cells to injury from respiratory viruses (Boscia et al., 1999).

Experimental chamber studies have noted increases in airway responsiveness to carbochol or methacholine following brief, high-level (1,500 or 2,000 ppb) NO_2 exposures in normal subjects (Frampton et al., 1991; Mohsenin, 1988; Utell et al., 1991)—exposure levels that have not typically elicited direct effects on lung function. Studies in asthmatics report enhanced airway responses to histamine or methacholine challenges at concentrations as low as 500–600 ppb, suggesting that the airways of asthmatics, already more responsive to nonspecific stimuli, are more sensitive to the enhancing effects of NO_2 (Mohsenin, 1987; Salome et al., 1996). While the mechanism for these effects remains uncertain, proinflammatory effects of NO_2 (Blomberg et al., 1999) may be involved. Although the relevance of these findings to typical indoor NO_2 concentrations encountered by the general public remains unclear, they do raise concerns for persons, such as mothers and infants, who may spend large amounts of time in kitchens where gas stoves are being used, especially in conjunction with low rates of home ventilation.

In addition to the enhancement of airway responses to nonspecific stimuli, NO_2 exposure at 400 ppb has been shown in experimental chamber studies to enhance the lung response, measured by a drop in forced expiratory volume in one second (FEV_1), to house dust mite aerosol inhalation by asthmatic adults (Rusznak et al., 1996; Tunnicliffe et al., 1994). Devalia and colleagues (1994) reported that exposure to a combination of SO_2 and NO_2 in concentrations that could be encountered in heavy traffic areas produced a statistically significant decrease in the concentration of allergen required to produce a 20% decrease in FEV_1 ($PD_{20}FEV_1$) of adult asthmatics challenged with *Dermatophagoides pteronyssinus* dust mite allergen. Again, no direct effects of NO_2 on lung function decline were noted in these studies.

Because of the importance of acute respiratory infections as triggers of asthma symptoms, an effect of NO_2 exposure on increased risk of respiratory infections might represent an indirect mechanism linking NO_2 with asthma exacerbations. In the late 1970s, Melia and colleagues reported an increased risk of respiratory infections among children living in homes with gas stoves in a large British cross-sectional survey (Melia et al., 1977, 1985). However, other studies have not confirmed this finding (Samet et al., 1993; Ware et al., 1984). In a prospective cohort study of infants, no association was found between NO_2 exposure or stove type and the incidence rate or duration of respiratory infections (Samet et al., 1993). Samet and colleagues (1987) extensively reviewed the historical literature on NO_2 and respiratory infections, concluding that "the findings on NO_2 exposure and respiratory illnesses indicate that the magnitude of the NO_2 effect at concentrations encountered in most U.S. homes is likely to be small." Although there is insufficient new evidence to alter this conclusion, it is worth noting that a 1999 longitudinal panel study reported enhanced severity of RSV infections in association with high indoor NO_2 concentrations (Chauhan et al., 1999).

Inconsistent associations between gas appliance use or NO_2 concentrations and declines in lung function have been reported in epidemiologic studies (Brunekreef et al., 1990; Dijkstra et al., 1990; Dodge, 1982; Fischer et al., 1985; Garrett et al., 1998; Hackney et al., 1992; Hasselblad et al., 1981; Hosein et al., 1989; Jarvis et al., 1996, 1998; Jedrychowski et al., 1995; Kattan, 1999; Speizer et al., 1980; Viegi et al., 1991; Ware et al., 1984), with significant associations reported in less than half of the studies. Where present, the nature of the lung function associations has not been consistent across studies, and the possibility of confounding exists in some cases.

As noted earlier, little evidence of acute lung function impacts of brief exposure to high concentrations of NO_2 has been observed in experimental chamber studies (Frampton et al., 1991; Hackney et al., 1992; Mohsenin, 1987, 1988; Salome et al., 1996; Tunnicliffe et al., 1994; Utell et al., 1991), except for four-hour exposures to very high concentrations (i.e., 2,000 ppb) (Blomberg et al., 1999). The committee concludes that lung function is not markedly affected either acutely or chronically by NO_2 at typical indoor concentrations.

Conclusions: Asthma Exacerbation and Development

• There is sufficient evidence of an association between brief high-level exposures to NO_2 and increased airway responses to both nonspecific chemical irritants and inhaled allergens among asthmatic subjects. These effects have been observed in human chamber studies at concentrations (400–700 ppb) that may occur only in poorly ventilated kitchens with gas appliances in use.

• There is limited or suggestive evidence of an association between the use of gas appliances and increased risk of respiratory symptoms, increased risk of respiratory infections, and to a lesser extent, decreased lung function. Data supporting this conclusion derive from epidemiologic studies.

• There is limited or suggestive evidence of no association between brief NO_2 exposures and acutely decreased lung function. This evidence comes from chamber studies of human subjects.

• There is inadequate or insufficient evidence to determine

whether or not an association exists between emissions from gas appliances and asthma development. However, the association observed between the use of gas appliances and the diagnosis of childhood asthma in two, large cross-sectional population surveys indicates that this topic should be examined more carefully in future research. As noted above, few studies have simultaneously measured NO_2 and associated co-pollutants and it is thus usually not possible to attribute health effects associated with gas appliance usage to NO_2 or other combustion by-product exposures per se.

Evidence and Conclusions:
Exposure Mitigation and Prevention

Indoor NO_2 mitigation has received relatively little attention in the published literature. Samet (1990) notes that general control options for pollutants emitted by indoor combustion appliances include source modification (removal, substitution, or emission reduction), ventilation (exhaust or dilution), or pollutant removal (filtration or reactivity). Source modification is usually the most effective approach. For example, that study recommended that "unvented combustion space heaters should not be used, particularly in cold climates where they may be on for prolonged periods." On the other hand, while removal of gas stoves would in theory represent an effective exposure reduction strategy, it may not be practical or economically feasible in most cases. Continuous pilot lights add between 10 and 20 ppb of NO_2 to background indoor levels and should be turned off or eliminated (Samet, 1990). Kitchen ventilation has the potential to be effective in reducing the impact of gas appliance emissions. However, exhaust hoods should be vented outdoors and must be used consistently while the appliance is on to be effective. Evidence suggests that only a small fraction of kitchens with gas stoves achieve these objectives (Fuhlbrigge and Weiss, 1997). For NO_2, pollutant removal via air cleaning is not a feasible approach. The committee did not identify any studies that addressed whether lowering indoor NO_2 levels had an effect on asthma outcomes.

Research Needs

Future research into the relationship between indoor NO_2 and asthma should target population subgroups that are likely to be most exposed. These include women and infants, especially those at the lower end of the socioeconomic spectrum, who may spend more time in kitchens during cooking. Further research is needed on the distributions of peak and mean personal exposures, on the relationship between exposure and asthma, and on the exacerbation of asthma among those with preexisting asthma. Research is also needed into the development of practical, economical ventilation methods for kitchens.

PESTICIDES

Definition of the Agent and Means of Exposure

"Pesticide" is a general term for an agent used to kill an undesirable organism. Among the categories of agents covered by this umbrella term are fungicides, herbicides, insecticides, and rodenticides.

Residential pesticides represent a broad class of chemicals and applications aimed at controlling flies, ants, moths, cockroaches, fleas, ticks, infectious organisms, fungi, plants and other unwanted species in and around the residential environment. Pesticides that have been measured in residential indoor air include chlordane, heptachlor, aldrin, dieldrin, diazinon, propoxur, dichlorvos, naphthalene, p-dichlorobenzene, pentachlorophenol, chlorpyrifos, malathion, and carbaryl (Baker and Wilkinson, 1990).

Factors Influencing Exposure

For members of the general population, the home represents the principal setting for pesticide exposures. It has been reported that approximately 90% of U.S. homes use pesticides, and ~84% use them inside the house (Baker and Wilkinson, 1990; U.S. EPA, 1979). Eskenazi and colleagues (1999) note that younger children may have higher exposure than others in a contaminated environment because of the greater amount of time spent indoors and closer contact with contaminated surfaces. They also identify farmworkers and their families as a vulnerable population through agents that penetrate or are tracked indoors. Children who live in poverty may be at particular risk of exposure through greater use of legal and illegal pesticides in urban environments (Landrigan et al., 1999).

In addition to diet and dermal absorption, inhalation of airborne pesticides, either in vapor form or adsorbed on particles, is an important route of exposure in the residential setting. Surveys of residential air concentrations of a range of pesticides were summarized by Baker and Wilkinson (1990, Table 2-11). Chlorpyrifos, which is currently one of the most commonly used residential pesticides, ranged in concentration from 2 to 37 $\mu g/m^3$ in a survey of U.S. homes published in 1983 (U.S. EPA, 1983).

Evidence Regarding Asthma Exacerbation and Development

Although there is a great deal of interest in the possibility of an association between pesticide exposure and asthma, relatively little research has been done on the topic to date. There is evidence suggesting that *high-level* exposures to some pesticides may elicit, via irritative mechanisms, persistent asthma. This consists primary of anecdotal and case reports.

A case report (Deschamps et al., 1994) gives an account of a 26-year-old male who developed persistent asthma after working in a closed room that was heavily treated with the organophosphate insecticide dichlorvos. Occupational exposure to the fungicides metam sodium (Cone et al., 1994), captafol (Royce et al., 1993), tetrachloroisophthalonitrile (chlorthalonil) (Honda et al., 1992), and tributyl tin oxide (Shelton et al., 1992) have also been associated with new asthma onset in case reports. Etzel (1995) cites additional examples. Exposure to pyrethrins or pyrethroids—insecticides commonly used in over-the-counter flea and insect sprays and powders—may cause allergic rhinitis (O'Malley, 1997). Individuals who are allergic to ragweed may also exhibit an allergic reaction to pyrethrins (Wax and Hoffman, 1994). Although it is not known whether and how these findings are relevant to the general population, they do raise concern about the possibility of asthmogenic effects of prolonged low-level exposures.

There is little relevant epidemiologic information in the literature. Senthilselvan and colleagues (1992) surveyed 1,939 male farmers in Saskatchewan on work, respiratory health, and exposure to chemicals. The prevalence of self-reported asthma was significantly associated with the use of carbamite insecticides. Other studies in agricultural and rural settings list pesticides along with numerous other exposures (animal, arthropod, and insect allergens; endotoxins; molds; plant proteins; pollen) potentially associated with asthma or other respiratory health outcomes (ATS, 1998).

The response noted in many of these reports appears to be similar to the reactive airways dysfunction syndrome (RADS), which is an asthma-like response that occurs in some people when exposed, even briefly, to a single high concentration of irritative chemicals, usually in the occupational setting. Whether a similar response may occur in a subset of the general population following lower-level but repeated exposures to irritating chemicals including pesticides is not known. This is an area that deserves further study.

Although some pesticides may affect the immune system, there is currently a lack of information on whether pesticide exposures can have an impact on the allergic asthma response. It has also been suggested that dysregulation of parasympathetic and sympathetic autonomic control of the airways through pesticide exposure may provide a mechanism (Eskenazi et al., 1999). Chan-Yeung and Lam (1986) note that acute asthma has been described in farm workers who use organophosphate insecticides, which act as an anticholinesterase and probably precipitate airflow obstruction on a pharmacologic basis without the requirement of an underlying predisposition to asthma.

Conclusions: Asthma Exacerbation and Development

• There is inadequate or insufficient evidence to determine whether or not an association exists between pesticide exposures at the levels typically encountered in nonoccupational or residential settings and the development or exacerbation of asthma. As noted in Chapter 5, proper use of some pesticides as part of an exposure control program may yield benefits for asthmatics through elimination or reduction of sources of allergen.

Evidence and Conclusions:
Exposure Mitigation and Prevention

There is considerable potential for reducing indoor residential exposures to pesticides. The methods of integrated pest management (IPM), which emphasize habitat modification in conjunction with minimal and selective use of pesticides, should be useful in this regard. Although strategies exist that may be effective in reducing indoor pesticide concentrations, it is not known whether reducing these concentrations has any significant impact on the risk of asthma development or exacerbation.

Research Needs

The most immediate need for research in this area is for information on whether prolonged exposures to low to moderate concentrations of airborne pesticides can elicit an irritative asthma response in susceptible individuals. Animal models may be helpful in this regard. Koren and O'Neill (1998) report that a presently unpublished study on a microbial biopesticide used against cockroaches suggests it can induce asthma in mice. It has been suggested that Flinders Sensitive Line rats, which exhibit increased responses to an agent similar to commonly used organophosphate pesticides, may be useful in the study of asthma outcomes.

VOLATILE ORGANIC COMPOUNDS

Definition of Agent and Means of Exposure

Volatile organic compounds are any of a large number of organic molecules that exist either as free vapors or adsorbed onto particles in air. More than 300 VOCs have been measured in indoor air (Wallace, 1987). Table 6-1 lists some of the more common VOCs and their sources.

TABLE 6-1 Common VOCs and Their Sources

Chemical	Major Sources of Exposure
1,1,1-Trichloroethane	Wearing or storing dry-cleaned clothes; aerosol sprays; fabric protectors
Aliphatic hydrocarbons (octane, decane, undecane)	Paints, adhesives, gasoline, combustion sources
Aromatic hydrocarbons (toluene, xylenes, ethyl-benzene, trimethylbenzene)	Paints, adhesives, gasoline, combustion sources
Benzene	Smoking, auto exhaust, passive smoking, driving, pumping gas
Carbon tetrachloride	Industrial-strength cleaners
Methylene chloride	Paint stripping, solvent usage
p-Dichlorobenzene	Room deodorizers, moth cakes
Terpenes (limonene, α-pinene)	Scented deodorizers, polishes, fabrics, fabric softeners, cigarettes, food, beverages
Tetrachloroethylene	Wearing or storing dry-cleaned clothes; visiting dry cleaners
Trichloroethylene	Unknown (cosmetics, electronic parts)

SOURCE: Adapted from Samet. *Indoor Air Pollution: A Health Perspective.* pp. 253. © 1991, The Johns Hopkins University Press.

Factors Influencing Exposure

Most available data suggest that personal exposures to VOCs are dominated by indoor exposures, even for persons living in urban areas near major outdoor VOC sources such as the petrochemical industry (Wallace, 1991; Wallace et al., 1987). Numerous VOC sources exist indoors, including cigarette smoking, combustion appliances, solvents, printed materials, photocopying machines, chlorinated water, dry-cleaned clothes, pesticides, silicone caulk, floor adhesive, particleboard, moth crystals, floor wax, wood stain, paint, furniture polish, floor finish, carpet shampoo, room deodorizer, and vinyl flooring and tiles (Gold, 1992; Samet, 1990; Wallace, 1991). Cleaning agents used in homes either may be a direct source of VOC emissions or may release VOCs through chemical decomposition of the surface being washed (Wolkoff et al., 1998).

Indoor VOC concentrations and time spent indoors are the two main factors influencing personal exposures to VOCs. Factors that modify indoor concentrations, including emission rates from sources and removal rates due to building ventilation, thus have a strong impact on personal exposures. Depending on the type of building material and the VOC, temperature and relative humidity may influence emission rates (Wolkoff, 1998). Activity patterns, personal habits, and hobbies also help determine exposure. Many VOCs can adsorb onto clothing and interior surfaces, which can then serve as reemission sources.

Table 6-2 shows average and maximum concentrations of 18 VOCs observed in a U.S. study of 24-hour and 12-hour personal exposures and in a German study of two-week average indoor concentrations. Average values ranged from 2 to 84 $\mu g/m^3$, with most concentrations falling below 10–15 $\mu g/m^3$. Wallace and colleagues (1987) reported data on simultaneous personal and outdoor VOC measurements in randomly selected groups of subjects living in two urban areas of New Jersey, in a moderately sized city in North Carolina, and in a small town in North Dakota. Personal exposures were consistently higher than simultaneous outdoor measurements in all three populations, presumably due to the dominant influence of indoor exposures. While outdoor concentrations increased with degree of urbanization across the three locations, personal exposures showed much less interlocation variability.

In contrast to combustion pollutants such as NO_2 and particulate matter, many VOCs tend to be source specific (Wallace et al., 1987). For example, the major source of chloroform in indoor air is hot water that has undergone chlorination. The principal sources of benzene in indoor air are smoking and gasoline vapors from attached garages. Tetrachloroethylene (PERC) is a VOC associated with dry cleaning; personal exposures are driven by time spent in laundromats and proximity to clothes that have been dry cleaned. p-Dichlorobenzene is a principal ingredient in both moth crystals and room deodorizers, two materials whose presence indoors is strongly predictive of variations in human exposure. Because human exposures to many VOCs are associated with spe-

TABLE 6-2 Concentrations of VOCs ($\mu g/m^3$) in Indoor Air in Germany Compared to Personal Exposures in the United States

Compound	Arithmetic Mean		Maximum	
	West Germany	United States	West Germany	United States
Chlorinated				
Chloroform	NM	3		NM
1,1,1-Trichloroethane	9	52	260	8,300
Trichloroethylene	11	6	120	350
Tetrachloroethylene	14	16	810	250
p-Dichlorobenzene	14	25	1,260	1,600
Aromatic				
Benzene	10	16	90	510
Styrene	2	3	41	76
Ethylbenzene	10	9	160	380
o-Xylene	7	9	45	750
m + p-Xylene	23	26	NM	300
Toluene	84	NM	1,710	NM
Aliphatic				
Octane	5	4	92	122
Nonane	10	12	140	177
Decane	15	6	240	161
Undecane	10	8	120	385
Dodecane	6	4	72	72
Terpenes				
α-Pinene	10	4	120	208
Limonene	28	43	320	2,530

NOTE: NM = not measured.
SOURCE: Adapted from Samet. *Indoor Air Pollution: A Health Perspective.* p. 259. © 1991, The Johns Hopkins University Press.

cific indoor source materials, exposure mitigation can be directed at reducing contact with specific sources.

Evidence Regarding Asthma Exacerbation and Development

A series of studies examined associations between asthma or asthma-related symptoms and indoor environmental factors including VOCs in Uppsala, Sweden. Norbäck and colleagues (1995) reported results from a small case-control study nested within a stratified population-based survey of adults 20–45 years old in this community: 47 cases with doctor-diagnosed asthma or asthma symptoms in the past year were compared with 41 subjects lacking both a diagnosis and symptoms. The symptom nocturnal shortness of breath was associated with the presence of carpets and indoor VOC, formaldehyde, and CO_2 concentrations. Significant VOC odds ratios ranged from 4.9 (95% CI 1.1–22.8) for toluene to 9.9 (95% CI 1.7–58.8) for total VOCs. The authors also reported associations between AHR and limonene (a terpene). Variability in peak flow rates was related to two other terpenes, α-pinene and δ-carene. A subsequent analysis of data from the same underlying population survey reported a marginally significant cross-sectional association (OR = 1.5, 95% CI 1.0–2.4) between asthma prevalence and the presence of newly painted interior surfaces in the home (Wieslander et al., 1997). Blood eosinophil concentrations were also elevated in subjects from recently painted dwellings. In a separate cross-sectional survey of 627 school children age 13–14 years, associations were observed between current asthma and school VOC concentrations in a logistic regression analysis that controlled for atopy, food allergies, and day care histories (Smedje et al., 1997). Many other environmental factors besides VOCs were associated with asthma, including school size, the presence of open shelves in classrooms, lower room temperature, higher relative humidity, and formaldehyde concentrations. This makes it difficult to attribute the association to VOCs alone. The VOC association was not robust to different VOC measurement methods; only passive, but not active, VOC monitoring data were associated with asthma. A mouse bioassay performed by Wolkoff and colleagues (1999) found that a mixture of (+)-α-pinene and ozone at concentrations close to their no-effect levels (NOELs) produced reaction products including formaldehyde and induced strong airway irritation. The authors determined that after accounting for sensitivity differences between mice and humans, the measured concentrations of formaldehyde and ozone in the reaction mixture were "not unrealistic for indoor settings."

While these studies suggest a variety of associations between indoor factors and measures of asthma status, because of the multiple correlations observed among exposure factors, it is not possible to clearly implicate VOCs as the causative agents. The possibility of residual confounding also clouds any causal interpretation. Despite these limitations, the studies reviewed here have generated several intriguing hypotheses regarding the possible influences of VOCs and other indoor environmental factors on asthma.

Small, acute impacts of high-level VOC exposures on decreased lung function and increased nasal inflammation have been suggested in controlled human chamber studies (Harving et al., 1991; Koren et al., 1992). Harving and colleagues (1991) reported a significant 9% pre–post exposure drop in FEV_1 among 11 asthmatics with bronchial reactivity who were exposed for 85 minutes to 25-mg/m^3 of a VOC mixture. However, this FEV_1 decline was not statistically different from the decline observed after sham exposure. Koren and colleagues (1992) exposed 16 adults without asthma or allergy to a 25 mg/m^3 VOC mixture for four hours. Significant increases in inflammatory cells (neutrophils) were observed in nasal washings performed immediately and 18 hours after exposure, adjusting for changes observed following sham exposure. The relevance of these high-exposure findings to the hundredfold lower concentrations observed in indoor residential settings in uncertain.

Conclusions: Asthma Exacerbation and Development

There is inadequate or insufficient evidence to determine whether or not an association exists between indoor residential VOC exposures and the development or the exacerbation of asthma, although elevated indoor concentrations of VOC mixtures are suspected to play a role in the constellation of symptoms known as sick building syndrome (e.g., headaches, fatigue, eye and upper respiratory irritation).

Evidence and Conclusions: Exposure Mitigation and Prevention

Although there is not clear evidence for a role of VOC exposures in asthma, it is deemed prudent to limit indoor exposures where practical. Approaches include indoor source removal or control (e.g., eliminating or reducing indoor smoking, air fresheners, moth crystals, use of solvent-based paints and solvent storage), source avoidance (e.g., altering activity patterns and habits to minimize exposures), and increased ventilation (e.g., increasing the overall home air exchange rate, installing ventilation fans in spaces containing sources).

Research Needs

With the advent of small, unobtrusive passive diffusion monitors capable of measuring microgram quantities of VOCs in 48-hour samples, it should be possible to incorporate personal VOC exposure assessment into future epidemiologic studies addressing environmental factors and asthma, yielding an expanded data base on which to judge the possible role of VOCs in asthma development and exacerbation.

Prospective cohort studies that characterize time-averaged personal VOC exposures of study subjects using passive badges and then follow subjects to assess the development and/or exacerbation of asthma could generate information that would allow a more confident assessment of any potential risk.

FORMALDEHYDE

Definition of the Agent and Means of Exposure

Although formaldehyde is a member of the VOC family, it is usually discussed separately from other VOCs because of the large body of scientific literature that specifically addresses formaldehyde exposures and human health. A colorless but characteristically pungent-smelling gas, formaldehyde is the simplest of the aldehydes, with the chemical formula HCHO. Formaldehyde is a component of many building materials, consumer products, and combustion gases. Potential indoor sources include urea–formaldehyde foam insulation (UFFI); glues used in plywood and pressed-board products; paper products including tissues, towels, and bags; cosmetics and detergents; and emissions from gas stoves and tobacco smoking.

Because of its polar structure, formaldehyde is highly soluble in water and thus is rapidly absorbed in the mucous membranes of the upper respiratory system. Because formaldehyde is highly reactive, it exerts its irritant effects primarily at or near the site of deposition.

Factors Influencing Exposure

Formaldehyde pollution is present both outdoors and indoors. Outdoor sources include motor vehicle exhaust, especially when oxygenates are used as fuel additives. Indoor concentrations are determined by the presence, number, and age of sources and by the degree of ventilation with fresh air. Formaldehyde off-gassing by materials decays over time, with the highest emission rates occurring when materials are new. Thus, formaldehyde concentrations will typically be highest in newly constructed or renovated interior spaces or in those containing new furniture and/or carpets, to the extent that these materials contain formaldehyde resins. Because mobile homes often contain a high percentage of such materials, as well as being relatively airtight, such structures will often contain higher formaldehyde concentrations than other types of residences.

Indoor residential formaldehyde concentrations in large numbers of conventional and mobile homes have been reported from several studies (Dally et al., 1981; Hanrahan et al., 1984; Ritchie and Lehnen, 1985; Sexton et al., 1986; Stock and Mendez, 1985) and were summarized by Marbury and Kreiger (1991). In monitoring surveys carried out in response to physician or resident requests, formaldehyde concentrations ranged from 0.01 to 5.62 parts per million (ppm), with mean concentrations falling below 0.50 ppm (Dally et al., 1981; Ritchie and Lehnen, 1985). In other studies, where populations were chosen either at random or by convenience, lower levels ranging from 0.01 to 0.80 ppm were observed, with means falling below 0.20. According to an EPA review (U.S. EPA, 1987), average concentrations in mobile homes (0.2 to 0.5 ppm) usually exceed those in conventional homes (0.03 to 0.09 ppm).

Evidence Regarding Asthma
Exacerbation and Development

Several excellent reviews of formaldehyde exposures and health have been published (Marbury and Krieger, 1991; Samet, 1990). The known health effects of formaldehyde at high exposure levels include nasal cancers in laboratory animals (Kerns et al., 1983) and irritation of the mucous membranes of the eyes, nose, and throat in humans. The influence of formaldehyde on asthma is much less clear, except for relatively rare cases of occupational asthma among workers exposed to high-level formaldehyde concentrations (Hendrick and Lane, 1977). With respect to occupational asthma, it remains unclear to what extent formaldehyde acts through immunologic mechanisms involving specific sensitization as opposed to irritant mechanisms (Nordman et al., 1985).

Although several surveys have reported high rates of upper-respiratory symptoms among residents of mobile homes and/or homes containing UFFI (Breysse, 1980; Dally et al., 1981; Norsted et al., 1985; Sardinas et al., 1979), interpretation is often difficult due to potential selection bias, the lack of adequate control groups, or the absence of exposure data (Marbury and Krieger, 1991). One large population-based study compared symptom rates between 1,396 residents of UFFI-containing homes and 1,395 residents of non-UFFI homes (Thun et al., 1982). Reports of wheezing were unusually low overall but were more frequent in residents of UFFI-containing homes (0.6 versus 0.1%). No airborne formaldehyde measurements were carried out however. Broder and colleagues (1988) reported elevated rates of a wide range of symptoms, including wheezing among 1,726 residents of UFFI-containing homes compared with 720 subjects from non-UFFI homes. Measured formaldehyde levels did not predict symptoms in a consistent way, suggesting the possible influence of other air contaminants besides, or in addition to, formaldehyde.

Several more recent population-based surveys suggest possible associations between indoor formaldehyde levels, along with many other indoor factors, and measures of asthma prevalence and/or symptoms in children and adults based on studies in the city of Uppsala, Sweden (Norbäck et al., 1995; Smedje et al.,

1997; Wieslander et al., 1997) (see earlier discussion of VOCs). None of these studies were able to isolate formaldehyde specifically as the causative agent in the reported associations. However, they do demonstrate the strong associations that may be found between a variety of measures of indoor air quality and health outcomes related to asthma.

Conclusions: Asthma Exacerbation and Development

• There is limited or suggestive evidence of an association between formaldehyde exposure and wheezing and other respiratory symptoms.
• There is inadequate or insufficient evidence to determine whether or not an association exists between formaldehyde exposure and asthma development.

Evidence and Conclusions: Exposure Mitigation and Prevention

Options for indoor formaldehyde mitigation include source control, air cleaning, and dilution (Samet, 1990). As is generally the case, source control is the option of first choice since it has the greatest potential for significant exposure reductions. The use of UFFI was banned in 1982 by the Consumer Products Safety Commission, eliminating this important source of indoor formaldehyde (Ashford et al., 1983). Interestingly, studies in which UFFI has been removed from houses have not demonstrated substantial decreases in formaldehyde levels. Of the remaining indoor sources, particleboard and hardwood plywood are the most significant. While removing these products from interior spaces would in theory represent an effective strategy for further exposure reduction, the resulting asthma risk reductions, if any, are considered unlikely to justify the costs associated with substitute materials. An additional consideration is the age of the material, since formaldehyde emission rates from these products diminish rapidly in the first year after they are manufactured (Samet, 1990). In-situ control of emissions can be achieved through the application of sealant coatings or coverings over the formaldehyde-containing materials, although VOC emissions from liquid coatings are likely to result. Other approaches to source reduction that have demonstrated effectiveness include ammonia fumigation and indoor temperature and humidity reduction.

Conclusions Regarding Means of Source Mitigation or Prevention

Although strategies exist that may be effective in reducing indoor formaldehyde concentrations, there is inadequate or insufficient evidence to determine whether or not an association exists between the implementation of these strategies and a decrease in asthma development or exacerbation.

Research Needs

No specific research is recommended.

FRAGRANCES

Definition of Agent and Means of Exposure

Fragrances are constituents of many personal (perfume, cologne, soap, shampoo, hairspray, shaving cream, deodorant, hand lotion, nail color and enamel) and household (detergent, fabric softener, dishwashing liquid, dishwasher detergent, air freshener) products. A 1991 survey identified several classes of polar VOCs in emissions from such products, including alcohols, esters, and aldehydes (Wallace et al., 1991). It is estimated that more than 3,000 chemicals are used in fragrance manufacture and that synthetic organic chemicals constitute the vast majority of the constituent raw materials (Fisher, 1998).

Evidence Regarding Asthma Exacerbation and Development

There is limited and largely anecdotal epidemiologic evidence suggesting that exposure to fragrances may induce asthma-like symptoms in some sensitive individuals and that some patients with asthma are sensitive to strong scents. A small number of controlled clinical challenge studies have been carried out, with some but not all of these demonstrating respiratory effects following fragrance exposure in sensitive population groups.

A study by Shim and Williams (1986) examined the effect of cologne and a saline placebo challenge on the expiratory volume of four patients with sensitivity to cologne. The researchers found an ~20–60% decline from baseline in FEV_1 during the 10-minute cologne exposure. Occlusion of the nostrils prevented the decline in one of the subjects. Kumar and colleagues (1995) studied the effect of exposure to commercial perfume-scented strips on 29 asthmatic adults and 13 nonasthmatic controls. They reported that perfume inhalation challenges produced significant declines in FEV_1 in asthmatic patients compared to control subjects, with greater declines noted among the more severely asthmatic subjects and among atopic (versus nonatopic) asthma subjects. A study by Millqvist and Löwhagen (1998) examined 10 asthmatic patients who complained of sensitivity to strong scents. The patients were challenged with methacholine, perfume (with and without a nose clip to block smell perception), and a placebo. They found no changes in lung function after the perfume provocation, although all of the patients were very sensitive to provocation with methacholine.

Millqvist and colleagues (1999) conducted a single-blinded, placebo-controlled randomized study of perfume exposure, using face masks or nose clips to block smell perception. They found that a group of 11 patients with a history of hyperreactivity to scents exhibited asthma-like symptoms that increased over the 30-minute exposure period when provoked. An earlier study by Millqvist and Löwhagen (1996) found that breathing through a carbon filter had no protective effect for nine patients with respiratory symptoms (but without immunoglobulin E [IgE] mediated allergy) subjected to perfume provocation, suggesting that the symptoms were not transmitted via the olfactory nerve. Mechanisms were also examined in an animal study by Hilton and colleagues (1996). They concluded that the fragrances eugenol and isoeugenol did not cause sensitization of the respiratory tract of mice or guinea pigs. Anderson and Anderson (1998) found that one-hour exposure to emissions of five commercial colognes or toilet water elicited pulmonary irritation and decreases in expiratory airflow velocity in male Swiss-Webster mice.

A clinic-based occupational study (Ross et al., 1998) found that workers in the perfume industry were among the groups with the highest incidence rates of occupational asthma in 1997. However, the specific workplace exposures that might account for the observation were not identified.

Conclusions: Asthma Exacerbation and Development

It is difficult to draw conclusions concerning a possible direct role of chemical fragrances in eliciting respiratory symptoms because many studies fail to control for the possible influence of odor perception. The studies that do account for this factor provide limited or suggestive evidence of an association between exposure to certain fragrances and the manifestation of respiratory symptoms in asthmatics sensitive to such exposures. The committee did not identify any carefully controlled studies that address the relationship between fragrance exposure and the development of asthma. In summary:

• There is limited or suggestive evidence of an association between exposure to certain fragrances and the manifestation of respiratory symptoms in asthmatics sensitive to such exposures.
• There is inadequate or insufficient evidence to determine whether or not an association exists between exposure to fragrances and asthma development.

Evidence and Conclusions: Exposure Mitigation and Prevention

There is no scientific literature on the effectiveness of fragrance exposure prevention measures for sensitive asthmatics. However, avoidance is clearly the most straightforward means of addressing problematic exposures for sensitive individuals.

Research Needs

Future research on fragrance exposures has to carefully control for confounding factors such as odor perception and to focus on objective measures of respiratory health. Additional research elucidating the mechanism or mechanisms underlying adverse respiratory reactions to nonacute exposure to fragrances or their component chemicals would also be helpful.

PLASTICIZERS

Definition of Agent and Means of Exposure

Chemicals that soften or enhance the flexibility of resins are called "plasticizers." These chemicals, which are also VOCs, are used primarily to process polyvinyl chloride (PVC) resin into useful forms. Plasticizer residues may be found in household items such as sheet vinyl flooring, wall coverings, vinyl upholstery, and shower curtains.

Evidence Regarding Asthma
Exacerbation and Development

There is a small literature addressing the possible connection between plasticizer exposure and asthma. Di(2-ethylhexyl) phthalate (DEHP), a plasticizer widely used in building materials, has been implicated in airway inflammation, and it has been suggested that residential exposure to DEHP may have a role in asthma pathogenesis (Øie et al., 1997) and bronchial obstruction (Øie et al., 1999). Doelman and colleagues (1990) report that mono(2-ethylhexyl) phthalate (MEHP), a breakdown product of DEHP, induces bronchial hyperreactivity in rats. Anderson and Anderson (1999) found that one-hour exposure to emissions of six brands of waterproof crib mattress covers, five of which were manufactured with PVC cores, induced asthma-like symptoms in male Swiss-Webster mice.

There are also reports of asthma exacerbation and development related to occupational exposures to plasticizers (Cipolla et al., 1999) and plastics manufacturing (Kogevinas et al., 1999), although their relevance to nonoccupational environments is open to question.

Conclusions: Asthma Exacerbation and Development and Research Needs

While the reports described above have attracted some interest in the research and building trades communities, there is inadequate or insufficient evidence to determine whether or not an association exists between nonoccupational exposure to plasticizers and the development or exacerbation of asthma. Research characterizing residential exposure to such compounds and evaluating asthma outcomes while controlling for known confounders would help resolve the question of their influence.

OTHER CHEMICAL EXPOSURES IN THE INDOOR ENVIRONMENT

Some outdoor air pollutants potentially associated with asthma penetrate the indoor environment and therefore present opportunities for indoor exposure. The following section addresses three of these pollutants: ozone, non-biologic particulate matter with sources other than tobacco smoke, and sulfur dioxide. Since the committee's mandate was to address indoor air pollutants, the discussion of these is less detailed than others in the chapter and no conclusions are drawn concerning indoor exposures and asthma outcomes.

Ozone

Ozone (O_3) is a natural part of the atmosphere that is also a by-product of the chemical reactions that create smog. Outdoor ozone penetrates dwellings only partially, depending on ventilation rate and reactions with indoor surfaces. In buildings without indoor sources, average ratios of indoor to outdoor (i/o) concentrations are typically around 0.5, although i/o ratios as low as 0.1 are observed in tightly sealed homes with air conditioners[1]. The

---

[1] Chapter 10 discusses the relationship of ventilation and air cleaning to pollutant exposures associated with asthma and asthma symptoms.

primary indoor residential source of ozone is an appliance called an ionizer or ozone generator, which is sold as an air freshening or air cleaning device. The EPA report *Ozone Generators That Are Sold as Air Cleaners* (U.S. EPA, 1999) details potential health problems from ozone exposure associated with the use of these devices. Xerographic copying machines—found in offices, schools, and some other indoor environments—also produce ozone.

Time-series epidemiologic studies have demonstrated significant associations between daily asthma hospitalizations and/or emergency room visits and daily outdoor ozone concentrations (U.S. EPA, 1996). Although outdoor ozone concentrations usually exceed indoor levels, it is likely that relevant exposures in these studies occurred predominantly indoors. Other studies indicate that higher ozone levels cause coughing and shortness of breath in asthmatics and nonasthmatics, and exacerbation of symptoms in asthmatics (Bielory and Deener, 1998). Peden and colleagues' (1995) study of 11 asthmatics who were allergic to dust mites (*Dermatophagoides farinae*) found that ozone exposure had both a priming effect on allergen-induced responses and an intrinsic inflammatory action in the nasal airways. A mixture of (+)-α-pinene and ozone yielded reaction products including formaldehyde and induced strong airway irritation in male mice (Wolkoff et al., 1999). The authors determined that after accounting for sensitivity differences between mice and humans, the measured concentrations of formaldehyde and ozone in the reaction mixture were "not unrealistic for indoor settings." There are no data suggesting that ozone exposure is associated with the development of asthma.

Chapter 10 contains an extended discussion of the impact of ventilation on the indoor concentrations of gaseous pollutants. As noted in that chapter, ozone is also removed from indoor air at a significant rate by deposition on or reaction with indoor surfaces. Rates of removal depend on the intensity of indoor air motion and other factors.

Particulates (Nonbiologic Particles)

"Particulate matter" (PM) is the name given to solid and liquid particles suspended in the air. Aside from outdoor infiltrate, the primary indoor sources of nonbiologic PM in indoor environments are combustion sources and tobacco smoke (which is addressed separately and in greater detail in Chapter 7). Unvented or poorly vented coal stoves and wood-burning stoves and fireplaces—where present—may be significant sources of indoor PM. For spaces without significant sources, indoors is a protective environment.

Studies consistently report an association between exposure to high outdoor levels of air pollutants, including PM, and adverse respiratory health effects (Koren, 1995). Evidence suggests that for fine particles (i.e., those with aerodynamic diameters less than 2.5 μm), outdoor PM often penetrates readily indoors. The literature specifically addressing asthma suggests an association between PM exposure and asthma exacerbation (e.g., Pope and Dockery, 1992; Roemer et al., 1993; Sheppard et al., 1999). Ostro and colleagues (1998) list three classes of possible mechanisms for this:

1. reflex bronchoconstriction via nonspecific irritant effects;
2. direct toxicity to the airway epithelium and resident immune cells, augmenting preexisting inflammation and airway hyperresponsiveness; and
3. induction of an inflammatory immune response, either because the particles themselves are allergenic or by permitting access of other allergens to the underlying tissues.

Aside from studies of the health effects of environmental tobacco smoke, where PM is part of a more complex exposure, data have not shown an association between PM and asthma development. Ongoing research is addressing this topic, including studies using animal models.

Limiting or eliminating sources is clearly the most straightforward means of addressing indoor PM exposures. Chapter 10 includes a discussion of the use of high-efficiency particulate air (HEPA) filters in lowering concentrations of indoor particulates.

Sulfur Dioxide

Sulfur dioxide (SO_2) is one of a family of gases called sulfur oxides (SO_x) formed when fuel containing sulfur—primarily coal and oil—is burned. Outdoor levels of SO_2 have diminished significantly since the 1960s in the United States due to the elimination of high-sulfur coal and oil as primary fuels for power generation and heating. Indoor sources include fossil fuel appliances and furnaces. However, these are not significant in most indoor environments, where outdoor infiltrate is the primary source.

Sensitive asthmatics breathing at elevated ventilation rates (during exercise, for example) experience bronchoconstriction and other airway responses in reaction to brief exposure to SO_2. These effects, which are relatively transitory, exhibit a dose–response relationship. Sulfur dioxide may both have a direct irritant effect and, possibly in combination with other air pollutants, potentiate the effect of antigens (Bielory and Deener, 1998). Exposure to a combination of SO_2 and NO_2 in concentrations that could be encountered in heavy-traffic areas produced a statistically significant decrease in the concentration of dust mite allergen required to produce a 20% decrease in FEV_1 of adult asthmatics (Devalia et al., 1994). There is no established mechanism for the effects of SO_2 on airways, although candidate mechanisms have been proposed (Peden, 1997). The committee did not identify any studies addressing indoor exposures to SO_2 and asthma development.

The exposure mitigation and prevention strategies discussed above for indoor sources of NO_2 are also relevant for SO_2.

REFERENCES

Anderson RC, Anderson JH. 1998. Acute toxic effects of fragrance products. Archives of Environmental Health 53(2):138–146.

Anderson RC, Anderson JH. 1999. Respiratory toxicity in mice exposed to mattress covers. Archives of Environmental Health 54(3):202–209.

Ashford NA, Ryan CW, Caldart CC. 1983. Law and science policy in federal regulation of formaldehyde. Science 222(4626):894–900.

ATS (American Thoracic Society). 1998. Respiratory health hazards in agriculture. Official Conference Report. American Journal of Respiratory and Critical Care Medicine 158(5 Pt 2):S1–S76.

Baker SR and Wilkinson CF Eds. 1990. The Effects of Pesticides on Human Health. Princeton, NJ: Princeton Scientific Publishing Co.

Bielory L, Deener A. 1998. Seasonal variation in the effects of major indoor and outdoor environmental variables on asthma. Journal of Asthma 35(1):7–48.

Blomberg A, Krishna MT, Helleday R, Soderberg M, Ledin MC, Kelly FJ, Frew AJ, Holgate ST, Sandstrom T. 1999. Persistent airway inflammation but accommodated antioxidant and lung function responses after repeated daily exposure to nitrogen dioxide. American Journal of Respiratory and Critical Care Medicine 159(2):536–543.

Boscia JA, Utell MJ, Torres A, Azadniv M, Roberts NJ Jr, Nichols J, Speers DM, Chalupa DC, Frasier LM, Frampton MW. 1999. NO_2 exposure in vivo predisposes airway epithelial cells to injury from respiratory viruses in vitro. American Journal of Respiratory and Critical Care Medicine 159(3):A699.

Braun-Fahrlander C, Ackermann-Liebrich U, Schwartz J, Gnehm HP, Rutishauser M, Wanner HU. 1992. Air pollution and respiratory symptoms in preschool children. American Review of Respiratory Disease 145(1):42–47.

Breysse PA. 1980. Formaldehyde Exposure in Mobile Homes: Occupational Safety and Health Symposia, 1979. Washington, DC: Government Printing Office. DHHS Publication no. (NIOSH) 80–139.

Broder I, Corey P, Cole P, Lipa M, Mintz S, Nethercott JR. 1988. Comparison of health of occupants and characteristics of houses among control homes insulated with urea formaldehyde foam. II. Initial health and house variables and exposure–response relationships. Environmental Research 45(2):156–178.

Brunekreef B, Houthuijs D, Dijkstra L, Boleij JS. 1990. Indoor nitrogen dioxide exposure and children's pulmonary function. Journal of the Air and Waste Management Association 40(9):1252–1256.

Chan-Yeung M, Lam S. 1986. Occupational asthma. American Review of Respiratory Disease 133(4):686–703.

Chauhan AJ, Linaker CH, Inskip H, Smith S, Schreiber J, Johnston SL, Holgate ST. 1999. Personal exposure to nitrogen dioxide (NO_2) and the risk of virus related asthma morbidity in children. American Journal of Respiratory and Critical Care Medicine 159:A699.

Cipolla C, Belisario A, Sassi C, Auletti G, Nobile M, Raffi GB. 1999. Asma occupazionale da dioctil-ftalato in una addetta alla produzione di tappi per bottiglia [Occupational asthma caused by dioctyl-phthalate in a bottle cap production worker]. Medicina del Lavoro 90(3):513–518.

Cone JE, Wugofski L, Balmes JR, Das R, Bowler R, Alexeeff G, Shusterman D. 1994. Persistent respiratory health effects after a metam sodium pesticide spill. Chest 106(2):500–508.

Dally KA, Hanrahan LP, Woodbury MA, Kanarek MS. 1981. Formaldehyde exposure in nonoccupational environments. Archives of Environmental Health 36(6):277–284.

Dekker C, Dales R, Bartlett S, Brunekreef B, Zwanenburg H. 1991. Childhood asthma and the indoor environment. Chest 100(4):922–926.

Deschamps D, Questel F, Baud FJ, Gervais P, Dally S. 1994. Persistent asthma after acute inhalation of organophosphate insecticide. Lancet 344(8938):1712.

Devalia JL, Rusznak D, Herdman MJ, Trigg CJ, Tarraf H, Davies RJ. 1994. Effect of nitrogen dioxide and sulfur dioxide on airway response of mild asthmatic patients to allergen inhalation. Lancet 344(8938):1668–1671.

Dijkstra L, Houthuijs D, Brunekreef B, Akkerman I, Boleij JS. 1990. Respiratory health effects of the indoor environment in a population of Dutch children. American Review of Respiratory Disease 142(5):1172–1178.

Dodge R. 1982. The effects of indoor pollution on Arizona children. Archives of Environmental Health 37(3):151–155.

Doelman CJ, Borm PJ, Bast A. 1990. Plasticisers, another burden for asthmatics? Agents and Actions Supplements 31:81–84.

Drye EE, Özkaynak H, Burbank B, Billick IH, Baker P, Spengler JD, Ryan PB, Colome SD. 1989. Development of models for predicting the distribution of indoor nitrogen dioxide concentrations. Journal of the Air and Waste Management Association 39(9):1169–1177.

Eskenazi B, Bradman A, Castorina R. 1999. Exposures of children to organophosphate pesticides and their potential adverse health effects. Environmental Health Perspectives 107(Supplement 3): 409–419.

Etzel RA. 1995. Indoor air pollution and childhood asthma: effective environmental interventions. Environmental Health Perspectives 103 (Supplement 6):55–58.

Fischer P, Remijn B, Brunekreef B, van der Lende R, Schouten J, Quanjer P. 1985. Indoor air pollution and its effect on pulmonary function of adult non-smoking women: II. Associations between nitrogen dioxide and pulmonary function. International Journal of Epidemiology 14(2):221–226.

Fisher BE. 1998. Focus: scents and sensitivity. Environmental Health Perspectives 106(12):A594–599.

Frampton MW, Morrow PE, Cox C, Gibb FR, Speers DM, Utell MJ. 1991. Effects of nitrogen dioxide exposure on pulmonary function and airway reactivity in normal humans. American Review of Respiratory Disease 143(3):522–527.

Fuhlbrigge A, Weiss S. 1997. Domestic gas appliances and lung disease. Thorax 52(Suppl 3):S58–S62.

Garrett MH, Hooper MA, Hooper BM, Abramson MJ. 1998. Respiratory symptoms in children and indoor exposure to nitrogen dioxide and gas stoves. American Journal of Respiratory and Critical Care Medicine 158(3):891–895.

Gold DR. 1992. Indoor air pollution. Clinics in Chest Medicine 13(2):215–229.

Goldstein IF, Andrews LR, Hartel D. 1988. Assessment of human exposure to nitrogen dioxide, carbon monoxide and respirable particulates in New York inner-city residences. Atmospheric Environment 22(10):2127–2139.

Hackney JD, Linn WS, Avol EL, Shamoo DA, Anderson KR, Solomon JC, Little DE, Peng RC. 1992. Exposures of older adults with chronic respiratory illness to nitrogen dioxide. A combined laboratory and field study. American Review of Respiratory Disease 146(6):1480–1486.

Hanrahan LP, Dally KA, Anderson HA, Kanarek MS, Rankin J. 1984. Formaldehyde vapor in mobile homes: a cross sectional survey of concentrations and irritant effects. American Journal of Public Health 74(9):1026–1027.

Harving H, Dahl R, Molhave L. 1991. Lung function and bronchial reactivity in asthmatics during exposure to volatile organic compounds. American Review of Respiratory Disease 143(4 Pt 1):751–754.

Hasselblad V, Humble CG, Graham MG, Anderson HS. 1981. Indoor environmental determinants of lung function in children. American Review of Respiratory Disease 123(5):479–485.

Hendrick DJ, Lane DJ. 1977. Occupational formalin asthma. British Journal of Industrial Medicine 34(1):11–18.

Hilton I, Dearman RJ, Fielding I Basketter DA, Dimber I. 1996. Evaluation of the sensitizing potential of eugenol and isoeugenol in mice and guinea pigs. Journal of Applied Toxicology 16(5):459–464.

Honda I, Kohrogi H, Ando M, Araki S, Ueno T, Futatsuka M, Ueda A. 1992. Occupational asthma induced by the fungicide tetrachloroisophthalonitrile. Thorax 47(9):760–761.

Hosein HR, Corey P, Robertson JM. 1989. The effect of domestic factors on respiratory symptoms and FEV_1. International Journal of Epidemiology 18(2):390–396.

Infante-Rivard C. 1993. Childhood asthma and indoor environmental risk factors. American Journal of Epidemiology 137(8):834–844.

Jarvis D, Chinn S, Luczynska C, Burney P. 1996. Association of respiratory symptoms and lung function in young adults with use of domestic gas appliances. Lancet 347(8999):426–431.

Jarvis D, Chinn S, Sterne J, Luczynska C, Burney P. 1998. The association of respiratory symptoms and lung function with the use of gas for cooking. European Community Respiratory Health Survey. European Respiratory Journal 11(3):651–658.

Jedrychowski W, Maugeri U, Gomola K, Tobiasz-Adamczyk B, Bianchi I. 1995. Effects of domestic gas cooking and passive smoking on chronic respiratory symptoms and asthma in elderly women. International Journal of Occupational Medicine and Environmental Health 1(1):16–20.

Kattan M. 1999. Environmental exposure to tobacco smoke and nitrogen dioxide in inner-city children with asthma. American Journal of Respiratory and Critical Care Medicine 159:A775.

Kerns WD, Pavkov KL, Donofrio DJ, Gralla EJ, Swenberg JA. 1983. Carcinogenicity of formaldehyde in rats and mice after long-term inhalation exposure. Cancer Research 43(9):4382–4392.

Kogevinas M, Antó JM, Sunyer J, Tobias A, Kromhout H, Burney P. 1999. Occupational asthma in Europe and other industrialised areas: a population-based study. Lancet 353(9166):1750–1754.

Koo LC, Ho JH, Ho CY, Matsuki H, Shimizu H, Mori T, Tominaga S. 1990. Personal exposure to nitrogen dioxide and its association with respiratory illness in Hong Kong. American Review of Respiratory Disease 141(5 Pt 1):1119–1126.

Koren HS, Graham DE, Devlin RB. 1992. Exposure of humans to a volatile organic mixture. III. Inflammatory response. Archives of Environmental Health 47(1):39–44.

Koren HS. 1995. Associations between criteria air pollutants and asthma. Environmental Health Perspectives 103(Supplement 6):235–242.

Koren HS, O'Neill M. 1998. Experimental assessment of the influence of atmospheric pollutants on respiratory disease. Toxicology Letters 28(102–103)317–321.

Kumar P, Caradonna-Graham VM, Gupta S, Cai X, Rao PN, Thompson J. 1995. Inhalation challenge effects of perfume scent strips in patients with asthma. Annals of Allergy, Asthma, and Immunology 75(5):429–433.

Landrigan PJ, Claudio L, Markowitz SB, Berkowitz GS, Brenner BL, Romero H, Wetmur JG, Matte TD, Gore AC, Godbold JH, Wolff MS. 1999. Pesticides and inner-city children: exposures, risks, and prevention. Environmental Health Perspectives 107(Supplement 3):431–437.

Leaderer BP, Zagraniski RT, Berwick M, Stolwijk JA. 1986. Assessment of exposure to indoor air contaminants from combustion sources: methodology and application. American Journal of Epidemiology 124(2):275–289.

Marbury MC, Kreiger RA. 1991. Formaldehyde. In Samet JM and Spengler JD, Eds. Indoor Air Pollution: A Health Perspective. Baltimore, MD: Johns Hopkins University Press.

Matti SJ, Chauhan AJ, Clough JB, Holgate ST. 1999. Distribution of personal exposure in the home to nitrogen dioxide in 8–12 year old children and their mothers. American Journal of Respiratory and Critical Care Medicine 159:A774.

Melia RJ, Florey CD, Altman DG, Swan AV. 1977. Association between gas cooking and respiratory disease in children. British Medical Journal 2 (6080):149–152.

Melia RJ, Florey CD, Chinn S, Morris RW, Goldstein BD, John HH, Clark D. 1985. Investigations into the relations between respiratory illness in children, gas cooking and nitrogen dioxide in the U.K. Tokai Journal of Experimental and Clinical Medicine 10(4):375–378.

Millqvist E, Löwhagen O. 1996. Placebo-controlled challenges with perfume in patients with asthma-like symptoms. Allergy 51(6):434–439.

Millqvist E, Löwhagen O. 1998. Methacholine provocations do not reveal sensitivity to strong scents. Annals of Allergy, Asthma, and Immunology 80(5)381–384.

Millqvist E, Bengtsson U, Löwhagen O. 1999. Provocations with perfume in the eyes induce airway symptoms in patients with sensory hyperreactivity. Allergy 54(5):495–499.

Mohsenin V. 1987. Airway responses to nitrogen dioxide in asthmatic subjects. Journal of Toxicology and Environmental Health 22(4):371–380.

Mohsenin V. 1988. Airway responses to 2.0 ppm nitrogen dioxide in normal subjects. Archives of Environmental Health 43(3):242–246.

Neas LM, Dockery DW, Ware JH, Spengler JD, Speizer FE, Ferris BG Jr. 1991. Association of indoor nitrogen dioxide with respiratory symptoms and pulmonary function in children. American Journal of Epidemiology 134(2):204–219.

Norbäck D, Björnsson E, Janson C, Widström J, Boman G. 1995. Asthmatic symptoms and volatile organic compounds, formaldehyde, and carbon dioxide in dwellings. Occupational and Environmental Medicine 52(6):388–395.

Nordman H, Keskinen H, Tuppurainen M. 1985. Formaldehyde asthma—rare or overlooked? Journal of Allergy and Clinical Immunology 75(1 Pt 1):91–99.

Norsted SW, Kozinetz CA, Annegers JF. 1985. Formaldehyde complaint investigations in mobile homes by the Texas Department of Health. Environmental Research 37(1):93–100.

Øie L, Hersoug LG, Madsen JØ. 1997. Residential exposure to plasticizers and its possible role in the pathogenesis of asthma. Environmental Health Perspectives 105(9):972–978.

Øie L, Nafstad P, Botten G, Magnus P, Jaakkola JK. 1999. Ventilation in homes and bronchial obstruction in young children. Epidemiology 10(3):294–299.

O'Malley M. 1997. Clinical evaluation of pesticide exposure and poisonings. Lancet 349(9059):1161–1166.

Ostro BD, Lipsett MJ, Mann JK, Wiener MB, Selner J. 1994. Indoor air pollution and asthma. Results from a panel study. American Journal of Respiratory and Critical Care Medicine 149(6):1400–1406.

Ostro BD, Lipsett MJ, Das R. 1998. Particulate matter and asthma: a quantitative assessment of the current evidence. Applied Occupational and Environmental Hygiene 13(6):453–460.

Peden DB, Setzer RW Jr, Devlin RB. 1995. Ozone exposure has both a priming effect on allergen-induced responses and an intrinsic inflammatory action in the nasal airways of perennially allergic asthmatics. American Journal of Respiratory and Critical Care Medicine 151(5):1336–1345.

Peden DB. 1997. Mechanisms of pollution-induced airway disease: in vivo studies. Allergy 52(38 Suppl):37–44, discussion 57–58.

Pope CA III, Dockery DW. 1992. Acute health effects of PM_{10} pollution on symptomatic and asymptomatic children. American Review of Respiratory Disease 145(5):1123–1128.

Quackenboss JJ, Spengler JD, Kanarek MS, Letz R, Duffy CP. 1986. Personal exposure to nitrogen dioxide: Relationship to indoor/outdoor air quality and activity patterns. Environmental Science and Technology 20(8):775–783.

Ritchie IM, Lehnen RG. 1985. An analysis of formaldehyde concentrations in mobile and conventional homes. Journal of Environmental Health 47(6):300–305.

Roemer W, Hoek G, Brunekreef B. 1993. Effect of ambient winter air pollution on respiratory health of children with chronic respiratory symptoms. American Review of Respiratory Disease 147(1):118–124.

Ross DJ, Keynes HL, McDonald JC. 1998. SWORD '97: surveillance of work-related and occupational respiratory disease in the UK. Occupational Medicine (London) 48(8):481–485.

Royce S, Wald P, Sheppard D, Balmes J. 1993. Occupational asthma in a pesticides manufacturing worker. Chest 103(1):295–296.

Rusznak C, Devalia JL, Davies RJ. 1996. Airway response of asthmatic subjects to inhaled allergen after exposure to pollutants. Thorax 51(11):1105–1108.

Ryan PB, Soczek ML, Treitman RD, Spengler JD, Billick IH. 1988. The Boston residential NO_2 characterization study. II. Survey methodology and population concentration estimates. Atmospheric Environment 22(10): 2115–2125.

Salome CM, Brown NJ, Marks GB, Woolcock AJ, Johnson GM, Nancarrow PC, Quigley S, Tiong J. 1996. Effect of nitrogen dioxide and other combustion products on asthmatic subjects in a home-like environment. European Respiratory Journal 9(5):910–918.

Samet J. 1990. Environmental controls and lung disease. American Review of Respiratory Disease 142:915–939.

Samet JM, Marbury MC, Spengler JD. 1987. Health effects and sources of indoor air pollution. Part I. American Review of Respiratory Disease 136(6):1486–1508.

Samet JM, Lambert WE, Skipper BJ, Cushing AH, Hunt WC, Young SA, McLaren LC, Schwab M, Spengler JD. 1993. Nitrogen dioxide and respiratory illnesses in infants. American Review of Respiratory Disease 148(5):1258–1265.

Sardinas AV, Connecticut State Department of Health Services, Most, Randi S, Giulietti, Mark A, Honchar P. 1979. Health effects associated with urea–formaldehyde foam insulation in Connecticut. Journal of Environmental Health 41(5):270–272.

Senthilselvan A, McDuffie HH, Dosman JA. 1992. Association of asthma with use of pesticides. Results of a cross-sectional survey of farmers. American Review of Respiratory Disease 146(4):884–887.

Sexton K, Liu KS, Petreas MX. 1986. Formaldehyde concentrations inside private residences: a mail-out approach to indoor air monitoring. Journal of the Air Pollution Control Association 36(6):698–704.

Shelton D, Urch B, Tarlo SM. 1992. Occupational asthma induced by a carpet fungicide—tributyl tin oxide. Journal of Allergy and Clinical Immunology 90(2):274–275.

Sheppard L, Levy D, Norris G, Larson TV, Koenig JQ. 1999. Effects of ambient air pollution on nonelderly asthma hospital admissions in Seattle, Washington, 1987–1994. Epidemiology 10(1):23–30.

Shim C, Williams MH. 1986. Effect of odors in asthma. American Journal of Medicine 80(1):18–22.

Smedje G, Norbäck D, Edling C. 1997. Asthma among secondary schoolchildren in relation to the school environment. Clinical and Experimental Allergy 27(11):1270–1278.

Smith B, Nitschke M, Ruffin R, Pilotto L, Wilson K, Pisaniello D. 1999. Cohort study of indoor nitrogen dioxide and respiratory symptoms in asthmatics. American Journal of Respiratory and Critical Care Medicine 159:A128.

Speizer FE, Ferris B Jr, Bishop YM, Spengler J. 1980. Respiratory disease rates and pulmonary function in children associated with NO_2 exposure. American Review of Respiratory Disease 121(1):3–10.

Spengler JD, Duffy CP, Letz R, Tibbitts TW, Ferris BG. 1983. Nitrogen dioxide inside and outside 137 homes and implications for ambient air quality standards and health effects research. Environmental Science and Technology 17(8):164–168.

Spengler JD, Schwab M, Ryan PB, Colome S, Wilson AL, Billick I, Becker E. 1994. Personal exposure to nitrogen dioxide in the Los Angeles Basin. Journal of the Air and Waste Management Association 44(1):39–47.

Spengler JD, Schwab M, McDermott A. Lambert WE, Samet JM. 1996. Nitrogen dioxide and respiratory illness in children. Part IV: Effects of housing and meteorologic factors on indoor nitrogen dioxide concentrations. Research Report/Health Effects Institute (58), 1–29; discussion 31–36.

Stock TH, Mendez SR. 1985. A survey of typical exposures to formaldehyde in Houston area residences. American Industrial Hygiene Association Journal 46(6):313–317.

Strachan DP, Carey IM. 1995. Home environment and severe asthma in adolescence: a population based case-control study. British Medical Journal 311(7012):1053–1056.

Thun MJ, Lakat MF, Altmen R. 1982. Symptom survey of residents of homes insulated with urea formaldehyde foam. Environmental Research 29(2):320–334.

Tunnicliffe WS, Burge PS, Ayres JG. 1994. Effect of domestic concentrations of nitrogen dioxide on airway responses to inhaled allergen in asthmatic patients. Lancet 344(8939–8940):1733–1736.

U.S. EPA (U.S. Environmental Protection Agency). 1979. National Household Pesticide Usage Survey, 1976–1977. EPA Report No. 540/0-80-002. Washington DC: U.S. Environmental Protection Agency, Office of Pesticides and Toxic Substances.

U.S. EPA. 1983. Analysis of Risks and Benefits of Seven Chemicals Used in Subterranean Termite Control. Washington DC: U.S. Environmental Protection Agency, Office of Pesticides and Toxic Substances.

U.S. EPA. 1987. Assessment of Health Risks to Garment Workers and Certain Home Residents from Exposure to Formaldehyde. Washington, DC: U.S. Government Printing Office.

U.S. EPA. 1996. Air Quality Criteria for Ozone and Related Photochemical Oxidants. Volume 3. EPA/600/P-93/004cF. Office of Research & Development, Research Triangle Park, North Carolina.

U.S. EPA. 1999. Ozone Generators That Are Sold as Air Cleaners. URL: http://www.epa.gov/iaq/pubs/ozonegen.html. Accessed September 6, 1999.

Utell MJ, Frampton MW, Roberts NJ Jr, Finkelstein JN, Cox C, Morrow PE. 1991. Mechanisms of nitrogen dioxide toxicity in humans. Research Report/Health Effects Institute Report (43):1–33.

Viegi G, Paoletti P, Carrozzi L, Vellitini M, Ballerin L, Biavati P, Nardini G, DiPede F, Sapigni T, Lebowitz MD, et al. 1991. Effects of home environment on respiratory symptoms and lung function in a general population sample in north Italy. European Respiratory Journal 4(5):580–586.

Viegi G, Carrozzi L, Paoletti P, Vellutini M, DiViggiano E, Baldacci S, Modena P, Pedreschi M, Mammini U, di Pede C, et al. 1992. Effects of the home environment on respiratory symptoms of a general population sample in middle Italy. Archives of Environmental Health 47(1):64–70.

Volkmer RE, Ruffin RE, Wigg NR, Davies N. 1995. The prevalence of respiratory symptoms in South Australian preschool children. II. Factors associated with indoor air quality. Journal of Paediatrics and Child Health 31(2):116–120.

Wallace LA. 1987. The Total Exposure Assessment Methodology (TEAM) Study: Summary and analysis. Volume 1. Office of Research and Development, USEPA, Washington, DC, EPA/600/6-87/002a.

Wallace LA. 1991. Volatile organic compounds. In Indoor Air Pollution: A Health Perspective. Samet JM and Spengler JD, Eds. Baltimore, MD: Johns Hopkins University Press.

Wallace LA, Pellizzari ED, Hartwell TD, Sparacino C, Whitmore R, Sheldon L, Zelon H, Perritt R. 1987. The TEAM (Total Exposure Assessment Methodology) Study: personal exposures to toxic substances in air, drinking water, and breath of 400 residents of New Jersey, North Carolina, and North Dakota. Environmental Research 43(2):290–307.

Wallace LA, Nelson WC, Pellizzari E, Raymer J, Thomas K. 1991. Identification of Polar Volatile Organic Compounds in Consumer Products and Common Microenvironments. U.S. Environmental Protection Agency report no. EPA/600/D-91/074. Washington DC: U.S. Government Printing Office.

Ware JH, Dockery DW, Spiro A III, Speizer FE, Ferris BG Jr. 1984. Passive smoking, gas cooking, and respiratory health of children living in six cities. American Review of Respiratory Disease 129(3):366–374.

Wax PM, Hoffman RS. 1994. Fatality associated with the inhalation of a pyrethrin shampoo. Journal of Toxicology – Clinical Toxicology 32(4):457-460.

Wieslander G, Norbäck D, Björnsson E, Jandon C, Boman G. 1997. Asthma and the indoor environment: the significance of emission of formaldehyde and volatile organic compounds from newly painted surfaces. International Archives of Occupational and Environmental Health 69(2):115–124.

Wolkoff P. 1998. Impact of air velocity, temperature, humidity, and air on long-term VOC emissions from building products. Atmospheric Environment 32(14–15):2659–2668.

Wolkoff P, Schneider T, Kildesø J, Degerth R, Jaroszewski M, Schunk H. 1998. Risk in cleaning: chemical and physical exposure. The Science of the Total Environment 215(1998):135–156.

Wolkoff P, Clausen PA, Wilkins CK, Hougaard KS, Nielsen GD. 1999. Formation of strong airway irritants in a model mixture of (+)-α-pinene/ozone. Atmospheric Environment 33(5):693–698.

7
EXPOSURE TO ENVIRONMENTAL TOBACCO *Smoke*

Involuntary exposure to environmental tobacco smoke (ETS), or "passive smoking," has been extensively investigated with respect to its potential health effects, particularly on respiratory health. There is a significant body of research on its potential effects regarding the incidence, prevalence, and exacerbation of established asthma. While attention has focused upon possible associations with childhood asthma, associations with asthma in adults also have been investigated. The following analysis relies heavily on several very detailed and comprehensive reviews, including those of the U.S. Environmental Protection Agency (EPA) (U.S. EPA, 1992), the California EPA's Office of Environmental Health Assessment (California EPA, 1997), the World Health Organization (WHO) International Consultation on Environmental Tobacco Smoke (ETS) and Child Health (WHO, 1999), the report of the United Kingdom's Scientific Committee on Tobacco and Health (SCOTH, 1998), and the series of ten meta-analyses (to date) of the health effects of ETS by Cook, Strachan, and colleagues (Anderson and Cook, 1997; Cook et al., 1998; Cook and Strachan, 1997, 1998, 1999; Strachan and Cook, 1997, 1998a–1998c).

DEFINITION OF ENVIRONMENTAL TOBACCO SMOKE (ETS)

Environmental tobacco smoke has been defined (Daisey et al., 1994) as:

> ... the smoke to which non-smokers are exposed when they are in an indoor environment with smokers. It is composed largely of sidestream tobacco smoke (SS), the smoke emitted by the smoldering end of a cigarette between puffs, with minor contributions from exhaled mainstream smoke (the smoke which is directly inhaled by the smoker) and any smoke that escapes from the burning part of the tobacco during puff-drawing by the smoker. ETS differs from SS in that it is highly diluted and dispersed within a room and it undergoes aging.

Tobacco smoke contains many chemical products with known or suspected adverse health effects. These products include eye and respiratory irritants, systemic toxicants, mutagens and carcinogens, and reproductive toxicants (California EPA, 1997). ETS consists of solid particulates, and semivolatile and volatile organic compounds (VOCs). The solid particulates have a mean diameter of 0.32 µm (National Research Council, 1986). "The aging process includes volatilization of nicotine, which is present in the particulate phase in mainstream smoke but is almost exclusively a component of the vapor phase of ETS" (U.S. EPA, 1992). The mean and standard deviation of the total emission factor for PM $_{2.5}$, determined for six commercial cigarettes and Kentucky reference cigarette 1R4F, is 8,100 ± 2,000 µg per cigarette. Bacterial endotoxin (lipopolysaccharide), previously associated with environmental lung diseases, has been reported to be a respirable constituent of both mainstream and sidestream smoke (Hasday et al., 1999).

Significant amounts of nearly 30 volatile organic compounds have been measured, including acetaldehyde, formaldehyde, nicotine, 3-vinylpyridine, toluene, pyridine, benzene, pyrrole, xylene, 2-butanone (methyl ethyl ketone [MEK]), phenol, and others. Many of the more volatile VOCs (such as aldehydes) remain in the air for prolonged periods of time following the smoking of a cigarette (at least four hours) and do not appear to undergo significant chemical reactions within this period. Some of the less volatile compounds and particulates appear to decrease over time due to deposition as well as ventilation effects. With the exception of nicotine, the emission factors for VOCs are significantly greater in ETS than in SS (U.S. EPA, 1992). Additional information on the physical and chemical properties of ETS and the biological activities can be found in the U.S. and California EPA reports (California EPA, 1997; U.S. EPA, 1992).

FACTORS CONTROLLING EXPOSURE TO ETS

Variations in Concentration of ETS in Indoor Environments

Exposure Assessment

Nicotine and particulate matter (PM), in addition to carbon monoxide, have been the constituents most extensively measured as a means of assessing ETS concentrations in indoor air. Nicotine is considered an adequate tracer for PM under certain conditions, and, possibly, for VOCs ranging from slightly to very volatile compounds (Daisey, 1999). Among the documented conditions influencing the concentration of nicotine are emission rates, ventilation, and (for VOCs/SVOCs) resorption and desorption from surfaces (Daisey, 1999). The EPA (1992) and Guerin et al. (1992) summarized more than 25 studies of nicotine concentration in more than 100 different indoor environments and found that the average concentrations of nicotine ranged from 0.3 to 30 μg/m^3, a hundredfold difference. In residences with one or more smokers, the typical range was from 2 to 10 μg/m^3, typically being higher in winter than in summer. Bars and smoking sections of commercial airplanes recorded the highest levels—up to 50–75 μg/m^3, although nonsmoking regulations and ordinances have significantly altered this. In general, the concentrations of nicotine have been found to increase with the number of smokers and number of cigarettes consumed in a given indoor environment (U.S. EPA, 1992).

One study involving personal monitor measurement of approximately 100 individuals in 16 metropolitan areas in the United States reported mean 24-hour time weighted average nicotine concentrations of 3.28 μg/m^3 for those exposed to ETS both at work and away from work; 1.41 µg/m^3 for those exposed away from work only; 0.69 µg/m^3 for those exposed at work only; and 0.05 µg/m^3 for those exposed at neither location (Jenkins et al., 1996; Jenkins and Counts, 1999). Particulate concentrations, unlike nicotine, are not specific to ETS as a source. However, although not unique to the combustion of tobacco, the quantity of respirable particulates produced by cigarette smoking, is large—significantly greater than the amounts produced by other common combustion sources within the home, such as wood-burning fireplaces, gas stoves, and kerosene space heaters (California EPA, 1997). Respirable suspended particles in homes with at least one smoker average about 20–100 µg/m^3 higher than the levels in similar nonsmoking homes. The highest concentrations have been reported in restaurants and bars—a maximum of 1,379 µg/m^3 and a range of average concentrations of 35–986 µg/m^3 (U.S. EPA, 1992). Ott et al. (1996) documented a 77% decrease in the average concentration of respirable suspended particles in a northern California tavern after a prohibition against smoking was instituted. In addition to the influence of the number of smokers and the amount smoked on the concentration of ETS in a given indoor environment, concentration is affected by the ventilation rate.

Long-term exposure to ETS has been of most concern from the standpoint of effects on lung development and cancers. However, ETS concentration varies over an extreme spatial and temporal range in indoor and outdoor environments, making it infeasible to comprehensively assess the ETS exposure history of an individual over their lifetime by direct exposure assessment or air sampling in all of the relevant environments. Critical aspects of this history can, however, be determined and more comprehensive and accurate assessment is often feasible for infants and very young children. Because of the difficulties involved, epidemiologists have tended to use questionnaires and interviews to determine individual history with regard to ETS exposure, classifying people into categorical groups to provide a semiquantitative measure of exposure. Direct measurement of exposure at or near the breathing zone is often done via personal monitors and can provide an assessment of integrated exposure, but this is feasible for monitoring only over a relatively limited period of time.

Biomarkers of Exposure

The most direct assessment of exposure involves the measurement of ETS constituents or their breakdown products in body fluids. To date, the most reliable of these biomarkers is cotinine, a metabolite of nicotine (Benowitz, 1999). Cotinine has an average half-life of approximately 16–19 hours (Benowitz and Jacob, 1994; Jarvis et al., 1988), making it highly useful for the assessment of integrated ETS exposure over the two to three days prior to the measurement. In infants and children, the half-life is appreciably longer, from approximately 40 hours in children more than 18 months old to approximately 65 hours in neonates (U.S. EPA, 1992). Because urinary cotinine excretion varies markedly among individuals as a result of renal function, urinary flow rate, and urinary pH (Benowitz et al., 1983), results often are expressed as nanograms of cotinine per milligram of creatinine, rather than simply in nanograms per milliliter of fluid. However, the production of creatinine is a function of muscle mass; hence excretion varies with age, sex, and other individual factors. In particular, the low level of creatinine produced in children means that the cotinine-to-creatinine ratios in children may fall into the range reported for active smokers (Watts et al., 1990).

The levels of exposure of nonsmokers to ETS are sufficient that nicotine and cotinine are detectable in their urine, blood, and saliva (Benowitz, 1996). Values are typically in the range of 0.5 to 10–15 ng/mL in the saliva and plasma, respectively, of nonsmokers, with urinary concentrations approximately three times higher—as much as 50 ng/mL or more (Guerin et al., 1992; U.S. EPA, 1992). A cutoff of 90 ng/mL has been used to distinguish active smokers from exposed and unexposed nonsmokers (Cummings et al., 1990), and studies consistently have been able to distinguish active smokers from exposed and unexposed nonsmokers (Jarvis et al., 1987). It has been more difficult to distinguish exposed from non-exposed non-smokers for a variety of reasons related to the validity of self-reported smoking status and ETS exposure, variability in nicotine metabolism, variability in sampling procedures, and the limits of sensitivity of the assay methods used (Idle, 1990). Increasing levels of cotinine have been generally found to be associated with increasing levels of self-reported ETS exposure (NRC, 1986; U.S. DHHS, 1986; U.S. EPA, 1992).

As would be expected from the results of measurement of ambient concentrations of nicotine, the maximum reported exposure levels have occurred in bars and restaurants and on commercial airline flights—approximately 30 ng/mg creatinine (Mattson et al., 1989). One study in which adults in an enclosed area were exposed to sidestream smoke from four cigarettes being smoked simultaneously and injected into the room continuously by machine, with ventilation conditions equivalent to those in the average office environment, found the air concentration of nicotine rapidly reached a stable level of 280 $\mu g/m^3$. Average nicotine concentration in saliva reached a maximum of 880 ng/mL after 60 minutes of exposure, and cotinine concentrations reached 3.4 ng/mL in serum and 55 ng/mg creatinine in urine, a little more than six hours after exposure.

A number of studies have compared biomarkers in active smokers with those in exposed and nonexposed nonsmokers. Jarvis and Russell (1984), for example, found mean urinary cotinine levels in these three groups of 1,390.0, 7.7, and 1.6 mg/mL, respectively ($p < .001$ between exposed and nonexposed nonsmokers). Cotinine concentrations of self-reported smokers and nonsmokers have generally been found to overlap.

In infants and children exposed to ETS, levels of cotinine have been found to be significantly higher in exposed than in nonexposed children. Direct exposure assessment has detected cotinine in the urine on the first day of life in neonates of both active smokers and ETS-exposed nonsmokers with significantly higher levels in the latter than in neonates of unexposed nonsmokers (Jordanov, 1990). Henderson et al. (1989) found that air nicotine concentration in the home was significantly associated with the average log urinary cotinine level ($r = 0.68$, $p = .006$). Greenberg et al. (1989) found a median concentration of 121 ng cotinine/mg creatinine (range 6–2,273 ng cotinine/mg creatinine) in children with any detectable cotinine. Chilmonczyk et al. (1990) found median levels of urinary cotinine of 1.6 mg/mL in nonsmoking households, 8.9 mg/mL where someone other than the mother smoked, 28 mg/mL where only the mother smoked, and 43 mg/mL where both the mother and someone else smoked.

Exposure Prevalence

In reviewing studies of ETS exposure prevalence, the California EPA (1997) concluded, "Taken as a whole, the various studies [at least 10 separate investigations including large representative sample surveys] . . . indicate that within California and the United States, exposure to ETS was widespread during the time period of the studies (1979 through 1992). Analysis of ETS exposure within California indicated that the workplace, home, and other indoor locations contributed significantly to the exposure of adults. For children, the home was the most important single location contributing to ETS exposure. In all studies using both self-reporting and a biological marker (cotinine level) as measures of exposure, prevalence was higher when determined using the biological marker." It further cited indirect evidence that "the prevalence of ETS exposure in the rest of the U.S. population is higher than that in California."

It is particularly noteworthy that despite aggressive antismoking education and regulation, and documented reductions in smoking rates (to 16.7% of the adult population in 1995 [CDHS, 1995]), in 1992 an estimated 9.4% of California women pregnant within the previous five years had smoked throughout pregnancy, and an estimated 19.6% of those 17 years of age may be exposed to ETS in their homes (Pierce et al., 1994). By inference from studies of adult smoking, it also would appear that the rates may be appreciably higher in specific subpopulations.

Influence of Activity Patterns on Exposure

The activity patterns of both children and adults have been studied in relation to exposure to ETS. For all ages, the home is the location in which the average person spends the most time— 921 minutes per day for adults and 1,078 minutes per day for children in California. Time within the home is spent primarily in the bedroom—an average of 524 minutes per day for adults and 674 minutes per day for children (Wiley et al., 1991). The next greatest amounts of time are spent by children in school or child care (an average of 109 minutes for all children and 330 minutes for those attending school), in other people's homes (80 minutes average and 251 minutes for those doing this), and in-transit (69 minutes overall and 83 minutes for those traveling). Overall, children spend an average of 1,230 min. (20.5 hours) each day indoors, 141 minutes outdoors, and 69 minutes in enclosed transit. Infants and other children ages 2 and under spend the most time indoors (an average of 21.6 hours), but somewhat less in enclosed transit (48 minutes). For adults, the times are 1,253 minutes indoors, 73 minutes outdoors, and 111 minutes in enclosed transportation, with time in the workplace replacing time spent in school or child care by children.

For children, the home is clearly the most likely source of exposure to ETS and the place that the child is most likely to sleep. While smoking is not permitted in schools or day care facilities and is prohibited in some states in licensed child care in private homes when children are present, the fact that many children are in nonlicensed child care arrangements or in states or communities where smoking prohibitions are not well enforced means that significant regular exposure may occur in home settings. Exposure during travel in the private automobile is another potential source of exposure.

For adults, research in California (Lum, 1994a, 1994b) has shown that exposure in the workplace is the most prevalent location for exposure of nonsmokers to ETS, with the home as the second most prevalent location. To the extent that workplaces adopt antismoking regulations, this exposure source may diminish in importance. The private automobile represents another potentially significant location for adult exposure.

It is possible for both adults and children to be exposed to ETS the majority of the time they are indoors, both during the day and at night. For the average preschool child, this could be virtually all of the time, for the school-aged child as many as 15.5 hours a day, and for adults anywhere from 12 hours (for those working in a nonsmoking environment) to 24 hours for those working as well as living in environments in which smoking is permitted. The only reliable exception would be time spent in school, public buildings, or public transit where smoking is prohibited. There is no reason to believe that the activity patterns of persons with asthma differ significantly from those of nonasthmatics, except for the possibility of their having lower activity levels that could result in more time spent indoors and hence greater exposure to any ETS present in indoor environments. Further, there are questions as to whether the sensitization of children to allergens (e.g., dust mites, cockroaches) in the home environment may be increased by the presence of ETS, as well as whether increased time spent in the indoor environment, if this occurs, results in greater exposure to ETS as well as to indoor allergens.

One study of children between the ages of 2 and 12 in Scotland, having at least one parent who smoked, found that salivary cotinine levels were nondetectable in only four children, all of whom had only a father who smoked (Irvine et al., 1997). In the remaining 493 children, the levels ranged from 0.5 ng/mL (barely detectable) to 21.2 ng/mL, with a mean of 4.35 ng/mL. The authors cite two studies in which levels of 14.3 ng/mL or higher have been taken as indicative of active smoking by a child. However, 13 of the 18 children who scored between 14.3 and 21.2 ng/mL were younger than 6 years of age and are presumed not to be active smokers. This study found that the age of the child, cotinine level and self-reported amount smoked in the home by the index parent, self-reported frequency of smoking in the same room as the child, whether the index parent's partner smoked, whether the child had contact with other smokers, the number of persons per room in the home, and whether the home had a yard or garden were all significantly and independently related to the child's cotinine level.

EVIDENCE OF A RELATIONSHIP BETWEEN ETS AND ASTHMA

Action of ETS on the Lungs

Tobacco smoke, whether mainstream, sidestream, or ETS, is a lung irritant. From a pathophysiologic point of view, active smoking is associated with significant structural changes in both the airways and the pulmonary parenchyma (U.S. DHHS, 1984), including hypertrophy and hyperplasia of the upper airway mucous glands, leading to an increase in mucous production with associated increased prevalence of cough and phlegm. Chronic inflammation of the smaller airways also occurs, leading to bronchial obstruction. In addition, airway narrowing may occur consequent to destruction of the alveolar walls, decreased lung elasticity, and development of centrilobular emphysema (U.S. EPA, 1992). Smoking also may increase mucosal permeability to allergens, increasing total and specific immunoglobulin E (IgE) levels (Zetterstrom et al., 1981) and blood eosinophil counts (Halonen et al., 1982).

The adverse health effects and pathophysiologic changes associated with active smoking have been observed at low-dose exposures, suggesting that ETS might have similar adverse effects, a suspicion that was heightened by the fact that ETS contains some volatile substances in greater quantities than are found in mainstream smoke (U.S. EPA, 1992). In addition, since large proportions of the population are involuntarily exposed to ETS, including more susceptible infants and children, the index of suspicion for adverse effects of ETS is high. Exposures early in life, when the lung is undergoing significant growth and remodeling, could plausibly alter lung development and increase the risk of various respiratory illnesses, including asthma. It is also plausible that, in addition to the marked susceptibility of young lungs, there is variable individual susceptibility in other respects, including genetic predisposition, lung injury such as bronchopulmonary dysplasia consequent to premature birth, and greater contact with a primary caregiver who smokes.

Maternal Active Smoking During Pregnancy

Exposure of the fetus to the products of maternal tobacco smoking is a form of "environmental" exposure to tobacco smoke, although not in the same proportions as in airborne ETS and not to all constituents of ETS (notably, not the particulates). It is plausible that virtually all products of active maternal smoking that enter the bloodstream of the mother, including products arising from mainstream and sidestream smoke, cross into the fetus through the placenta with a diffusion gradient. This has been confirmed in the case of carbon monoxide (Longo, 1970) and cotinine. A biomarker for nicotine exposure, cotinine has been detected in the amniotic fluid of ETS-exposed women and the urine of their neonates in significantly higher concentrations than in nonexposed nonsmoking women (Jordanov, 1990). Transplacental passage of the bloodborne products of *passive* maternal ETS exposure also would be expected, although at lower levels and with a different chemical composition than if the mother were an active smoker.

Active maternal smoking has been associated with reduced size of the placental arteries (Asmussen, 1979), a reduction in average birthweight of 75–400 gm. (Abell et al., 1991; Asmussen, 1979; Lodrup Carlsen et al., 1997; Milner et al., 1999; Sherwood et al., 1999; Wang et al., 1997), and altered lung function measured shortly after birth (Lodrup Carlsen et al., 1997). Small but statistically significant deficits in forced expiratory volume in one second (FEV_1) and other spirometric indices (forced vital capacity [FVC], mid expiratory flow [MEF], and end expiratory flow [EEF]) have been fairly consistently demonstrated in school-aged children (data reviewed in Cook and Strachan, 1999) and as early as three days after birth (Lodrup Carlsen et al., 1997), thereby strongly implicating maternal smoking during pregnancy as the cause of these deficits. However, in Turkey, where there is heavy smoking by men and virtually none by women, exposure of children also has been associated with significant deficits in lung function (e.g., Bek et al., 1999). Experimental studies in animals have demonstrated that ETS exposure of pregnant rats is associated with reduced lung volume, number of saccules and septal crests, and elastin fibers in fetal lungs (Collins et al., 1985). More recently, Sekhon et al. (1999) reported that nicotine alone, when administered to pregnant rhesus monkeys, altered the expression of nicotine receptors in the developing fetal lung, leading to lung hyperplasia with structural alterations and reduced complexity of the gas-exchange surface.

ETS and Children's Respiratory Health

Recent reviews of an extensive body of cross-sectional, case-control, and longitudinal epidemiologic research on the effects of parental smoking on children's respiratory health have come to very similar, although not identical, conclusions. These reviews include both systematic, quantitative meta-analyses (Cook and Strachan, 1999) and narrative reviews (California EPA, 1997; U.S.

EPA, 1992; SCOTH, 1998; WHO, 1999). In updating their earlier quantitative meta-analysis to include additional studies conducted between April 1997 and June 1998, Cook and Strachan (1999) summarize their earlier general conclusions (Cook and Strachan, 1997, 1998; Cook et al., 1998; Strachan and Cook, 1997, 1998a–1998c) as follows:

> Overall, there is a very consistent picture with odds ratios for respiratory illnesses and symptoms and middle ear disease of between 1.2 and 1.6 for either parent smoking, the odds usually being higher in pre-school than school-aged children and higher for maternal smoking than for paternal smoking.

Virtually all of the evidence with regard to the effects of chronic ETS exposure in children comes from epidemiologic research, with very limited investigation of acute exposures. True experimental investigations of controlled acute exposure in chambers has been limited to adults.

Chronic ETS Exposure and Asthma Incidence, Prevalence, and Severity in Infants and Children

With respect specifically to the prevalence of asthma and respiratory symptoms in school-aged children, both the previously analyzed and the newer studies reviewed by Cook and Strachan (1999) supported the conclusion that parental smoking is associated with "increased prevalence of asthma and respiratory symptoms in school children" and that "among children with established asthma, parental smoking was associated with more severe disease." Indicators of disease severity for which such an association has been documented include emergency room visits, life-threatening attacks, and symptoms.

As indicated in Table 7-1, among children ages 5–16, pooled odds ratios (ORs) for asthma prevalence in studies reported through April 1997 were 1.21 (95% confidence interval [CI] 1.10–1.34, 21 studies) for either parent smoking from cross-sectional studies and 1.37 (1.15–1.64, 14 studies) from case-control studies, 1.36 (1.20–1.55, 11 studies) for maternal smoking only, 1.07 (0.92–1.24, 9 studies) for paternal smoking only, and 1.50 (1.29–1.73, 8 studies) for both parents smoking. Maternal smoking was associated with an OR of 1.31 (1.22–1.41, 4 studies) for asthma incidence under age 6 and with an OR of 1.13 (1.04–22, 4 studies) over age 6.

In younger children (0–2 years of age), the odds ratio for wheezing illness was 1.55 (95% CI 1.16–2.08, 5 studies) for either parent smoking and 2.08 (1.59–2.71, 7 studies) for mother smoking. These data suggest that parental smoking is more influential as a cause of "wheezy bronchitis" in infants and toddlers than of later-onset asthma.

There is, at present, some inconsistency with regard to the interpretation of studies that have attempted to separate the influence of maternal smoking during pregnancy from postnatal maternal smoking. Separation of the effects is difficult since those who smoke during pregnancy are very likely to continue to do so after the birth of the child, although some smokers may quit during the first trimester and abstain for the remainder of the pregnancy, often resuming thereafter. One U.S. study of 705 Chicago fifth graders (Hu et al., 1997) found that maternal smoking during pregnancy was more strongly related to doctor-diagnosed asthma than was current maternal smoking. Similarly, a Scandinavian study of nearly 16,000 children 6–12 years of age (Forsberg et al., 1997) found that asthma attacks, dry cough, and asthma treatment in the preceding year were inversely associated with current smoking in the home but positively associated with smoking in the first two years of life. The inverse relationship with current smoking suggests that parents (at least in Scandinavia) may modify their smoking behavior as a result of the child's asthma.

Several observations may be relevant in understanding the lower odds ratios for asthma prevalence and incidence in school-aged children than for wheeze in younger children, especially where the data come from cross-sectional studies and relate to current smoking in the home. ETS exposure of older children may be lessened by virtue of the greater amounts of time spent outside the home and may not reflect their smoke exposure at a younger age. Cotinine, a marker for smoke exposure, has been shown to be lower in school-aged than in younger children, among children with comparable levels of smoking in the home (Irvine et al., 1997).

As already noted, maternal antenatal smoking has been asso-

TABLE 7-1 Summary of Effects of Parental Smoking on the Respiratory
Health of Children

Outcome	Either Parent OR (95% CI)
Lower respiratory illnesses	
(LRI) at age 0–2	
All studies	1.57 (1.42–1.74) [27]
Community studies of wheeze	1.55 (1.16–2.08) [5]
Community studies of LRI,	1.54 (1.31–1.80) [11]
bronchitis and/or pneumonia	
Hospital admission for LRI,	1.71 (1.21–2.40) [8]
bronchitis, bronchiolitis, or	
pneumonia	
Prevalence rates at age 5–16	
Wheeze	1.24 (1.17–1.31) [30]
Cough	1.40 (1.27–1.53) [30]
Phlegm	1.35 (1.13–1.62) [6]
Breathlessness	1.31 (1.08–1.59) [6]
Asthma (cross-sectional studies)	1.21 (1.10–1.34) [21]
Asthma (case-control studies)	1.37 (1.15–1.64) [14]
Bronchial reactivity	
Skin prick positivity	
Incidence of asthma	
Under age 6	
Over age 6	
Middle-ear disease	
Acute otitis media	Range 1.0–1.6 [8]
Recurrent otitis media	1.48 (1.08–2.04) [7]
Middle-ear effusion	1.38 [c] (1.23–1.55) [4]
Referral for glue ear	1.21 [c] (0.95–1.53) [7]
Sudden infant death[d]	

NOTE: Numbers in square brackets are numbers of studies on which pooled odds ratios
(OR) are based.

[a]Relates largely, but not entirely to maternal smoking.
[b]Results relate to maternal smoking during pregnancy or exposure to ETS in infancy.
Data for ETS exposure during later childhood are too heterogeneous for meta-analysis.
[c]Based on fixed-effects estimate.

Mother Only OR (95% CI)	Father Only OR (95% CI)	Both parents OR (95% CI)
1.72 (1.55–1.91) [27] 2.08 (1.59–2.71) [7] 1.57 (1.33–1.86) [7]	1.29 (1.16 – 1.44) [16]	
1.53 (1.25–1.86) [9]	1.32 (0.87 – 2.00) [6}	
1.28 (1.19–1.38) [18] 1.40 (1.20–1.64) [14]	1.14 (1.06–1.23) [10] 1.21 (1.09–1.34) [9]	1.47 (1.14–1.90) [11] 1.67 (1.48–1.89) [16] 1.46 (1.04–2.05) [5]
1.36 (1.20–I.55) [11]	1.07 (0.92–1.24) [9]	1.50 (1.29–1.73) [8]
1.29 [a] (I.10–1.50) [10] 0.87[b] (0.64–1.24) [8]		
1.31[c] (1.22–1.41) [4] 1.13 [c] (1.04–1.22) [4]		
2.13 (1.86–2.43) [18]		

[d]Estimates and confidence limits differ due to exclusion of the study by Bulterys et al. (1993) (see Erratum at the end of Cook and Strachan, 1999).

SOURCE: Cook and Strachan, 1999; based on studies published through April 1997. Reprinted with permission of BMJ Publishing Group.

ciated with reduced size of the placental vessels and decreased blood flow to, if not oxygenation of, the fetus, a reduction in birthweight of infants carried to term, and decreased airflow. In addition, the risks of prematurity, neonatal respiratory distress syndrome, and bronchopulmonary dysplasia (BPD) are greater in children of mothers who smoke during pregnancy. Antenatal smoke exposure is associated with decreased airflow, which is considered likely to be related to airway size (Hanrahan and Halonen, 1998), and it has been suggested that postnatal exposure may induce or augment airway inflammation, both of which could contribute to the observed greater likelihood of development of wheezing and respiratory infections in young children (Cook and Strachan, 1999; U.S. EPA, 1992). Arguably, this may also increase the likelihood of both respiratory infections and sensitization to aeroallergens. All of these factors may, especially in an infant genetically predisposed to allergen sensitization and asthma, increase the likelihood that a persistent inflammatory condition will be established in the airways, thus promoting the development of asthma and perhaps hastening its manifestation. However, since asthma clearly occurs in children from nonsmoking homes with little or no ETS exposure, the gradual addition of such children to the pool of "cases" might tend to weaken the observed association between asthma and ETS exposure among older children, whether they are considered cross-sectionally or as a birth cohort followed longitudinally. A possibly more delayed development of asthma in some non-ETS-exposed children would not dilute the observed relationship between ETS exposure and early wheezing.

Dose–Response Relationship Between ETS Exposure and Asthma

As summarized in Table 7-1 and noted above, the OR for asthma prevalence when both parents smoke tends to be higher than when only the mother smokes, which in turn is higher than when only the father smokes. The presumed explanation is that, in general, fathers have less intense contact with the child (and/ or that a nonsmoking mother may exert some influence in protecting the child against ETS exposure due to the father's smoking). The relative risks associated with these four situations with respect to tobacco smoke in the home (i.e., both parents smoke, only the mother smokes, only the father smokes, neither parent smokes) suggests a dose–response relationship to asthma prevalence as well as to wheeze, cough, phlegm, and breathlessness. This further suggests that a reduction in exposure, short of total avoidance, confers some benefit. However, the OR for either parent smoking and for paternal smoking is still greater than 1, indicating that this level of exposure is not without risk. Moreover, once asthma is established, the evidence supports the conclusion that ETS exposure is associated with more frequent asthma exacerbations. Although a threshold may exist, there is no evidence as to what, if any, level of ETS exposure of a child, especially a child with asthma, could be said to be "risk free."

ETS Exposure and Asthma Incidence in Adults

There do not appear to be any studies linking chronic adult ETS exposure to adult onset asthma or any findings of an increased prevalence of asthma in adults exposed to ETS compared to those not exposed. In fact, if adults with asthma purposely avoided such exposure, a negative association might be observed. However, one study has shown an increased likelihood of new onset of wheezing in young adults, as well as children, attributable to maternal smoking during pregnancy, even after controlling for exposure in the home and other risk factors (Strachan et al., 1996).

Acute ETS Exposure and Asthma Exacerbations

Assessing the contribution of acute ETS exposure to asthma exacerbations is difficult since, for a significant proportion of exposed individuals, exposure is likely to be chronic (although variable). The evidence from studies comparing reported recent ETS exposure and cotinine levels in children seen for acute asthma versus similar children seen for well-child visits is somewhat equivocal (Ehrlich et al., 1992; Ogborn et al., 1994). These studies suffer from small sample sizes and low power. One large study of adults correlated asthma symptoms with reported daily ETS exposure and reported an OR of 1.61 (95% CI 1.06–2.46) for restricted activity days in relation to ETS exposure level, with a somewhat higher ratio (2.05; 95% CI = 1.78–2.40) for the level of asthma symptoms associated with having a smoker in the home, suggesting an effect of chronic as well as acute ETS exposure (Ostro et al., 1994).

ETS exposure in a chamber under controlled conditions has been investigated predominantly in adults with asthma, rather than in children. These studies, which were reviewed in detail by the California EPA (1997), have shown slight to moderate transient effects on lung function in at least a portion of participants but have not demonstrated a consistent effect. The studies had significant design limitations, including exclusion of participants who had recently been ill or had brittle asthma and, in many cases, the use of exposures of an hour or less in duration. Although participants in some of these studies may have been vulnerable to the effects of psychological suggestion because researchers did not disguise the concentration of ETS delivered, others with effective "blinding" of participants had observed effects.

CONCLUSIONS REGARDING THE HEALTH IMPACTS OF ETS WITH RESPECT TO ASTHMA

The evidence cited above permits the following conclusions with regard to the relationships between ETS exposure and asthma:

• There is sufficient evidence to conclude that there is a causal relationship between ETS exposure and exacerbations of asthma in preschool-aged children.
• There is sufficient evidence to conclude that there is an association between ETS exposure and the development of asthma in younger children. In the limited number of studies that have been able to separate the effects of maternal active smoking during pregnancy from the effects of ETS exposure after birth, evidence suggests that—while both exposures are detrimental—maternal smoking during pregnancy has the stronger adverse effect.
• There is limited or suggestive evidence of a relationship

between chronic ETS exposure and exacerbations of asthma in older children and adults. Limited or suggestive evidence of an association between acute ETS exposure and exacerbation also exists for asthmatics sensitive to this exposure.

• There is inadequate or insufficient evidence to determine whether or not an association exists between ETS exposure and the development of asthma in school-age children.

EVIDENCE REGARDING MEANS OF SOURCE MITIGATION OR PREVENTION

Ventilation and Air Cleaning

At present, source control appears to be the only reliably effective means of preventing ETS exposure. As discussed in Chapter 10, ventilation and air cleaning measures are available that have the technical capability of reducing the particulate components of ETS in indoor environments. However, these measures would be unlikely to appreciably reduce exposure of the fetus of a pregnant woman who smokes. Further, there is currently no direct evidence as to how much a reduction in the concentration of ETS particulates in a home, if achieved, would reduce the demonstrated adverse effects of ETS exposure on asthma. Also, there is no evidence regarding the degree of reduction in ETS particulate concentration that actually would be achieved through ventilation and air cleaning in the homes of smokers who continue to smoke indoors, even if these were introduced by an aggressive educational intervention. Any changes in ventilation that smokers did implement might also vary in effectiveness as a function of season and weather conditions. Nor is it known how such measures would affect the actual exposure of the residents, particularly children, and how this might vary as a function of who smokes and how many smokers are in the home. A more thorough discussion of ventilation and air-cleaning technologies is contained in Chapter 10.

Gas-phase air cleaning systems are available and potentially effective for some gas-phase constituents of ETS; however, no proven, reliable, and cost-effective means of air cleaning currently exists of removing the broad range of gaseous components of ETS from the indoor air. Until the components of ETS that affect asthma are better characterized, including the role, if any, of specific VOCs, the importance of developing a means for the removal of ETS-related VOCs as a means of addressing the asthma problem will remain unclear.

Source Control

If all ETS exposure were eliminated for fetuses, infants, and children, and for persons of any age who have already developed asthma, it is reasonable to assume that the population risk of developing wheezing with respiratory infections and the risk of asthma exacerbations would decrease to the levels currently observed among similar persons who are not exposed. This conclusion, however, is inferred primarily from the epidemiologic data comparing persons from homes with smokers to those living in homes with no smoker. No demonstration has been reported showing that exposure can be totally eliminated by an educational intervention, much less that doing so achieves beneficial asthma outcomes. However, Eisner et al. (1998) have reported an association of asthma severity, health status, and health care initiation with ETS exposure in 451 nonsmoking adults. They also reported that cessation of ETS exposure at follow-up was associated with an improvement in the severity of asthma scores and reduced health care utilization.

Even in accepting the likelihood that a benefit would result from truly effective elimination of exposure, questions remain about the extent to which this can be achieved in practice. Successful elimination of exposure is dependent on the extent to which the initiation of smoking can be prevented, especially in young women of childbearing age; that women who do smoke can be induced to cease smoking during and following pregnancy; that all persons, particularly parents but also other caregivers and frequent visitors, can be induced not to smoke at all or not in the environment of a child with asthma; and that adults with asthma will actually eliminate their exposure to ETS. The evidence that these changes can be induced by regulatory or educational means is reviewed below.

Regulatory Strategies

Where they exist and are enforced, regulatory strategies pro-hibiting smoking in public buildings, schools and child care fa-cilities, on public transit, and in the workplace (e.g., offices, plants, commercial airplanes, restaurants, bars) have clearly been associ-ated with decreased population exposure to ETS (California EPA, 1997). Within private homes, however, regulatory strategies have been deemed unacceptable, except where licensed child care is being provided. This leaves source control in the major environ-ment where ETS exposure occurs—the home—to be addressed by indirect regulatory forces (e.g., increased cigarette taxes, con-trols on cigarette advertising, etc.) and by educational or behav-ioral change methods.

The overall prevalence of cigarette smoking in the United States declined substantially from 40% in 1965 to 29% in 1987, but the decline has leveled off and has not reached the public health goal of 15% set for the year 2000 (U.S. DHHS, 1991). In 1995, the overall prevalence of cigarette smoking was 25% (CDC, 1997). The decline has been marginal among those with low education aspi-rations. Of every five persons who use tobacco, four begin before age 18 (CDC, 1989). After several years of substantial decline among adolescents in four ethnic minority groups, smoking prevalence increased during the 1990s among African-American and Hispanic youth (CDC, 1998). These trends and the success of efforts at smoking prevention and cessation among young women in particular are especially relevant to the issue of avoiding ante-natal and postnatal exposure of children to maternal smoking.

Adolescent Smoking Prevention and Cessation

School-based programs to prevent the initiation of smoking can be successful if they include social reinforcement and other strategies demonstrated to promote behavioral change (Bruvold, 1993). Moreover, properly designed school smoking policies (i.e., multiple components including a greater emphasis on prevention and less emphasis on cessation) are associated with lower amounts of smoking in adolescents (Pentz et al., 1989). It also has been shown that certain strategies directed at adolescents can have an effect opposite from that intended (McKenna and Williams, 1993).

Success of Smoking Cessation Efforts Directed at Adults

Public health strategies to prevent initiation of smoking and encourage cessation have clearly been associated with a decline in smoking prevalence. However, smoking is an addictive behavior with many personal and social factors that support its continuation. Many unsuccessful strategies to get smokers to quit have been attempted, notably those based on simply providing information and/or those directed at the general population of smokers who have not evidenced an interest in quitting (e.g., Gritz et al., 1992). Programs directed at smokers who are highly addicted or who initiated smoking earlier in their lives have been less successful than those directed at shorter-term, less addicted smokers (Chen and Millar, 1998; Killen et al., 1988; Senore et al., 1998; Smith et al., 1999). As overall smoking cessation rates in the United States have decreased, those who continue to smoke tend to be heavier smokers (COMMIT, 1995). However, the comparison of less and more successful programs has enabled a distillation of the components of the more successful approaches. It has been clearly demonstrated that well-designed smoking cessation programs, delivered by trained counselors, can be effective in achieving smoking cessation in adult men and women, including ethnic and minority groups (AHCPR, 1996). Such programs are associated with greater and more sustained short- and longer-term quit rates than the rates among persons who quit on their own, without the benefit of such assistance. Cessation programs are more successful to the extent that they are more intensive, take account of the varying motivations and level of addiction of participants, and are attuned to the individual's readiness to consider and initiate cessation attempts.

With regard to smoking cessation attempts in the clinical setting, strong cessation messages from clinicians, structured in relation to the readiness and personal needs of the patients and utilizing nicotine replacement therapy and supplementary educational and behavioral interventions, have been associated with an increase in both initial and sustained quit rates in controlled trials

(Law and Tang, 1995; Ockene and Zapka, 1997). The Agency for Health Care Policy and Research (AHCPR) recently reviewed more than 3,000 scientific articles that addressed the assessment and treatment of tobacco dependence, nicotine addiction, and clinical practice in order to develop guidelines for smoking cessation for primary care and specialist physicians (AHCPR, 1996; Fiore et al., 1997). These guidelines emphasize the importance of several components: nicotine replacement therapy (NRT), social support from the clinician, and skills training or problem solving based on practical advice and techniques to help individuals adapt to life as a nonsmoker. The inclusion of NRT is associated with pooled ORs of smoking cessation at six months, compared with a placebo, of from ~1.6 to ~3.0, depending on the method of delivery. Odds ratios are lowest for gum, rising to ~2.0 for the transdermal patch, and 2.92 and 3.05 for nasal spray and inhaled nicotine (Cepeda-Benito, 1993; Fiore et al., 1994; Law and Tang, 1995; Li Wan Po, 1993; Silagy et al., 1994; Tang et al., 1994; Viswesvaran and Schmidt, 1992).

Smoking Cessation Interventions in Pregnant Women

As discussed above, maternal smoking, in particular, has been associated with adverse respiratory and asthma outcomes. In the United States in 1994, 23.1% of all women and 14.6% of pregnant women smoked (Kendrick and Merritt, 1996). Special efforts to obtain cessation in women, particularly pregnant women and mothers, appear to be warranted. A meta-analysis of randomized trials of prenatal smoking cessation interventions that measured effects between the sixth and ninth months of pregnancy concluded that "prenatal smoking cessation interventions increase rates of smoking cessation during pregnancy" (Dolan-Mullen et al., 1994). Haddow et al. (1991), not included in the review, reported only modest success in getting pregnant women to cease smoking during pregnancy using a cotinine-assisted intervention. The relative success of such interventions with women of various ages, ethnicities, and education has not been analyzed, although most of the reported studies took place with patients seen in public clinic settings, suggesting that the results are not limited to middle- or upper-income and education groups. There is evidence that many women who quit smoking during pregnancy resume soon after delivery (McBride and Pirie, 1990; Mullen et al., 1997).

Three studies were reviewed that included low birthweight and other pregnancy outcome measures in addition to smoking cessation risk ratios. Reduced risk of low birthweight was found in studies that achieved higher rates of smoking cessation. Although these studies are suggestive of a potential beneficial effect on respiratory outcomes, no studies to date appear to have investigated these outcomes directly. Long-term follow-up is also needed to determine the effectiveness in sustaining cessation after delivery.

Reduction of ETS Exposure in Children with Asthma by Source Control Methods Other Than Smoking Cessation

Efforts to reduce the exposure of children, with or without asthma, by getting family members who smoke and others to limit their smoking to outside the home or even to certain well-ventilated areas within the home appear to face significant challenges due to the inherent inconvenience to the smoker and the limitations posed by inclement weather and building characteristics. Nevertheless, it is useful to consider what is known about the effectiveness of educational programs in achieving the goal of protecting children from exposure.

Intervention attempts to reduce passive smoking of infants and children, with or without asthma, have had mixed success. Greenberg et al. (1994) reported on an intervention designed to assist families in reducing infants' ETS exposure. The intervention was based on social learning theory and was delivered during four nurse home visits within the first 6 months of life. There was a tendency for nonparticipants to include higher proportions of mothers who smoked, as well as black, younger, and less educated mothers. Intervention effects were considered separately for families where the mother smoked and families where the mother did not smoke. Among those randomized, when the mother smoked the intervention was associated with significantly lower self-reported exposure of the infant to tobacco smoke from the mother and from nonmaternal household members. Infants whose mother did not smoke had low reported exposure from the outset of the study, and no intervention effect was observed. This differential self-reported exposure of infants of maternal smokers was not, however, accompanied by a significant differential in the cotinine-to-creatinine ratios of the intervention and control children. In fact, the proportion with detectable urine cotinine levels tended to increase over the year of follow-up in both groups. The incidence of all acute lower respiratory illnesses (ALRIs) and of severe acute respiratory illnesses did not decrease in the intervention group, and in fact, there was a small but statistically significant difference in all ALRIs favoring the control group. There was a significant difference in the frequency of persistent lower-respiratory symptoms in the maternal smoking subsample, but only where the head of household had a high school education or less. The authors interpret the results as indicating that mothers took steps to protect the infant from exposure by removing them from the vicinity of the smoker and that the infants were nevertheless subsequently exposed to residual nicotine but not to other ETS products, "which may be more likely than nicotine to have acute and chronic toxicity for passive smokers." The authors did not discuss whether parental report could have been biased in the direction of reduced reporting of exposure, and the unplanned subgroup analysis means that the positive results with regard to persistent lower-respiratory symptoms are merely suggestive.

Chilmonczyk et al. (1992) reported an unsuccessful physician's office-based intervention strategy that used feedback from the physician to the parent on infant urine cotinine measurements in an attempt to reduce the infant's exposure to ETS. The 6% reduction of urine cotinine levels for the intervention group at follow-up two months later was not statistically significant. This lack of success was in contrast to the investigator's previous success in getting women to stop smoking during pregnancy based on feedback on their own urine cotinine levels (Haddow et al., 1991), suggesting there may be greater motivation and ability of women to cease smoking and eliminate exposure of their fetus than to prevent exposure of infants and older children. An earlier unsuccessful attempt to reduce passive smoking in infancy was reported by Woodward et al. (1987).

Hovell et al. (1994) and Wahlgren et al. (1997) have reported that among children with asthma, a preventive medicine counseling intervention was associated with a greater reduction in self-reported and air monitor-verified ETS exposure than a monitored or usual care control condition. McIntosh et al. (1994) did not report a significant benefit of a cotinine-assisted, minimal-contact intervention.

Where positive results and promising interventions have been reported, there is a need for replication and, if possible, extension to other populations. Extensions of interventions should be made to populations including those who tend to be more resistant to cessation efforts and may be more typical of those whose children are being exposed to significant levels of ETS and are at risk for poor asthma outcomes for a variety of reasons. Wilson et al. (1996) have found that both adults with asthma who smoke and smoking parents of children with asthma are less likely than nonsmokers to attend an asthma education program, making it less likely that they will modify the child's exposure or experience the other benefits of such asthma education programs.

None of the studies to date that have investigated educational interventions to reduce ETS exposure have extended this to include asthma outcomes—either doctor-diagnosed asthma or wheezing illness incidence, or the prevalence or exacerbations of established asthma. Until this is done, it leaves unanswered the question of whether any ETS exposure reduction that may be achieved is sufficient to alter these disease outcomes, as well as whether there is any safe ETS exposure level. This is particularly important when the intervention aims to reduce infant exposure by means other than cessation of smoking by all caregivers and others in the child's environment. For this reason it also is impossible to directly answer questions regarding the cost-effectiveness of mitigation and prevention strategies.

CONCLUSIONS REGARDING ETS SOURCE CONTROL OR MITIGATION: FEASIBILITY AND BENEFITS

Conclusions Regarding the Effects of Complete Avoidance of ETS Exposure

Based on reasoning from the epidemiologic evidence presented above, the following conclusions can be reached regarding the potential benefits of essentially complete avoidance of ETS exposure, if this could be achieved:

• There is sufficient evidence to conclude that complete avoidance of ETS exposure would be associated with a lower likelihood of exacerbations of asthma in preschool children with established asthma.

• There is limited evidence suggesting that complete avoidance of ETS exposure would be associated with a lower likelihood of exacerbations of asthma in older children and adults.

• There is sufficient evidence to conclude that complete avoidance of ETS exposure, if this could be achieved, would reduce the probability of the development of wheezing with respiratory illness in younger children.

• There is limited or suggestive evidence that complete avoidance of ETS exposure, if this could be achieved, would reduce the likelihood of the persistence of asthma or of new-onset asthma in children and adults.

Conclusions Regarding Mitigation Through Source Control

• There is sufficient evidence to conclude that increased ventilation and air-cleaning methods are technologically capable of reducing the concentration of ETS particulates in indoor air.

• There is no evidence as to how readily the necessary ventilation and air-cleaning methods or technologies would be adopted and how effectively they actually would be used to reduce ETS concentration.

• There is no evidence of whether interventions designed to encourage the use of the requisite ventilation and air-cleaning methods would be associated with a reduction in ETS concentration, in the exposure of persons with asthma to ETS, or in asthma prevalence or exacerbations.

• There is inadequate evidence to conclude that interventions intended to establish smoke-free homes where a family member has asthma and to require smokers to smoke only outdoors are associated with a reduction in ETS exposure or asthma exacerbations.

RESEARCH NEEDS

A better understanding is needed of the mechanism(s) by which ETS and its individual constituents may

- impair the normal development of the airways in the fetus,
- promote allergic sensitization,
- promote respiratory infections,
- promote early wheezing illness, and
- (possibly) induce pathophysiologic changes that may promote the establishment of asthma.

Research is also needed to understand the nature of the interactions, both at the population or epidemiologic level and at the molecular and cellular levels, between the genetic predispositions to allergic sensitization and bronchial hyperresponsiveness and ETS exposure as they relate to the development of asthma. The respective roles of antenatal and postnatal exposure to ETS in the pathophysiologic changes associated with asthma and other respiratory illnesses are in need of further investigation.

Behavioral research also is needed to better understand the factors that lead to the initiation of smoking in adolescents, especially young women, and to the maintenance of smoking in pregnant women and mothers. Additionally, there is a need to develop more effective interventions to achieve sustained pre- and postnatal smoking cessation in pregnant women and mothers, especially those whose children are at higher risk of developing asthma due to their family history, socioeconomic status, and place of residence. Since ETS exposure of children at greatest risk for adverse asthma outcomes (especially low-income and minority children of African-American ancestry) may come from other caregivers as well as the mother or parents (i.e., other family members with whom the mother and child live and from day care providers), interventions must be developed that will be effective in reducing the child's exposure from all sources. The effectiveness of ETS exposure reduction interventions in actually improving asthma outcomes should be evaluated as well.

REFERENCES

Abell TD, Baker LC, Ramsey CN Jr. 1991. The effects of maternal smoking on infant birth weight. Family Medicine 23(2):103–107.

AHCPR (Agency for Health Care Policy and Research). 1996. Smoking Cessation: Clinical Practice Guideline (No. 18). DHHS Publication No. (AHCPR) 96-0892. U.S. Department of Health and Human Services, Public Health Service. Washington, DC.

Anderson HR, Cook DG. 1997. Health effects of passive smoke. 2. Passive smoking and sudden infant death syndrome. Review of the epidemiological evidence. Thorax 52(11):1003–1009. [Published erratum appears in Thorax 1999. 54(4):365–366.]

Asmussen I. 1979. Fetal cardiovascular system as influenced by maternal smoking. Clinical Cardiology 2(4):246–256.

Bek K, Tomac N, Delibas A, Tuna F, Tezic HT, Sungur M. 1999. Department of Pediatric Allergy, Dr Sami Ulus Children's Hospital, Ankara, Turkey. Postgraduate Medicine Journal 75(884):339–341.

Benowitz NL, Jacob P III. 1994. Metabolism of nicotine to cotinine studied by a dual stable isotope method. Clinical Pharmacology Therapeutics 56(5):483–493.

Benowitz JL, Kuyt F, Jacob P III, Jones RT, Osman AL. 1983. Cotinine disposition and effects. Clinical Pharmacology Therapeutics 34(5):604–611.

Benowitz NL. 1996. Cotinine as a biomarker of environmental tobacco smoke exposure. Epidemiologic Reviews 18(2):188–204.

Benowitz NL. 1999. Biomarkers of environmental tobacco smoke exposure. Environmental Health Perspectives 107(Suppl 2):349–355.

Bruvold WH. 1993. A meta-analysis of adolescent smoking prevention programs. American Journal of Public Health 83(6):872–880.

Bulterys M. 1993. Passive tobacco exposure and sudden infant death syndrome. Pediatrics 92(3):505–506.

California EPA (California Environmental Protection Agency). 1997. Health Effects of Exposure to Environmental Tobacco Smoke. Office of Environmental Health Hazard Assessment. Sacramento, CA.

CDC (Centers for Disease Control and Prevention). 1989. Reducing the Health Consequences of Smoking: 25 Years of Progress—A Report of the Surgeon General. U.S. Department of Health and Human Services, Public Health Service, CDC, DHHS publication no. (CDC) 89-8411. Washington, DC.

CDC. 1997. Cigarette smoking among adults—United States, 1995. Morbidity and Mortality Weekly Report 46(51):1217–1220.

CDC. 1998. Tobacco Use Among U.S. Racial/Ethnic Minority Groups—African Americans, American Indians and Alaska Natives, Asian American and Pacific Islanders, and Hispanics: A Report of the Surgeon General. U.S. Department of Health and Human Services, National Center for Chronic Disease Prevention and Health Promotion, Office on Smoking and Health. Atlanta, GA.

CDHS (California Department of Health Services). 1995. Are Californians protected from environmental tobacco smoke? A summary of the findings on work site and household policies. California adult tobacco study. CDHS Tobacco Control Section, Sacramento, CA.

Cepeda-Benito A. 1993. Meta-analytical review of the efficacy of nicotine chewing gum in smoking treatment programs. Journal of Consulting Clinical Psychology 61(5):822–830.

Chen J, Millar WJ. 1998. Age of smoking initiation: implications for quitting. Health Reports 9(4):39–46.

Chilmonczyk BA, Knight GJ, Palomaki GE, Pulkkinen AJ, Williams J, Haddow JE. 1990. Environmental tobacco smoke exposure during infancy. American Journal of Public Health 80(10):1205–1208.

Chilmonczyk BA, Palomaki GE, Knight GJ, Williams J, Haddow JE. 1992. An unsuccessful cotinine-assisted intervention strategy to reduce environmental tobacco smoke exposure during infancy. American Journal of Diseases of Children 146(3):357–360.

Collins MH, Moessinger AC, Kleinerman J, Bassi J, Rosso P, Collins AM, James LS, Blanc WA. 1985. Fetal lung hypoplasia associated with maternal smoking: a morphometric analysis. Pediatric Research 19(4):408–412.

COMMIT. 1995. Community Intervention Trial for Smoking Cessation (COMMIT): I. Cohort results from a four-year community intervention. American Journal of Public Health 85(2):183–192.

Cook DG, Strachan DP. 1997. Health effects of passive smoking. 3. Parental smoking and prevalence of respiratory symptoms and asthma in school age children. Thorax 52(12):1081–1094.

Cook DG, Strachan DP. 1998. Health effects of passive smoking. 7. Parental smoking, bronchial reactivity and peak flow variability in children. Thorax 53(4):295–301.

Cook DG, Strachan DP, Carey IM. 1998. Health effects of passive smoking. 9. Parental smoking and spirometric indices in children. Thorax 53(10):884–893.

Cook DG, Strachan DP. 1999. Health effects of passive smoke. 10. Summary of effects of parental smoking on the respiratory health of children and implications for research. Thorax 54(4):357–366.

Cummings KM, Markello SJ, Mahoney M, Bhargava AK, McElroy PD, Marshall JR. 1990. Measurement of current exposure to environmental tobacco smoke. Archives of Environmental Health 45(2):74–79.

Daisey JM, Mahanama KRR, Hodgson AT. 1994. Toxic volatile organic compounds in environmental tobacco smoke: Emission factors for modeling exposures of California populations. A133–186; California Air Resources Board, Sacramento, CA.

Daisey JM. 1999. Tracers for assessing exposure to environmental tobacco smoke: what are they tracing? Environmental Health Perspectives 107(Suppl 2): 319–327.

Dolan-Mullen P, Ramirez G, Groff JY. 1994. A meta-analysis of randomized trials of prenatal smoking cessation interventions. American Journal of Obstetrics and Gynecology 171(5):1328–1334.

Ehrlich R, Kattan M, Godbold J, Saltzberg DS, Grimm KT, Landrigan PJ, Lilienfeld DE. 1992. Childhood asthma and passive smoking. Urinary cotinine as a biomarker of exposure. American Review of Respiratory Disease 145(3):594–599.

Eisner MD, Yelin EH, Henke J, Shiboski SC, Blanc PD. 1998. Environmental tobacco smoke and adult asthma. The impact of changing exposure status on health outcomes. American Journal of Respiratory and Critical Care Medicine 158(1):170–175.

Fiore MC, Smith SS, Jorenby DE, Baker TB. 1994. The effectiveness of the nicotine patch for smoking cessation. Journal of the American Medical Association 271(24):1940–1947.

Fiore MC, Jorenby DE, Baker TB. 1997. Smoking cessation: principles and practice based upon the AHCPR Guideline, 1996. Agency for Health Care Policy and Research. Annals of Behavioral Medicine 19(3):213–219.

Forsberg B, Pekkanen J, Clench-Aas J, Martensson MB, Stjernberg N, Bartonova A, Timonen KL, Skerfving S. 1997. Childhood asthma in four regions in Scandinavia: risk factors and avoidance effects. International Journal of Epidemiology 26(3):610–619.

Greenberg RA, Bauman KE, Glover LH, Strecher VJ, Kleinbaum DG, Haley NJ, Stedman HC, Fowler MG, Loda FA. 1989. Ecology of passive smoking by young infants. Journal of Pediatrics 114(5):774–780.

Greenberg RA, Strecher VJ, Bauman KE, Boat BW, Fowler MG, Keyes LL, Denny FW, Chapman RS, Stedman HC, LaVange LM, et al. 1994. Evaluation of a home based intervention program to reduce infant passive smoking and lower respiratory illness. Journal of Behavioral Medicine 17(3):273–290.

Gritz ER, Berman BA, Bastani R, Wu M. 1992. A randomized trial of a self-help smoking cessation intervention in a nonvolunteer female population: testing the limits of the public health model. Health Psychology 11(5):280–289.

Guerin MR, Jenkins RA, Tomkins BA. 1992. The Chemistry of Environmental Tobacco Smoke: Composition and Measurement. Boca Raton, FL: Lewis Publishers.

Haddow JE, Knight GJ, Palomaki GE, Wald NJ. 1991. Cotinine-assisted intervention in pregnancy to reduce smoking and low birthweight delivery. British Journal of Obstetrics and Gynaecology 98(9):859–865.

Halonen M, Barbee RA, Lebowitz MD, Burrows B. 1982. An epidemiologic study of interrelationships of total serum immunoglobulin E, allergy skin-test reactivity, and eosinophilia. Journal of Allergy and Clinical Immunology 69(2):221–228.

Hanrahan JP, Halonen M. 1998. Antenatal interventions in childhood asthma. European Respiratory Journal 12(Suppl 27):46s–51s.

Hasday JD, Bascom R, Costa JJ, Fitzgerald T, Dubin W. 1999. Bacterial endotoxin is an active component of cigarette smoke. Chest 115(3):829–835.

Henderson FW, Reid HF, Morris R, Wang OL, Hu PC, Helms RW, Forehand L, Mumford J, Lewtas J, Haley NJ, et al. 1989. Home air nicotine levels and urinary cotinine excretion in preschool children. American Review of Respiratory Disease 140(1):197–201.

Hovell MF, Meltzer SB, Zakarian JM, Zakarian JM, Wahlgren DR, Emerson JA, Hofstetter CR, Leaderer BP, Meltzer EO, Zeiger RS, O'Connor RD, et al. 1994. Reduction of environmental tobacco smoke exposure among asthmatic children: a controlled trial. Chest 106(2):440–446. [Published erratum appears in Chest 1995. 107(5):1480.]

Hu FB, Persky V, Flay BR, Zelli A, Cooksey J, Richardson J. 1997. Prevalence of asthma and wheezing in public schoolchildren: association with maternal smoking during pregnancy. Annals of Allergy, Asthma, and Immunology 79(1):80–84.

Idle JR. 1990. Titrating exposure to tobacco smoke using cotinine—a minefield of misunderstandings. Journal of Clinical Epidemiology 43(4):313–317.

Irvine L, Crombie IK, Clark RA, Slane PW, Goodman KE, Feyerabend C, Cater JI. 1997. What determines levels of passive smoking in children with asthma? Thorax 52(9):766–769. [Comment in Thorax 1998. 53(3):233–234.]

Jarvis MJ, Russell MAH. 1984. Measurement and estimation of smoke dosage to non-smokers from environmental tobacco smoke. European Journal of Respiratory Disease Supplement 133:68–75.

Jarvis MJ, Tunstall-Pedoe H, Feyerabend C, Vesey C, Saloojee Y. 1987. Comparison of tests used to distinguish smokers from nonsmokers. American Journal of Public Health 77:1435–1438.

Jarvis MJ, Russell MA, Benowitz NL, Feyerabend C. 1988. Elimination of cotinine from body fluids: implications for noninvasive measurement of tobacco smoke exposure. American Journal of Public Health 78(6):696–698.

Jenkins RA, Palausky A, Counts RW, Bayne CK, Dindal AB, Guerin MR. 1996. Exposure to environmental tobacco smoke in sixteen cities in the United States as determined by personal breathing zone air sampling. Journal of Exposure Analysis and Environmental Epidemiology 6(4):473–502.

Jenkins RA, Counts RW. 1999. Personal exposure to environmental tobacco smoke: salivary cotinine, airborne nicotine, and nonsmoker misclassification. Journal of Exposure Analysis and Environmental Epidemiology 9(4):352–363.

Jordanov JS. 1990. Cotinine concentrations in amniotic fluid and urine of smoking, passive smoking and non-smoking pregnant women at term and in the urine of their neonates on 1st day of life. European Journal of Pediatrics 149(10):734–737.

Kendrick JS, Merritt RK. 1996. Women and smoking: an update for the 1990s. American Journal of Obstetrics and Gynecology 175(3 Pt 1):528–535.

Killen JD, Fortmann SP, Telch MJ, Newman B. 1988. Are heavy smokers different from light smokers? A comparison after 48 hours without cigarettes. Journal of the American Medical Association 260(11):1581–1585.

Law M, Tang JL. 1995. An analysis of the effectiveness of interventions intended to help people stop smoking. Archives of Internal Medicine 155(18):1933–1941.

Li Wan Po A. 1993. Transdermal nicotine in smoking cessation. A meta-analysis. European Journal of Clinical Pharmacology 45(6):519–528.

Lodrup Carlsen KC, Jaakkola JJ, Nafstad P, Carlsen KH. 1997. In utero exposure to cigarette smoking influences lung function at birth. European Respiratory Journal 10(8):1774–1779.

Longo LD. 1970. Carbon monoxide in the pregnant mother and fetus and its exchange across the placenta. Annals of the New York Academy of Sciences 174(1):312–341.

Lum S. 1994a. Duration and location of ETS exposure for the California population. Memorandum from S. Lum, Indoor Exposure Assessment Section, Research Division, California Air Resources Board, to L. Haroun, Reproductive and Cancer Hazard Assessment Section, Office of Environmental Health Hazard Assessment, February 3.

Lum S. 1994b. Corrections to the table of duration and location of ETS exposure for kids 6–11 years old transmitted February 3, 1994. Memorandum from S. Lum, Indoor Exposure Assessment Section, Research Division, California Air Resources Board, to L. Haroun, Reproductive and Cancer Hazard Assessment Section, Office of Environmental Health Hazard Assessment, July 19.

Mattson ME, Boyd G, Byar D, Brown C, Callahan JF, Corle D, Cullen JW, Greenblatt J, Haley N, Hammond K, Lewtas J, Reeves W. 1989. Passive smoking on commercial airline flights. Journal of the American Medical Association 261(6):867–872.

McBride CM, Pirie PL. 1990. Postpartum smoking relapse. Addictive Behaviors; 15(2):165–168.

McIntosh NA, Clark NM, Howatt WF. 1994. Reducing tobacco smoke in the environment of the child with asthma: a cotinine-assisted, minimal-contact intervention. Journal of Asthma 31(6):453–462.

McKenna JW, Williams KN. 1993. Crafting effective tobacco counter-advertisements: lessons from a failed campaign directed at teenagers. Public Health Report 108 (Suppl 1):85–89.

Milner AD, Marsh MJ, Ingram DM, Fox GF, Susiva C. 1999. Effects of smoking in pregnancy on neonatal lung function. Archives of Disease in Childhood. Fetal and Neonatal Edition 80(1):F8–F14.

Mullen PD, Richardson MA, Quinn VP, Ershoff DH. 1997. Postpartum return to smoking: who is at risk and when. American Journal of Health Promotion 11(5):323–330.

National Research Council. 1986. Environmental Tobacco Smoke: Measuring Exposures and Assessing Health Effects. Washington, DC: National Academy Press.

Ockene JK, Zapka JG. 1997. Physician-based smoking intervention: a rededication to a five-step strategy to smoking research. Addictive Behaviors 22(6): 835–848.

Ogborn CJ, Duggan AK, DeAngelis C. 1994. Urinary cotinine as a measure of passive smoke exposure in asthmatic children. Clinical Pediatrics 33(4): 220–226.

Ostro BD, Lipsett MJ, Mann JM, Weiner M, Selner JS. 1994. Indoor air pollution and asthma: results from a panel study. American Journal of Respiratory and Critical Care Medicine 149(6):1400–1406.

Ott WR, Switzer P, Robinson J. 1996. Particle concentration inside a tavern before and after prohibition of smoking: evaluating the performance of an indoor air quality model. Journal of the Air and Waste Management Association 46:1120–1134.

Pentz MA, Brannon BR, Charlin VL, Barrett EJ, MacKinnon DP, Flay BR. 1989. The power of policy: the relationship of smoking policy to adolescent smoking. American Journal of Public Health 79(7):857–862.

Pierce JP, Evans N, Farkas SJ, Cavin SW, Berry C, Kramer M, Kealey S, Rosbrook B, Choi W, Kaplan RM. 1994. Tobacco use in California: an evaluation of the tobacco control program, 1989–1993. La Jolla, CA. Cancer Prevention and Control, University of California, San Diego.

Scientific Committee on Tobacco and Health (SCOTH). 1998. Report of the Scientific Committee on Tobacco and Health. Her Majesty's Stationery Office (United Kingdom). URL: http://www.official-documents.co.uk/document/doh/tobacco/contents.htm. Accessed December 10, 1999.

Sekhon HS, Jia Y, Rab R, Kuryatov A, Pankow JF, Whitsett JA, Lindstrom J, Spindel ER. 1999. Prenatal nicotine increases pulmonary 7 nicotinic receptor expression and alters fetal lung development in monkeys. Journal of Clinical Investigation 103(5):637–647.

Senore C, Battista RN, Shapiro SH, Segnan N, Ponti A, Rosso S, Aimer D. 1998. Predictors of smoking cessation following physicians' counseling. Preventive Medicine 27(3):412–421.

Sherwood RA, Keating J, Kavvadia V, Greenough A, Peters TJ. 1999. Substance misuse in early pregnancy and relationship to fetal outcome. European Journal of Pediatrics 158(6):488–492.

Silagy C, Mant D, Fowler G, Lodge M. 1994. Meta-analysis on efficacy of nicotine replacement therapies in smoking cessation. Lancet 343(8890):139–142.

Smith PM, Kraemer HC, Miller NH, DeBusk RF, Taylor CB. 1999. In-hospital smoking cessation programs: who responds, who doesn't? Journal of Consulting and Clinical Psychology 67(1):19–27.

Strachan DP, Cook DG. 1997. Health effects of passive smoking. 1. Parental smoking and lower respiratory illness in infancy and early childhood. Thorax 52(10):905–914.

Strachan DP, Cook DG. 1998a. Health effects of passive smoking. 4. Parental smoking, middle ear disease and adenotonsillectomy in children. Thorax 53(1):50–56.

Strachan DP, Cook DG. 1998b. Health effects of passive smoking. 5. Parental smoking and allergic sensitisation in children. Thorax 53(2):117–123. [Published erratum appears in Thorax 1999. 54(4):366.]

Strachan DP, Cook DG. 1998c. Health effects of passive smoking. 6. Parental smoking and childhood asthma: longitudinal and case-control studies. Thorax 53(3):204–212.

Strachan DP, Butland BK, Anderson HR. 1996. Incidence and prognosis of asthma and wheezing illness from early childhood to age 33 in a national British cohort. British Medical Journal 312(7040):1195–1199.

Tang JL, Law M, Wald N. 1994. How effective is nicotine replacement therapy in helping people to stop smoking? British Medical Journal 308(6920):21–26.

U.S. DHHS (U.S. Department of Health and Human Services). 1984. The Health Consequences of Smoking: Chronic Obstructive Lung Disease. A Report of the Surgeon General. U.S. DHHS, Public Health Service, Office of the Assistant

Secretary for Health, Office of Smoking and Health, Washington, DC. DHHS Pub. No. (PHS) 84-50205.

U.S. DHHS. 1986. The Health Consequences of Involuntary Smoking. A Report of the Surgeon General. DHHS (CDC) Pub. No. 87-8398. Washington, DC: U.S. Government Printing Office.

U.S. DHHS. 1991. Healthy People 2000: National Health Promotion and Disease Prevention Objectives. DHHS Pub. No. (PHS) 91-50212. Washington, DC: Office of the Assistant Secretary for Health.

U.S. EPA (U.S. Environmental Protection Agency). 1992. Respiratory Health Effects of Passive Smoking: Lung Cancer and Other Disorders. EPA/600/6-90/006F. Washington, DC.

Viswesvaran C, Schmidt FL. 1992. A meta-analytic comparison of the effectiveness of smoking cessation methods. Journal of Applied Psychology 77(4):554–561.

Wahlgren DR, Hovell MF, Meltzer SB, Hofstetter CR, Zakarian JM. 1997. Reduction of environmental tobacco smoke exposure in asthmatic children. A 2-year follow-up. Chest 111(1):81–88.

Wang X, Tager IB, Van Vunakis H, Speizer FE, Hanrahan JP. 1997. Maternal smoking during pregnancy, urine cotinine concentrations, and birth outcomes. A prospective cohort study. International Journal of Epidemiology 26(5):978–988.

Watts RR, Langone JJ, Knoght GJ, Lewtas J. 1990. Cotinine analytical workshop report: consideration of analytical methods for determining cotinine in human body fluids as a measure of passive exposure to tobacco smoke. Environmental Health Perspectives 84:173–182.

WHO (World Health Organization). 1999. Tobacco Free Initiative—Consultation Report. WHO/NCD/TFI/99.10. International Consultation on Environmental Tobacco Smoke (ETS) and Child Health proceedings, January 11–14, Geneva, Switzerland.

Wiley JA, Robinson JP, Cheng YT, Piazza T, Stork L, Pladsen K. 1991. Study of children's activity patterns. ARB/9-93/489; California Air Resources Board, Sacramento, CA.

Wilson SR, Latini DM, Starr-Schneidkraut N, Fish L, Loes L, Page A, Kubic P. 1996. Education of parents of infants and very young children with asthma: a developmental evaluation of the Wee Wheezers program. Journal of Asthma 33(4):239–254. [Published erratum appears in J Asthma 1997. 34(3):261.]

Woodward A, Owen N, Grgurinovich N, Griffith F, Linke H. 1987. Trial of an intervention to reduce passive smoking in infancy. Pediatric Pulmonology 3(3):173–178.

Zetterstrom O, Osterman K, Machado L, Johansson SG. 1981. Another smoking hazard: raised serum IgE concentration and increased risk of occupational allergy. British Medical Journal (Clinical Research Edition) 283(6301):1215–1217.

8

INDOOR DAMPNESS
AND *Asthma*

O ne of the environmental factors most commonly associated with respiratory disease is building dampness. This chapter presents an overview of the research on dampness and human health, and the nature, causes, and control of dampness problems. There is a large and detailed literature addressing the engineering and physics underlying moisture control in buildings (e.g., Trechsel, 1994). In addition, professional organizations such as the American Society for Testing and Materials and the American Society of Heating, Refrigerating, and Air Conditioning Engineers publish reference materials and promulgate consensus standards intended, in part, to limit indoor moisture problems. A task force formed by the International Society of Indoor Air Quality and Climate published guidelines for the control of moisture-related problems in 1996 (Flannigan and Morey, 1996).

A complete treatment of the science and art of controlling indoor moisture is beyond the scope of this report. As this overview suggests, moisture in buildings is an area of research that has the potential to significantly affect public health. The committee believes that better communication between health, engineering, and building professionals is likely to result in more informed studies on the health effects associated with moisture problems and the means to prevent or remediate these problems. It strongly encourages efforts to bring these groups together to educate one another on their areas of expertise and to establish collaborations aimed at improving the public's health.

INDOOR WATER SOURCES AND REMOVAL PROCESSES

Water is present in buildings as vapor within the indoor air, as a liquid reservoir, as a solid (ice or frost), as a layer of molecules adsorbed on the surface of building materials, or as condensation within the pores of these materials. A continuous process of moisture transfer occurs among the phases and indoor locations of water. The observed relationship of moisture problems to asthma is presumed to be a consequence of the influence of moisture on the growth of microorganisms on building materials; consequently, the moisture content of materials is of primary interest for asthma. However, the moisture content of building materials is influenced by the other phases and locations of moisture within the building.

Sources of water on or within building materials include leaks of liquid water from interior plumbing or from outdoors above or below grade. Other sources include melting of ice or frost (e.g., in attics), capillary transport from moist soil (e.g., through concrete foundations), water vapor condensation, and the water present in building materials (e.g., wood or concrete) at the time of building construction, which is particularly significant during the first year after construction. Additionally, water vapor from air surrounding a building material adsorbs on or in building materials. The equilibrium moisture content of a building material surrounded by air is primarily a function of the relative humidity of the air; however, the equilibrium relationship varies considerably among building materials (Kumaran et al., 1994).

Sources of water vapor in indoor air include the water vapor in incoming outdoor air, which is often a dominant source during hot humid weather. Air that is drawn into buildings from crawl spaces or after passing through soil and cracks in the building substructure may be particularly humid. Additional indoor water vapor sources include human respiration; evaporation that occurs from water-using activities such as cooking, bathing, washing, and drying; intentional humidification; evaporation from liquid water that originates from leaks or condensation; and the desorption of water from materials. Combustion products that enter the indoor air space due to the use of unvented gas or kerosene heaters or a failure of the venting of "vented" heating equipment can be a large source of water vapor.[1] Christian (1994) summarizes information on the rates of water vapor production from many of these sources.

Removal processes for indoor water vapor include the flow of air with water vapor from indoors to outdoors (i.e., ventilation), absorption on surfaces, the intentional water vapor condensation that occurs in air conditioning and dehumidification systems, and unintentional water vapor condensation on surfaces.

Water vapor condenses from air when the air temperature is cooled below the dew-point temperature. In buildings, unintentional water vapor condensation occurs when humid air contacts cool surfaces. In winter, the temperature of portions of the building envelope that exchange heat with outdoor air or soil may fall below the dew-point temperature of indoor air, leading to water vapor condensation. The condensation that occurs on windows is a familiar phenomenon; however, condensation can also occur within the building envelope as humid indoor air passes through the envelope and comes in contact with cool surfaces. In the summer, surface temperatures in air-conditioned buildings may be lower than the dew-point temperature of the outdoor air. Condensation may occur as the warm humid outdoor air flows into the building and contacts these surfaces. Vapor barriers (i.e., sheets of material with a low permeability to water vapor that also retard air flow) are commonly installed in building envelopes to limit moisture transport and the associated risks of condensation. Vapor barriers should be installed near the warm side of the building envelope—for example, near the inner surface of walls in a building located in a cold climate. Improper placement of vapor barriers or unintentional vapor barriers can lead to condensation. Impermeable vinyl wallpaper located on the inner sur-

---

[1]An estimated 0.5 million unvented gas heaters were sold in 1996 (Apte, 1996). Even a small 10,000 Btu-per-hour unvented gas heater, about one-tenth the capacity of a residential furnace system, would release water vapor equivalent to about 10 L per day of liquid water (Apte, 1996).

faces of walls of air-conditioned buildings has been associated with moisture problems behind the wallpaper (Lstiburek and Carmody, 1994; Rose, 1994).

In mechanically ventilated buildings, pressure differences between indoors and outdoors or between indoors and the surrounding soil are generated by mechanically produced airflows. These pressure differences modify the rates of moist air transport through the building envelope in a manner that may inhibit or increase indoor moisture problems. In cold climates, the design intent is often to maintain buildings depressurized relative to outdoors and to prevent humid indoor air from flowing outward through the building envelope. In warm, humid climates, pressurization of the building to limit infiltration of humid outdoor air is usually the design intent. Because houses rarely have continuously operating mechanical ventilation systems, outdoor air infiltrates into the building through portions of the building envelope and indoor air exfiltrates through other sections of the envelope.

Based on this discussion, the risk factors for moisture problems include water leaks from the building interior or exterior; unusually high rates of water vapor generation indoors; a high rate of moisture entry into buildings from moist soil (often associated with problems in water drainage around foundations); a low rate of ventilation particularly during the winter (see Chapter 10); the absence or improper location of vapor barriers; insufficient water vapor removal by dehumidifiers or air-conditioning systems; and an unusually cold or unusually humid climate. The prevalence of these different risk factors and the implications for microbiological growth are not well documented; however, water leaks are very common (see below) and cited in many case studies of building-related respiratory health problems.

Measures of Dampness

Moisture Measurements

Temperature and relative humidity measurements are the fundamental information regarding moisture in buildings. Humidity is usually measured psychometrically and is represented as relative humidity—the amount of water in the air relative to the amount of water the air can hold at saturation at the same temperature and pressure.

At present there are no accepted protocols for characterizing the moisture levels of buildings. However, wide variations in both temperature and humidity are to be expected from place to place and over time, daily and seasonally. Given this variation, monitoring may be necessary to track environmental changes over time. The design of a monitoring system depends on the type of information one desires. Building monitoring may be designed to disclose center-of-room moisture conditions, moisture conditions at particular (perhaps troublesome) locations, response of the building to exterior conditions such as the entry of rainwater, moisture movement within the building, and change in humidity conditions over time. However, monitoring cannot be used to discern the appearance of undesirable conditions away from where instruments are placed, nor can it be used to determine that all moisture-related aspects in a building are acceptable.

There are a wide variety of tools available for measuring specific moisture conditions (Lagus, 1994). Infrared pyrometers can be used to identify surface temperature anomalies, which may become moisture and mold sites. Pin-type moisture meters measure material moisture content, although calibration difficulties may limit their effectiveness. Smoke pencils can help identify envelope leakage sites and also the effectiveness of combustion product discharge. Pressure difference instruments (manometers and micromanometers) may be employed to estimate the role of air pressure and movement in the development of water problems in building envelopes.

Abe and colleagues (1996) developed a sensor for determining the ability of fungi to grow in microenvironments. The method uses an agar medium impregnated with spores of a xerophilic fungus and relies on water absorption from the environment to stimulate fungal growth. Microclimate differences and changes were documented in an apartment using the method.

Visible Signs of Dampness

Although humans are poor humidity sensors, there are signs

of inappropriate moisture conditions that can be directly perceived. Such perceptions are the basis for most epidemiologic studies in which moisture condition data are collected by questionnaire. Questions are typically formulated to seek information on whether conditions such as leaks, floods, wet basements, window condensation, visible fungal growth, or moldy odors are currently present or have been present in the past. It should be noted, however, that reporting bias is an important source of error in such studies. Dales and colleagues (1997) report that under some conditions, allergy patients may be more likely than nonallergic people to report visible fungal growth. Additionally, smokers may be less likely than nonsmokers to report such growth.

Moisture conditions in buildings are best discovered through direct observation and inspection. Home inspectors are known to rely on smell to supplement visual inspection. Among the items typically included in an inspection report are presence of mold, water stains, evidence of leaks or flooding, current leaks, crawl space conditions, attic sheathing condition, and overall stoutness or dilapidation of the building. Characterization of rainwater discharge and management is also necessary, given the importance and prevalence of foundation leakage to the overall moisture balance of a building.

Extent of the Home Dampness Problem

The reported percentage of homes with dampness problems varies widely depending on geography, the approach to home selection, and the types of questions or inspection procedures used to detect such problems. Selected study results are presented in Table 8-1.

The U.S. Census collects data on water leakage in homes. The results for 1973–1984 are reported in the Annual Housing Survey. Results after 1984 are reported in the American Housing Survey for the United States. Based on these data, Figure 8-1 shows the percentage of homes with water leakage from indoor and outdoor water sources. The overlap between these categories and the total percentage of homes with water leakage are not reported. The reported percentages of homes with water leaks are relatively constant, and most water leaks are from outdoors. Based on the

TABLE 8-1 Reported Prevalence of Home Dampness Problems

Author	Country	N	Indicator	Prevalence
Tsongas (1985) (as reviewed in Tsongas, 1994)	U.S. (Spokane, WA)	96	Inspection	59% condensation on windows 38% mold and mildew
Rose (1986) (as reviewed in Tsongas, 1994)	U.S. (Illinois)	670	Inspection	5.4% had major moisture problems 35% had visible mildew or surface stains or surface damage
Trechsel et al. (1987) (as reviewed in Tsongas, 1994)	U.S. (Florida)	86 (homes in naval station with masonry walls)	Inspection	30% current, past, or potential moisture problems 48% had mildew problems 66% had mildew or moisture problems
Dales et al. (1991)	Canada	403	Questionnaire	38%
Brunekreef (1992)	Netherlands	3,300	Questionnaire	25.4%
Pirhonen et al. (1996)	Finland	1,460	Questionnaire	23%
Leung et al. (1998)	Hong Kong	40	Questionnaire	27%
Norbäck et al. (1999)	Sweden	98 + 357	Questionnaire	27%
Schafer et al. (1999)	Germany	1,235	Questionnaire	8.8% (10.3% East, 1.9% West)

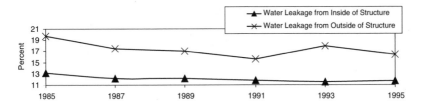

FIGURE 8-1 Prevalence of water leaks in U.S. housing, based on data from the U.S. Census.

data in Table 8-1 and Figure 8-1, a substantial portion of homes have moisture problems.

Underlying Causes of Moisture Problems

Although extensive guidance is available on means of preventing and remediating moisture problems in both residential and commercial buildings (e.g., Trechsel, 1994, chapters 15–22), moisture problems are very common. Economic and institutional barriers contribute to moisture problems. For example, water leaks will not be repaired if financial resources are inadequate. Also, the relevant features of building design, operation, and maintenance may be determined substantially by speculative builders or other decision-makers who are substantially unaffected by future moisture problems. Similarly, landlords who do not reside in the affected building may not be motivated to repair water leaks rapidly.

Despite the technical knowledge about prevention of moisture problems that is available in current scientific and engineering literature, the errors in building design and construction reported in case studies indicate that many architects, engineers, builders, and roofers have an inadequate knowledge of the means of preventing moisture problems. Additionally, these professionals and tradesmen, and the general public, may not fully understand the association between moisture problems and health problems. The lack of awareness of health consequences reduces the motivation to prevent or correct problems.

Moisture and Sources for Disease Agents

The relationship between dampness and respiratory disease is considered to be linked to allergens or other materials derived from fungal growth, or dust mites, or both, although none of these pathways has been clearly documented. Billings and Howard (1998) reviewed the literature on indoor dampness and respiratory disease and concluded that dampness increases allergen burden, leading to increased risk of developing asthma.

Fungi

Fungi are among the more versatile organisms with respect to water requirements and are able to grow with extremely little moisture under some conditions. Fortunately, conditions in buildings do not usually permit such growth, which occurs only under optimal substrate and temperature conditions for the fungus. The majority of fungal growth occurs on surfaces that are continuously wet for many days. Such wetness may occur on cold surfaces where condensation is continuously present, in materials that remain at or near the dew point (e.g., carpeting on uninsulated cold or damp surfaces), or in materials that get wet as a result of leaks or floods. Fungal growth can also occur without the presence of liquid water from leaks or condensation. High relative humidities (e.g., greater than approximately 80%) increase the risk of fungal growth on some surfaces (Foarde et al., 1996). It should be noted that indoor middle-of-the-room humidity is not a consistent predictor of the presence of fungal growth, and fungal control cannot be achieved just with ambient indoor air humidity control. Fungi grow on surfaces under microenvironmental conditions that may be very different from those in room air (Li and Hsu, 1997).

Data are equivocal on the relationship of airborne fungi to humidity or other dampness indicators. Verhoeff and colleagues (1992) found only a weak relationship between airborne fungi and dampness as characterized by a checklist. On the other hand, Garrett and colleagues (1998) found associations between high airborne fungal concentrations and musty odor, water intrusion, limited ventilation, and high indoor humidity. Relatively weak relationships have also been shown for dampness indicators and concentrations of some fungi in dust (Dales et al., 1997; Douwes et al., 1999). Visible fungal growth has been associated with high airborne concentrations of some fungi but not others. However, good studies examining the association between surface and airborne fungi are lacking.

Dust Mites

Dust mite populations have been shown experimentally to respond to increasing humidity (Arlian et al., 1998). Dust mite allergen concentrations have been associated with humidity and signs of dampness (Couper et al., 1998; Julge et al., 1998; Nicolai et al., 1998; van Strien et al., 1994), as well as with conditions known to lead to dampness such as increasing number of occupants and reduced ventilation (Couper et al., 1998; Sundell et al., 1995; van Strien et al., 1994). Chapter 5 contains a discussion of the influence of humidity levels on dust mites.

INDOOR DAMPNESS AND RESPIRATORY DISEASE

Overview

Table 8-2 summarizes studies that report odds ratios (ORs) for the association of dampness or visible mold with asthma.

Ronmark and colleagues (1998) ranked home dampness second only to family history as a risk factor for asthma in northern Sweden. Øie and colleagues (1999) report an association between ventilation rate and dampness in the home and bronchial obstruction during the first year of life. The OR for bronchial obstruction related to dampness in low-ventilation homes was 9.6 (confidence interval [CI] 1.05–87.4); in high ventilation homes the OR was 2.3 (CI 0.83–6.39).

Connections between animal allergens, dampness, and asthma have been made in several studies. Norbäck and colleagues (1995) report that symptoms related to asthma were more common in dwellings with house dust mites and visible signs of dampness or microbial growth. However, a later study by some of the same individuals (Norbäck et al., 1999) reports no connec-

TABLE 8-2 Odds Ratios (ORs) for Association of Dampness or Visible Mold with Asthma

Author/ Location	Population (N)	Dampness Indicator	OR (CI) for Asthma-Related Health Effect
Aldous et al., 1996 (Arizona)	Infants (936)	Presence or absence of evaporative cooler	1.8 (1.1–3.0)[a]
Nafstad et al., 1998 (Norway)	Children: 251 cases, 251 controls	Questionnaires, home visits	3.8 (2.0–7.2)[a]
Yazicioglu et al., 1998 (Turkey)	Children: 597 controls, 85 asthmatics	Questionnaire	2.62 (1.13–6.81)[b]
Yang et al., 1998b (Taiwan)	Children: 86 cases, 86 controls	Reported dampness	1.77 (1.24–2.53)[b]
Norbäck et al., 1999 (Sweden)	Adults: 98 asthmatics, 357 controls	Observed dampness	Overall: 1.8 (1.1–3.0)[b] Floor: 4.6 (2.0–10.5)[a]
Jaakkola et al., 1993 (Finland)	Children (2,568)	Any dampness or mold	1.10 (0.54–2.24)[b] 2.17–2.62[a]
Slezak et al., 1998 (U.S.)	Head Start children (1,085)	Signs of dampness, visible mold	4.5 (1.25–16.3)[b] 1.94 (1.23–3.04)[b]
Hu et al., 1997	Young adults (2,041)	Mold growth	2.0 (1.2–3.2)[b]
Brunekreef et al., 1989 (Netherlands)	Children (4,625)	Dampness and mold	1.27–2.12 [a]
Strachan, 1988 (Scotland)	Children (873)	Visible mold	3.0 (1.72–5.25)[a]
Yang et al., 1998a (Taiwan)	Children: 165 cases, 165 controls	Reported dampness	2.65[b]
Nicolai et al., 1998 (Germany)	Adolescents (155)	Observed dampness	16.14 (3.53–73.73)[a]
Jedrychowski and Flak, 1998 (Poland)	Children (1,129)	Reported dampness	1.6 (1.1–2.5)[a]

NOTE: CI = confidence interval.

[a]Wheeze, persistent cough, bronchial obstruction.
[b]Asthma.

tion between allergy to dust mites and either current asthma or home dampness in Sweden. In another study, positive skin tests to dust mite allergens were more prevalent among occupants of damp houses than among occupants of homes without reported signs of dampness (Iversen and Dahl, 1995). Lindfors and colleagues (1995) report that a combination of high-dose exposure to cat or dog allergen, environmental tobacco smoke, and damp housing was significantly associated with asthma (OR = 8.0; CI 1.9–34.1).

Williamson and colleagues (1997) report a correlation between mold growth indicators and asthma (196 age- and sex-matched subjects): $r = 0.23$, $p < .035$. Norbäck and colleagues (1999) report that allergy to fungi (*Cladosporium* or *Alternaria*) was more prevalent in damp homes (9.3% versus 3.9%) and fungal sensitivity was related to current asthma (OR = 3.4; CI 1.4–8.5). Brunekreef and colleagues (1989) studied 4,625 children relating symptoms and pulmonary function to reported dampness and molds in the homes. The OR for reported molds varied from 1.27 to 2.12. There was a drop in FEF_{25-75} of 1.6% associated with reported mold. Among a population sample of 873 children, Strachan (1988) studied respiratory symptoms, measured pulmonary function, and evaluated the home environment for reported dampness or mold. Wheezing in the past year was most closely associated with visible mold (adjusted OR = 3.0; CI 1.72–5.25). However, there was no association with the degree of bronchospasm in children measured in homes with and without mold; the association may be due to awareness of mold leading to increased reporting. The prevalence of lower-respiratory symptoms (any cough, phlegm, wheezing, or wheeze with dyspnea) was increased among those reporting dampness or mold compared with those not reporting dampness or mold as follows: 38% versus 27% among current smokers, 21% versus 14% among ex-smokers, and 19% versus 11% among nonsmokers (all p values $< .001$). Maier and colleagues (1997) indicate an association between reported household water damage and physician-diagnosed asthma in Seattle school children.

Conclusions Regarding Health Effects

1. Damp conditions are associated with the existence of doctor-diagnosed asthma and with the presence of symptoms considered to reflect asthma (i.e., dampness may lead to the development of asthma).

2. Symptom prevalence among asthmatics is also related to home dampness indicators (i.e., dampness may exacerbate existing asthma).

3. The factors related to dampness that actually lead to the development of disease and to disease exacerbation are not yet confirmed, but probably relate to dust mite and fungal allergens.

DAMPNESS CONTROL

Implementation of measures that are effective in reducing dampness problems should be a logical approach to lessening the indoor asthma problem. Several recommendations for prevention and remediation of moisture problems, including periodic inspections to prevent water leakage, are provided in a review article based on a workshop held by the American Thoracic Society (ATS, 1997). However, no intervention studies clearly document that any form of dampness control works effectively to reduce symptoms or to reduce the chances of asthma development. Rose (1994) provides recommendations for retrofitting existing buildings for dampness control. Peat and colleagues (1998) suggest that *retrofitting* for dampness control is expensive and unlikely to have long-term beneficial effects, and that houses be *designed to prevent* dampness problems.

Logical steps can be taken to reduce dampness problems, and most have been documented as effective. For houses, these steps include

1. powered mechanical ventilation to remove and/or dilute occupant-generated moisture (Harrje, 1994);

2. proper installation of vapor barriers (Lsteburek and Carmody, 1993);

3. channeling ground water away from foundations and seal-

ing below-ground walls to prevent water intrusion (Lstiburek and Carmody, 1994);

4. properly protecting ground-level concrete slabs from moisture intrusion (Lstiburek and Carmody, 1994); and

5. constructing crawl spaces to prevent water intrusion (Lstiburek and Carmody, 1994).

RESEARCH NEEDS

With respect to the association of dampness problems with asthma development and symptoms, research is needed to clearly identify the causative agents (e.g., molds, dust mite allergens) and to document more accurately the relationship between dampness and allergen exposure. Research is also needed to characterize and demonstrate the reductions in asthma morbidity from prevention or remediation of moisture problems.

Regarding characterization of moisture problems, research is needed to:

1. develop accurate, standardized protocols for assessing moisture problems in buildings;

2. develop and document the effectiveness of specific measures for dampness reduction in existing buildings; and

3. develop standardized, effective protocols for flood cleanup that will limit microbial growth.

In addition to these research needs, there is a need for improved education of the public about the consequences of moisture problems and for better education of building professionals regarding means of preventing moisture-related problems.

REFERENCES

Abe K, Nagao Y, Nakada T, Sakuma S. 1996. Assessment of indoor climate in an apartment by use of a fungal index. Applied and Environmental Microbiology 62(3):959–963.

Aldous MB, Holberg CJ, Wright AL, Martinez FD, Taussig LM. 1996. Evaporative cooling and other home factors and lower respiratory tract illness during the first year of life. American Journal of Epidemiology 143(5):423–430.

ATS (American Thoracic Society). 1997. Achieving Healthy Indoor Air. Report of the ATS Workshop: Santa Fe, New Mexico, November 16–19, 1995. American Journal of Respiratory and Critical Care Medicine 156(3):S33–S64.

Apte MG. 1996. Unvented gas space heaters: drainless sinks. Home Energy 13(5):9–10.

Arlian LG, Confer PD, Rapp CM, Vyszenski-Moher DL, Chang JC. 1998. Population dynamics of the house dust mites *Dermatophagoides farinae, D. pteronyssinus,* and *Euroglyphus maynei* (Acari: Pyroglyphidae) at specific relative humidities. Journal of Medical Entomology 35(1):46–53.

Billings CG, Howard P. 1998. Damp housing and asthma [Review]. Monaldi Archives for Chest Disease 53(1):43–49.

Brunekreef B. 1992. Damp housing and adult respiratory symptoms. Allergy 47(5):498–502.

Brunekreef B, Dockery DW, Speizer FE, Ware JH, Spengler JD, Ferris BG. 1989. Home dampness and respiratory morbidity in children. American Review of Respiratory Disease 140(5):1363–1367.

Christian JE. 1994. Moisture Sources. In Moisture Control in Buildings, Trechsel HR, Ed. American Society for Testing and Materials, Philadelphia.

Couper D, Ponsonby AL, Dwyer T. 1998. Determinants of dust mite allergen concentrations in infant bedrooms in Tasmania. Clinical and Experimental Allergy 28(6):715–723.

Dales RE, Burnett R, Zwanenburg H. 1991. Adverse health effects among adults exposed to home dampness and molds. American Review of Respiratory Disease 143(3):505–509.

Dales RE, Miller D, McMullen E. 1997. Indoor air quality and health: validity and determinants of reported home dampness and moulds. International Journal of Epidemiology 26(1):120–125.

Douwes J, van der Sluis B, Doekes G, van Leusden F, Wijnands L, van Strien R, Verhoeff A, Brunekreef B. 1999. Fungal extracellular polysaccharides in house dust as a marker for exposure to fungi: relations with culturable fungi, reported home dampness, and respiratory symptoms. Journal of Allergy and Clinical Immunology 103(3 Pt 1):494–500.

Flannigan B, Morey PR. 1996. Control of moisture problems affecting biological indoor air quality. ISIAQ guideline TFI-1996. Ottawa, Canada: International Society of Indoor Air Quality and Climate.

Foarde KK, Van Osdell DW, Chang JCS. 1996. Evaluation of fungal growth on fiberglass duct materials for various moisture, soil, use and temperature conditions. Indoor Air 6(2):83–92.

Garrett MH, Rayment PR, Hooper MA, Abramson MJ, Hooper BM. 1998. Indoor airborne fungal spores, house dampness and associations with environmental factors and respiratory health in children. Clinical and Experimental Allergy 28(4):459–467.

Harrje DT. 1994. Effect of air infiltration and ventilation. In: Moisture Control in Buildings, Trechsel HR, Ed. Philadelphia, PA: American Society for Testing and Materials, pp. 185–194.

Hu FB, Persky V, Flay BR, Richardson J. 1997. An epidemiological study of asthma prevalence and related factors among young adults. Journal of Asthma 34(1):67–76.

Iversen M, Dahl R. 1995. Characteristics of mold allergy. Journal of Investigational Allergology and Clinical Immunology 5(4):205–208.

Jaakkola JJ, Jaakkola N, Ruotsalainen R. 1993. Home dampness and molds as determinants of respiratory symptoms and asthma in pre-school children. Journal of Exposure Analysis and Environmental Epidemiology 3(Suppl 1):129–142.

Jedrychowski W, Flak E. 1998. Separate and combined effects of the outdoor and indoor air quality on chronic respiratory symptoms adjusted for allergy among preadolescent children. International Journal of Occupational Medicine and Environmental Health 11(1):19–35.

Julge K, Munir AK, Vasar M, Bjorksten B. 1998. Indoor allergen levels and other environmental risk factors for sensitization in Estonian homes. Allergy 53(4):388–393.

Kumaran MK, Mitalas GP, Bomberg MT. 1994. Fundamentals of transport and storage of moisture in building materials and components. In: Moisture Control in Buildings, Trechsel HR, Ed. Philadelphia, PA: American Society for Testing and Materials.

Lagus PL. 1994. Measurement techniques and instrumentation. In: Moisture Control in Buildings, Trechsel HR, Ed. Philadelphia, PA: American Society for Testing and Materials.

Leung R, Lam CW, Chan A, Lee M, Chan IH, Pang SW, Lai CK. 1998. Indoor environment of residential homes in Hong Kong—relevance to asthma and allergic disease. Clinical and Experimental Allergy 28(5):585–590.

Li CS, Hsu LY. 1997. Airborne fungus allergen in association with residential characteristics in atopic and control children in a subtropical region. Archives of Environmental Health 52(1):72–79.

Lindfors A, Wickman M, Hedlin G, Pershagen G, Rietz H, Nordvall SL. 1995. Indoor environmental risk factors in young asthmatics: a case-control study. Archives of Disease in Childhood 73(5):408–412.

Lstiburek J, Carmody J. 1993. Moisture Control Handbook. New York: Van Nostrand Reinhold.

Lstiburek J, Carmody J. 1994. Moisture control for new residential buildings. In: Moisture Control in Buildings, Trechsel HR, Ed. Philadelphia, PA: American Society for Testing and Materials, pp. 321–347.

Maier WC, Arrighi HM, Morray B, Llewellyn C, Redding GJ. 1997. Indoor risk factors for asthma and wheezing among Seattle school children. Environmental Health Perspectives 105(2):208–214.

Nafstad P, Øie L, Mehl R, Gaarder PI, Lodrup-Carlsen KC, Botten G, Magnus P, Jaakkola JJ. 1998. Residential dampness problems and symptoms and signs of bronchial obstruction in young Norwegian children. American Journal of Respiratory and Critical Care Medicine 157(2):410–414.

Nicolai T, Illi S, von Mutius E. 1998. Effect of dampness at home in childhood on bronchial hyperreactivity in adolescence. Thorax 53(12):1035–1040.

Norbäck D, Björnsson E, Janson C, Widstrom J, Boman G. 1995. Asthmatic symptoms and volatile organic compounds, formaldehyde, and carbon dioxide in dwellings. Occupational and Environmental Medicine 52(6):388–395.

Norbäck D, Björnsson E, Janson C, Palmgren U, Boman G. 1999. Current asthma and biochemical signs of inflammation in relation to building dampness in dwellings. International Journal of Tuberculosis and Lung Disease 3(5):368–376.

Øie L, Nafstad P, Botten G, Magnus P, Jaakkola JK. 1999. Ventilation in homes and bronchial obstruction in young children. Epidemiology 10(3):294–299.

Peat JK, Dickerson J, Li J. 1998. Effects of damp and mould in the home on respiratory health: a review of the literature. Allergy 53(2):120–128.

Pirhonen I, Nevalainen A, Husman T, Pekkanen J. 1996. Home dampness, moulds and their influence on respiratory infections and symptoms in adults in Finland. European Respiratory Journal 9(12):2618–2622.

Ronmark E, Lundback B, Jonsson E, Platts-Mills T. 1998. Asthma, type-1 allergy and related conditions in 7- and 8-year-old children in northern Sweden: prevalence rates and risk factor pattern. Respiratory Medicine 92(2):316–324.

Rose WB. 1986. Case study: moisture damage to homes in Champaign County, IL. Proceedings of the Building Thermal Envelope Coordinating Council Symposium on Air Infiltration, Ventilation, and Moisture Transfer, Dallas, TX.

Rose WB. 1994. Recommendations for remedial and preventive actions for existing residential buildings. In: Moisture Control in Buildings, Trechsel HR, Ed. Philadelphia, PA: American Society for Testing and Materials.

Schafer T, Kramer U, Dockery D, Vieluf D, Behrendt H, Ring J. 1999. What makes a child allergic? Analysis of risk factors for allergic sensitization in preschool children from East and West Germany. Allergy and Asthma Proceedings 20(1):23–27.

Slezak JA, Persky VW, Kviz FJ, Ramakrishnan V, Byers C. 1998. Asthma prevalence and risk factors in selected Head Start sites in Chicago. Journal of Asthma 35(2):203–212.

Strachan DP. 1988. Damp housing and childhood asthma: validation of reporting of symptoms. British Medical Journal 297(6658):1223–1226.

Sundell J, Wickman M, Pershagen G, Nordvall SL. 1995. Ventilation in homes infested by house-dust mites. Allergy 50(2):106–112.

Trechsel HR, et al. 1987. Field study on moisture problems in exterior walls of a masonry housing development on the coast of the Gulf of Mexico. In: Thermal Insulation: Materials and Systems, Powell FJ, Matthews SL, Eds. Philadelphia, PA: American Society for Testing and Materials.

Trechsel HR. Ed. 1994. Moisture Control in Buildings. ASTM Manual Series: MNL 18. Philadelphia, PA: American Society for Testing and Materials.

Tsongas GA. 1985. The Spokane wall insulation project: A field study of moisture damage in walls insulated without a vapor barrier. U.S. DOE/ Bonneville Power Administration DOE/BP-541.

Tsongas PF. 1994. The health care industry's challenge. Health Systems Review 27(3):7–10.

van Strien RT, Verhoeff AP, Brunekreef B, van Wijnen JH. 1994. Mite antigen in house dust: relationship with different housing characteristics in the Netherlands. Clinical and Experimental Allergy 24(9):843–853.

Verhoeff AP, van Wijnen JH, Brunekreef B, Fischer P, van Reenen-Hoekstra ES, Samson RA. 1992. Presence of viable mould propagules in indoor air in relation to house damp and outdoor air. Allergy 47(2 Pt 1):83–91.

Williamson IJ, Martin CJ, McGill G, Monie RD, Fennerty AG. 1997. Damp housing and asthma: a case-control study. Thorax 52(3):229–234.

Yang CY, Lin MC, Hwang KC. 1998a. Childhood asthma and the indoor environment in a subtropical area. Chest 114(2):393–397.

Yang CY, Tien YC, Hsieh HJ, Kao WY, Lin MC. 1998b. Indoor environmental risk factors and childhood asthma: a case-control study in a subtropical area. Pediatric Pulmonology 26(2):120–124.

Yazicioglu M, Saltik A, Ones U, Sam A, Ekerbicer HC, Kircuval O. 1998. Home environment and asthma in school children from the Edirne region in Turkey. Allergologia et Immunopathologia 26(1):5–8.

9
ASTHMA AND NONRESIDENTIAL
INDOOR *Environments*

Many cross-sectional epidemiologic studies document an association of asthma diagnoses or asthma symptoms with aspects of the residential environment, such as dampness or mold. In contrast to residences, nonindustrial work buildings often have large numbers of occupants in whom the epidemiology of asthma can be studied in relation to the built environment. Several lines of evidence suggest the efficiency of pursuing indoor environmental factors in relation to asthma among office workers, school staff, and students.

This chapter briefly reviews the scientific literature regarding asthma and nonresidential indoor environments—primarily office buildings and schools. Industrial environments, which may expose workers to a wide variety of allergens and irritants capable of inducing asthma (Chan-Yeung, 1995), are outside the scope of this chapter and report.

BUILDING-RELATED ASTHMA

New-onset asthma caused by specific building environments has been investigated infrequently. Case reports of office building-related asthma with a clear work-related pattern exist, and the causes were related to humidifiers or the biocides used in humidifiers (Finnegan and Pickering, 1986; Robertson and Burge,

1985). Epidemic asthma occurred in a printing factory in association with a contaminated humidifier (Burge et al., 1985). The most compelling description of office building-related asthma found that employees of a county social services agency in Denver had a 2.9-fold rate of physician-diagnosed asthma arising since building occupancy compared to employees of a comparable suburban social service agency (Hoffman et al., 1993). The Denver workers reported excess shortness of breath and chest tightness, and 36% of Denver employees with preexisting asthma reported exacerbation of their asthma in relation to building occupancy, in contrast to none of the preexisting asthma cases in the suburban county agency. Peak expiratory flow measurements of a sentinel case fell markedly over the course of time at work, with recovery within an hour outside the office building. Bronchial reactivity measurements from serial methacholine challenge tests showed improvement to the normal range within two to three months of assignment to another building; recurrence of bronchial hyperreactivity with reassignment to the implicated building; and partial resolution with removal for the second time. In addition to a 4.9-fold excess of asthma with onset or exacerbation since building occupancy, cases of hypersensitivity pneumonitis and other interstitial lung disease occurred among occupants of the implicated building. In this investigation, the suspected cause was bioaerosols associated with below-grade moisture incursion from the posterior wall that was built into an earthen bank.

Although the few publications discussed above document the phenomenology of building-related asthma, little information exists regarding the contribution of building-related asthma to the increasing asthma burden in the United States or elsewhere. Several state health departments solicit physician case reports of occupational asthma for the purpose of state- or federally funded surveillance. Data from four states reported to the National Institute for Occupational Safety and Health (NIOSH) for 1993–1995, documented 86 physician-diagnosed cases attributed to indoor air quality deficiencies, without other specific cause, accounting for 8% of all occupational asthma cases recognized by reporting physicians (Romero Jajosky et al., 1999). Michigan, Massachusetts, and Connecticut supplied data over a longer time, which document increasing proportions of all reported occupational asthma cases attributable to indoor environmental quality, up to 25% (unpublished reports from Carolyn Jean Dupuy, Connecticut Department of Public Health, Respiratory Disease and the Indoor Environment, sent to NIOSH by email on May 13, 1999; from Ruth Vanderwaals, Michigan Department of Public Health sent by email on March 22, 1999; and from Letitia Davis, Massachussetts Department of Health, sent by email on March 6, 1999).

Another source of physician case reporting is the data base of the Association of Occupational and Environmental Health Clinics (AOEC). An unpublished report indicates that from 1991 to 1996, 15% of the 542 occupational asthma cases seen in the reporting members of largely academic clinics were related to indoor air exposures, with the annual proportion increasing from 6% in 1991 to 30% in 1996 (Hunting, 1999). Half of the cases of reported asthma–reactive airways dysfunction syndrome (RADS) from primary and secondary schools and vocational schools in this data base were attributed to indoor air exposures. These data must be interpreted with caution because the number and location of reporting clinics varied from year to year and the numbers are likely influenced by the characteristics of the responding facilities.

This limited information from practitioners reporting to public health agencies or their occupational health association suggests that physicians are seeing patients with work-related patterns of asthma symptoms or objective measurements for which they are unable to identify specific causes apart from alleged poor indoor air quality. Investigations in response to these reports by public health agencies have been limited by lack of knowledge about what agents to measure in evaluating hazards for asthma or for less specific indoor air quality (IAQ) complaints.

Attribution of asthma and asthma symptoms to specific building environments by the lay public far exceeds physician recognition of building-related asthma. One source of data regarding public concerns about building contributions to asthma are the requests from employees or management for Health Hazard Evaluations, a mandated service program of NIOSH. Since the late 1970s, after the energy crisis and changes in ventilation codes, health hazard evaluations coded as "indoor air quality requests" from office and school employees have increased persistently in numbers and proportions of all requests. Of the 100–200 IAQ requests received each year since 1990, the proportion mentioning asthma in the written request increased during the 1990s to 14% in 1998; the proportion mentioning any chest symptom increased to 46% in 1997–98. Among IAQ requests mentioning asthma as a health concern to be investigated in the health hazard evaluation, a substantial minority—43%—come from school staff, management, or unions representing teachers.

Most office workers do not attribute chest symptoms to their building environments. Preliminary analyses of questionnaire information from occupants of 29 buildings surveyed in the U.S. Environmental Protection Agency's (EPA's) Building Assessment and Evaluation Study (Brightman et al., 1997), revealed that the median prevalence of occupants reporting frequent work-related shortness of breath was 2%. An unpublished NIOSH report evaluating a series of buildings selected without regard to indoor air quality complaints indicated a maximum prevalence of 8% (Sieber and Godwin, 1998). In contrast, one-third of 80 office buildings studied by NIOSH in 1993 in response to health hazard evaluation requests had prevalences of frequent work-related shortness of breath greater than 8%, ranging up to 24% (Malkin et al., 1996). In conclusion, most buildings in which occupants have IAQ complaints probably do not have excesses of respiratory disease, but a substantial subset of "complaint" buildings may have occupants who associate their chest symptoms with building occupancy.

The NIOSH experience with indoor air quality investigations in 1993 gives some epidemiologic leads regarding building-related asthma and its causes (Sieber et al., 1996). Among 2,435 occupants of 80 office buildings classified according to whether moisture was found in the heating, ventilation, and air-conditioning (HVAC) system, the prevalence of physician-diagnosed asthma arising since building occupancy was 61% higher in buildings with HVAC moisture, and the prevalence was doubled of having three of four chest symptoms (cough, shortness of breath, wheezing, and chest tightness) at least once per week in the last month that improved away from work (Table 9-1). In logistic analyses controlled for age and gender, postoccupancy asthma was associated with dirty HVAC filters, debris in the air intake system, and renovation with drywall within the previous three weeks; daily surface cleaning with wet methods appeared protec-

TABLE 9-1 Health Condition Prevalence (%), by Moisture Status of HVAC, Among 2,435 Occupants of 80 Office Buildings with IAQ Complaints

Health Condition	Moisture Present	Moisture Absent	Range
Any physician-diagnosed asthma	11.7	11.6	
Postoccupancy asthma	3.7	2.3	0–33.3
Chest symptoms[a]	4.1	1.8	0–33.3

[a]Three of four (cough, shortness of breath, wheezing, chest tightness) occurring ≥1/week in last month and improving away from work.

SOURCE: W. Karl Sieber, Division of Surveillance, Hazard Evaluations, and Field Studies, National Institute for Occupational Safety and Health, written communication to Kathleen Kreiss, March 8, 1999.

TABLE 9-2 Relative Risks by Health Condition for Environmental Factors

Environmental Variable	Postoccupancy Asthma	Multiple Chest Symptoms
Dirty filters	2.0 [a]	1.9 [a]
Air intake debris	2.0 [a]	3.1 [a]
Recent renovation with drywall	2.5 [a]	1.1
Ceiling panels	3.2	3.4
Daily surface cleaning	0.5 [a]	0.7

[a]$p < .05$.

SOURCE: W. Karl Sieber, Division of Surveillance, Hazard Evaluations, and Field Studies, National Institute for Occupational Safety and Health, written communication to Kathleen Kreiss, March 8, 1999.

tive in logistic regression models (Table 9-2). Similar risk factors existed for the outcome of frequent multiple work-related chest symptoms, which may represent asthma-like conditions for which medical consultation has not been sought or for which an asthma diagnosis was not made. Although these findings do not document specific agents causing asthma in office buildings, they suggest that bioaerosols associated with HVAC moisture, maintenance deficiencies, and water damage requiring renovation are promising hypotheses.

Commercial office building stock and schools are often constructed with flat roofs, which predispose to puddling and moisture incursion; other common sources of moisture problems in nonindustrial building stock are HVAC coils, drain pans, and duct liners; below-grade drainage; inadequate moisture barriers; flashing leaks; flooding; and plumbing mishaps. At present, little information is available regarding the relative health risks of these conditions. Absent risk assessment information, building occupants, tenants, and managers have little leverage to have these conditions fixed.

STUDIES OF SCHOOLS

School is an important indoor environment for many children. The possible influence on health of indoor air in schools and day care centers have been the subject of research. However, very little of this literature specifically addresses indoor air exposures and asthma. Of the available studies on the topic, few have done careful and complete exposure assessments.

Schools are subject to many of the same exposure problems found in homes, although not necessarily at the same levels. The level of some allergens such as cockroach and dust mite is thought to be generally lower in schools because of the relative scarcity of appropriate habitats for infestation (Perzanowski et al., 1999; Sarpong et al., 1997). The absence of a direct source does not, however, necessarily preclude its presence. Cat allergen—which clings to clothing and other belongings—has been measured in schools and other indoor environments where cats have never been (Chan-Yeung et al., 1999). Indeed, schools may be a major source of exposure to cat and dog allergen for individuals who do not have these pets at home (Almqvist et al., 1999; Lonnkvist et al., 1999).

A survey of research on indoor air quality in schools noted that the major building-related problem identified was inadequate outdoor air ventilation (Daisey and Angell, 1998). Water damage and concomitant mold growth constituted the second greatest problem.

Among the epidemiologic studies identified by the committee, Smedje and colleagues (1997) investigated air quality, environmental characteristics, and asthma outcomes in 762 students in 39 schools in Uppsala County, Sweden. They used a self-administered questionnaire for health status information and employed trained occupational hygienists to gather building information and measurements in the schools. Data gathered included air exchange rate, temperature, and relative humidity; airborne levels of volatile organic compounds (VOCs), NO_2, molds, and bacteria; and levels of endotoxin, and cat, dog and mite allergen in settled dust. They found that self-reported current asthma was more common in schools that were larger, and had more open shelves (a repository of settled dust), lower room temperature, higher relative humidity, higher concentrations of formaldehyde or other VOCs, viable mold or bacteria, and more cat allergen. The observations were drawn on the basis of 40 asthma cases.

Haverinen and colleagues (1999) investigated three school buildings that were suspected of causing occupant health problems. The buildings were subjected to extensive structural and microbial contamination surveys; questionnaires were used to collect data on health problems. Widespread moisture damage was observed. Fungi identified (*Aspergillus, Eurotium*) were typical of buildings with mold problems. The prevalence of self-identified asthma among upper secondary school students, but not elementary or high school students, was reported to be higher than typical for the age group.

Hunting (1999) noted that half (14 of 28) of the cases of reported "occupational asthma" or RADS from primary and secondary schools and vocational schools in AOEC's Occupational and Environmental Disease Surveillance Database over the years 1991–1996 were attributed to indoor air exposures. The specific exposures were not identified.

Nafsted and colleagues (1999) investigated the respiratory health of 3,853 children in Oslo, Norway. Parents completed a questionnaire concerning day care arrangements, the child's health, environmental conditions, and family characteristics. The study found that the lifetime risk of doctor-diagnosed asthma was higher in children who started attending day care centers during the first two years of life. The authors speculated that this was a consequence of the higher rate of early respiratory infections in children in day care. There were no data reported on other environmental exposures.

Two studies addressed respiratory outcomes in children in the United States. A study of 1,268 children in Minnesota by Marbury and colleagues (1997) found that day care attendance was associated with an increased risk of recurrent wheezing illnesses and lower-respiratory infections. Celedon and colleagues (1999) examined respiratory tract illnesses in 498 Boston-area children who had at least one parent with a history of allergy or asthma. Researchers found that day care attendance increased the risk of upper and lower respiratory tract illnesses in the first year of life for these children. (Chapter 5 includes a discussion of the state of the literature regarding exposure to infectious agents and asthma.)

Such findings are not, however, uniform. McCutcheon and Woodward (1996) studied the respiratory health of 445 Adelaide, South Australia, school children, using data from questionnaires completed by parents and school records. They found that children who had attended child care prior to commencing school experienced half as many episodes of asthma as children who had never attended child care. Among the explanations for the findings offered was protection against later respiratory illness as a result of early exposure, although selection of illness-prone children into home care might also have had an influence.

School exposures are an area of continuing research interest. One ongoing study of schools in the Chicago area is monitoring levels of a wide variety of potentially problematic agents—including molds, dust mites, animal and insect allergens, particulates and various chemicals—and other environmental characteristics such as ventilation and humidity. The study is also evaluating the effect of interventions including educational initiatives and integrated pest management. An interesting observation from this effort is the importance of involving engineering and janitorial staff, along with school management, in the planning and implementation of interventions (Persky, 1999).

CONCLUSION

Published case reports, public health surveillance of physician reporting, and cross-sectional studies of building occupants with indoor air quality complaints provide limited or suggestive evidence of an association between aspects of the nonindustrial indoor environment and the development of asthma with a building occupancy-related pattern of symptoms and, in some instances, objective abnormalities. What is lacking for the most part, however, is knowledge of specific etiologic agents in these nonindustrial indoor environments that might be responsible for these new work-related asthma cases. Epidemiologic data suggest that moisture and ventilation system problems are markers for the problematic agents when work-related asthma arises in particular office and school buildings. Similarly, one cross-sectional study comparing two building populations (with comparable IAQ discomfort complaints) provides evidence that one particular office environment caused exacerbation of asthma among more than a third of occupants with preoccupancy physician diagnoses of asthma, and the other office environment did not. However, the proportion of building environments that precipitate exacerbation of asthma and the specific etiologies of asthma exacerbation are unstudied.

RESEARCH NEEDS

The few available studies suggest the importance of building factors in relation to asthma, but further research is critical to assessing the attributable risks, remediable risk factors, and means of hazard assessment. Development of methods for representative quantitative assessment of bioaerosols of fungal and bacterial origin is a high priority for health outcome studies and hazard assessment. In addition, knowledge of the epidemiology of building-related asthma in problem buildings where there are excess chest complaints among occupants, in comparison to buildings where there are no complaints, can advance our understanding of specific bioaerosols in relation to asthma. Research should focus on exposure–response studies of many building environments and populations; clinical investigation of patients with building-related asthma; and intervention studies, even without knowing the specific etiology involved. This research agenda requires new partnerships among academic investigators, clinicians, public health agencies, industrial hygienists, and building scientists.

REFERENCES

Almqvist C, Larsson PH, Egmar AC, Hedren M, Malmberg P, Wickman M. 1999. School as a risk environment for children allergic to cats and a site for transfer of cat allergen to homes. Journal of Allergy and Clinical Immunology 103(6):1012–1017.

Brightman HS, Womble SE, Girman JR, Sieber WK, McCarthy JF, Buck RJ, Spengler JD. 1997. Preliminary comparison of questionnaire data from two IAQ studies: occupant and workspace characteristics of randomly selected buildings and "complaint" buildings. Healthy Buildings/IAQ Conference proceedings, September 1997, Washington, DC.

Burge PS, Finnegan M, Horsfield N, Emery D, Austwick P, Davies PS, Pickering CA. 1985. Occupational asthma in a factory with a contaminated humidifier. Thorax 40(4):248–254.

Celedon JC, Litonjua AA, Weiss ST, Gold DR. 1999. Day care attendance in the first year of life and illnesses of the upper and lower respiratory tract in children with a familial history of atopy. Pediatrics 104(3 Pt 1):495–500.

Chan-Yeung M. 1995. Occupational asthma. Environmental Health Perspectives 103(Supplement 6): 249–252.

Chan-Yeung M, McClean PA, Sandell PR, Slutsky AS, Zamei N. 1999. Sensitization to cat without direct exposure to cats. Clinical and Experimental Allergy 29(6):762–765.

Daisey JM, Angell WJ. 1998. A Survey and Critical Review of the Literature on Indoor Air Quality, Ventilation and Health Symptoms in Schools. Lawrence Berkeley National Laboratory. LBNL-41517.

Finnegan MJ, Pickering CAC. 1986. Building-related illness. Clinical Allergy 16(5):389–405.

Haverinen U, Husman T, Toivola M, Suonketo J, Pentti M, Lindberg R, Leinonen J, Hyvärinen A, Meklin T, Nevalainen A. 1999. An approach to management of critical indoor air problems in school buildings. Environmental Health Perspectives 107(Suppl 3):509–514.

Hoffman RE, Wood RC, Kreiss K. 1993. Building-related asthma in Denver office workers. American Journal of Public Health 83(1):89–93.

Hunting K. 1999. Reporting Trends in ASTHMA and RADS Related to Indoor Air Problems. Data from the AOEC Database. [unpublished report dated 3/18/99].

Lonnkvist K, Hallden G, Dahlen SE, Enander I, van Hage-Hamsten M, Kumlin M, Hedlin G. 1999. Markers of inflammation and bronchial reactivity in children with asthma, exposed to animal dander in school dust. Pediatric Allergy and Immunology 10(1):45–52.

Malkin R, Wilcox T, Sieber WK. 1996. The National Institute for Occupational Safety and Health indoor environmental evaluation experience. Part two: symptom prevalence. Applied Occupational and Environmental Hygiene 11(6):540–545.

Marbury MC, Maldonado G, Waller L. 1997. Lower respiratory illness, recurrent wheezing, and day care attendance. American Journal of Respiratory and Critical Care Medicine 155(1):156–161.

McCutcheon H, Woodward A. 1996. Acute respiratory illness in the first year of primary school related to previous attendance at child care. Australian and New Zealand Journal of Public Health 20(1):49–53.

Nafstad P, Hagen JA, Øie L, Magnus P, Jaakkola JJ. 1999. Day care centers and respiratory health. Pediatrics 103(4 Pt 1):753–758.

Persky VW. 1999. Asthma and Exposure in the School Environment. Presentation made at a public workshop before the Committee on the Assessment of Asthma and Indoor Air on March 22, 1999. Washington DC.

Perzanowski MS, Ronmark E, Nold B, Lundback B, Platts-Mills TA. 1999. Relevance of allergens from cats and dogs to asthma in the northernmost province of Sweden: schools as a major site of exposure. Journal of Allergy and Clinical Immunology 103(6):1018–1024.

Robertson AS, Burge PS. 1985. Building sickness. Practitioner 229(1404):531–534.

Romero Jajosky RA, Harrison R, Reinisch F, Flattery J, Chan J, Tumpowsky C, Davis L, Reilly MJ, Rosenman KD, Kalinowski D, Stanbury M, Schill DP, Wood J. 1999. Surveillance of work-related asthma in selected U.S. states using surveillance guidelines for state health departments—California, Massachusetts, Michigan, and New Jersey, 1993–1995. Morbidity and Mortality Weekly Report 48(SS03):1–20.

Sarpong SB, Wood RA, Karrrison T, Eggleston PA. 1997. Cockroach allergen (Bla g 1) in school dust. Journal of Allergy and Clinical Immunology 99(4):486–492.

Sieber WK, Stayner LT, Malkin R, Petersen MR, Mendell MJ, Wallingford KM, Crandall MS, Wilcox TG, Reed L. 1996. The National Institute for Occupational Safety and Health indoor environmental evaluation experience. Part three: associations between environmental factors and self-reported health conditions. Applied Occupational and Environmental Hygiene 11(12):1387–1392.

Sieber WK, Godwin KJ. 1998. Comparison of Prevalence of Medical Symptoms in Two Indoor Environmental Quality (IEQ) Surveys, 1998. Unpublished report of the National Institute for Occupational Safety and Health.

Smedje G, Norbäck D, Edling C. 1997. Asthma among secondary schoolchildren in relation to the school environment. Clinical and Experimental Allergy 29(6):762–765.

10
IMPACT OF VENTILATION AND AIR CLEANING ON *Asthma*

Indoor exposures to pollutants associated with the incidence or symptoms of asthma are affected by many aspects of building design, maintenance, and operation. Building features modify the indoor sources of pollutants, the rates of pollutant entry from outdoors, and the rates of pollutant removal from indoors. Building ventilation and air cleaning are the two primary processes used intentionally within buildings to remove pollutants from the indoor air and maintain acceptable indoor environmental conditions. This chapter provides an overview of the relationship of building ventilation and particle air cleaning to exposures to indoor-generated pollutants that are associated with asthma. The findings from experimental assessments of the effects of air cleaning on allergy and asthma symptoms are also summarized. Because the association of asthma with pollutants from outdoor air is not a primary focus of this report, even though the exposures may occur primarily indoors, the dependence of these exposures on ventilation and air cleaning is not addressed in this chapter.

THEORETICAL BACKGROUND

This section provides a very brief overview of theoretical considerations that are necessary to understand the influence of building ventilation and air cleaning on indoor pollutant concentrations. Emphasis is placed on indoor particles because the indoor-generated pollutants most clearly associated with asthma are particles. Appendix A provides a more detailed technical discussion of this topic along with the equations and parameter values used for the theoretical predictions later in this chapter.

From conservation of mass, the steady-state indoor air concentration[1] of a pollutant that is emitted indoors and absent from outdoor air equals the indoor pollutant generation rate divided by the *sum* of all pollutant removal rates. In the present context, the most important pollutant removal processes are (1) ventilation (i.e., the flow of indoor air containing pollutants to outdoors); (2) pollutant depositional losses on indoor surfaces; and (3) air cleaning (i.e., intentional removal of pollutants from indoor air by air filters and other types of air cleaners). The influence of changes in ventilation or air-cleaning rates on the indoor pollutant concentration depends on the magnitude of the other two pollutant removal processes.

Many of the indoor-generated pollutants important for asthma are particles with diameters ranging from a fraction of a micrometer (1 μm equals one-millionth of a meter) to approximately 20 μm. Table 10-1 provides information of the sizes (aerodynamic diameters)[2] of these particles. The available data are limited and sometimes contradictory. Many of the bioaerosols associated with asthma, particularly dust mite allergens, whole pollens, cockroach allergen, and many fungal spores are large particles greater than a few micrometers in diameter. There are contradictions among available data on the size of particles with cat allergen; however, a significant fraction of airborne cat allergen appears to be associated with particles smaller than a few micrometers. Environmental tobacco smoke is composed almost entirely of submicron-size particles (i.e., particles smaller than 1 μm). Droplet nuclei from coughs and sneezes, which often contain virus, are included in Table 10-1 because viral infections are strongly

---

[1]For this discussion, we have assumed perfect mixing of the indoor air. See Appendix A for more information.

[2]Except that the sources of the data for pollens and fungal spores do not indicate whether the sizes are physical or aerodynamic diameters.

TABLE 10-1 Reported Size Distributions of Particles Associated with Asthma

Type of Particle	Size Ranges Reported	References
Dust mite allergens	30% to 90% of allergen is in particles <11 μm; only a few percent are in particles <2 μm	Tsubata et al., 1996; Platts-Mills et al., 1986
Plant pollens	15–25 μm for intact pollen; however, much of the grass and birch allergen is present in much smaller particles, including particles <1 μm	American Thoracic Society, 1997; Holmquist and Vesterberg, 1999; Pehkonen and Rantio-Lehtimaki, 1994; Rantio-Lehtimaki et al., 1994; Schappi et al., 1997; Spieksma et al., 1990, 1991
Common fungal spores	2–10 μm	American Thoracic Society, 1997
Cat antigen	From <1 μm to >10 μm. Based on Custovic et al. 1998), (49% are >9 μm, 23% are <4.7 μm. Based on Luczynska et al. (1990), 75% are >5 μm and 25% (range 10–62%) are <2.5 μm	Custovic et al., 1998; Luczynska et al., 1990; Rantio-Lehtimaki et al., 1994
Cockroach antigen	~ 80% of allergen >10 μm	de Blay et al., 1997b
Environmental tobacco smoke (ETS)	For emitted particles, the mass median diameter is 0.22 μm. Virtually all particle mass is within 0.09- to 1.0-μm-size particles	Miller, 1996
Droplet nuclei from coughs and sneezes	~ 50–70% of particles are <1 μm; 20%–40% are between 1 and 2 μm; <1% of particles >4 μm ~ 25% of particle volume is in particles <2 μm; 20% of volume is in particles between 2 and 4 μm; ~ 50% of volume in particles >4 μm	Gerone et al., 1966

linked to exacerbation of asthma, at least in children (Johnston et al., 1995). There is evidence that rates of building ventilation and occupant density modify the rates of respiratory illness experienced by building occupants (Fisk, 1999; Fisk and Rosenfeld, 1997), presumably by changing exposures to infectious droplet nuclei. Data on the size distribution of droplet nuclei are extremely limited and the methods employed to obtain the data may have resulted in an undercounting of the larger particles. The available data indicate that most of these particles are submicron in size but most of the particle volume is associated with particles larger than 1 μm. It is not clear whether the number concentration or volume concentration of infectious droplet nuclei is more relevant for disease transmission.

The magnitude of two of the particle removal processes—deposition on surfaces and air cleaning—can vary dramatically with particle size. Particles deposit on indoor surfaces when indoor air motion, gravitational settling, electrostatic forces, and other phenomena cause them to collide with indoor surfaces. For particles, larger than a few micrometers in diameter, depositional losses are dominated by rates of gravitational settling. A 20-μm particle falls a distance of 1 m in about 80 seconds so it remains suspended indoors for only a short period. The deposition losses of such large particles tend to overwhelm normal rates of particle removal by ventilation or air cleaning. In contrast, a 0.2-μm particle falls a distance of 1 m in about five days. The rate of depositional removal of 0.2-μm particle from the indoor air, which is controlled by the indoor air motion, indoor surface roughness, and other factors, is almost a factor of 100 lower than the rate of depositional removal of a 20-μm particle.

Some gaseous pollutants such as nitrogen dioxide and ozone are also removed from indoor air at a significant rate by deposition (often called sorption) on or reaction with indoor surfaces. Rates of depositional removal depend on the chemical nature of the pollutant, the intensity of indoor air motion, and other factors. Gravitational settling is unimportant for gaseous pollutants.

Particles deposited on indoor surfaces can be resuspended in indoor air when the surfaces are disturbed by human activities (e.g., walking, vacuuming) or by high air velocities (e.g., air exiting a fan). Theory (Hinds, 1982) and limited empirical data

(Thatcher and Layton, 1995) indicate that resuspension occurs predominantly for particles larger than approximately 2 μm. Based on our current knowledge of the behavior of particles, exposures to some of the larger particles associated with asthma may be substantially influenced by the localized resuspension of particles that results from occupant activities.

BUILDING VENTILATION

Background

In this document, the term "ventilation" refers to the flow of outside air indoors, which is always accompanied by an equal flow of indoor air outdoors. Ventilation removes and dilutes indoor airborne pollutants, brings outdoor air pollutants into buildings, and often removes or supplies heat and water vapor. Ventilation is also needed to maintain oxygen concentrations inside buildings, although the quantity of ventilation needed to supply oxygen is very small relative to other ventilation requirements.

Increasing the rate of ventilation generally leads to overall improvements in indoor air quality; however, the indoor concentrations of some pollutants from outdoors, such as outdoor particles and ozone, can increase with the ventilation rate. Indoor humidity can increase or decrease with ventilation rate. When it is cold and dry outdoors, increased ventilation usually reduces the indoor humidity.

While increased ventilation rates are usually considered beneficial for health and for improving perceived air quality (e.g., odors), ventilation air must often be heated (and sometimes humidified) or cooled and dehumidified. Consequently, the ventilation rates selected for buildings must strike a balance between the benefits of energy savings with reduced ventilation and the known or suspected benefits to health with increased ventilation.

Several metrics are used to specify the rates of building ventilation. Generally, these metrics are flow rates of outside air normalized by the number of occupants, floor area, or indoor volume. Corresponding units of ventilation rates include the following: liters per second per person (L s^{-1} per person); liters per second per square meter of floor area (L s^{-1} per square meter); and air changes per hour (h^{-1}).

Municipalities typically adopt one of the several building design codes used in the United States. These codes, or state energy codes, include building design provisions intended to maintain ventilation rates above a minimum rate that varies with the building type. The American Society of Heating, Refrigerating, and Air Conditioning Engineers (ASHRAE) publishes a minimum ventilation standard that is the basis for the ventilation specifications in many codes. The current version of the ASHRAE standard is Standard 62-1999—Ventilation for Acceptable Indoor Air Quality (ASHRAE, 1999). Standard 62-1999 lists 0.35 h^{-1} as a minimum ventilation rate in residences,[3] 10 L s^{-1} per person (20 cubic feet per minute [cfm] per person) as a minimum ventilation rate in offices, and 8 L s^{-1} per person (15 cfm per person) as a minimum ventilation rate in schools. Due to a paucity of scientific data on the relationship of building ventilation rates with the health and well-being of occupants (Seppanen et al., 1999), the minimum ventilation rates in the ASHRAE standard are based substantially on professional judgment and on studies performed in laboratories with conditions quite different from those encountered in real buildings.

Building design codes and ASHRAE's minimum ventilation standard do not ensure that all buildings maintain the specified minimum ventilation rates. In most states and municipalities, there are no legal requirements to actually maintain ventilation rates at or above the levels in building design codes. Additionally, building ventilation rates are difficult to measure accurately, infrequently measured, and as discussed later, poorly controlled.

Ventilation systems, although intended to remove indoor pollutants, can also become sources of pollutants. Portions of ventilation systems, particularly components that become wet, can become colonized by microorganisms and produce bioaerosols that

---

[3]Standard 62-1999 states that the 0.35 h^{-1} of ventilation is normally satisfied by infiltration and natural ventilation but includes no technical specifications for the building to ensure that this ventilation rate is met continuously or on average. Standard 62-1999 also specifies installed mechanical exhaust capacities of 50 L s^{-1} (100 cfm) per kitchen and 25 L s^{-1} (50 cfm) per bathroom.

are transported by the airflow to the occupied space. In addition, particles, fibers, and odorous and potentially irritating volatile organic compounds (VOCs) may be emitted from synthetic materials, including fibrous insulation materials, from residual oils used in component production, from deposited dusts, and from microorganisms. Ventilation also affects the indoor humidity which in turn influences the growth or survival of microorganisms within buildings.

Heating, ventilating, and air conditioning (HVAC) is the more general process of thermally conditioning and ventilating buildings. In commercial buildings, these functions are usually integrated. The HVAC process employed in commercial buildings is reviewed here because HVAC features may influence exposure to pollutants that are known or thought to be associated with asthma.

Methods and Rates of Ventilation in U.S. Single-Family Residences

Diamond (1999) has summarized many of the basic physical characteristics of the U.S. residential building stock. In 1997, detached single-family units and row houses constituted 73% of the U.S. housing stock, 6% of the housing stock was mobile homes, and the remainder was apartments. The average heated floor space in all U.S. housing stock was 181 m^2 (1,950 square feet) and air conditioning was installed in 70% of these dwellings. The average conditioned floor area of mobile homes was 87 m^2 (940 square feet) and 70% of mobile homes had air conditioners. Fourteen percent of all housing units used humidifiers, and nine percent had dehumidifiers.

When windows are closed, the ventilation of single-family residences in the United States is almost exclusively an uncontrolled process. In air infiltration (or infiltration and exfiltration), air leaks through unintentional cracks and holes in the building envelope. The infiltration rate is driven by small pressure differences across the building envelope that are typically less than a few pascals in magnitude. These pressure differences arise due to the differences between the indoor and outdoor air temperatures, resulting in different indoor and outdoor air densities, and also as a consequence of wind. Unintentional air leakage in the ductwork of forced-air heating and air-conditioning systems located in attics and crawl spaces, also causes large increases in air infiltration. Even if the ducts do not leak, forced-air systems can pressurize or depressurize specific rooms relative to the outdoor pressure, forcing air leakage through the building envelope.

U.S. homes often have intermittently-operated exhaust fans in bathrooms and kitchens. When operated, these fans draw outdoor air into the building. Window and door opening by occupants, predominantly during mild weather, also has a large influence on residential ventilation rates.

A very small portion of single-family dwellings in the United States have mechanical ventilation systems (i.e., fans operating continuously or intermittently to provide ventilation). Mechanical ventilation is most common in the State of Washington because the state energy code now requires mechanical ventilation. The technologies used to mechanically ventilate residences are described in Roberson et al. (1998).

Ventilation rates in residences vary considerably over time. The lowest ventilation rates occur during mild weather with windows and doors closed. When weather is more severe, windows remain closed but ventilation rates are higher due to increased indoor-to-outdoor temperature differences and increased use of forced-air heating and air conditioning. The highest ventilation rates generally occur when windows or doors are open.

Present data on ventilation rates in U.S. single-family residences are limited and possibly not representative of the building stock. One source of information is measurements of the airtightness of building envelopes with windows and doors closed. Ventilation rates are predicted with semiempirical models, using measured values of building airtightness[4] combined with climate data and indicators of a building's shielding from wind, as model in-

---

[4]Airtightness is used here as a general term understandable to a broad audience. The actual measured parameter is the effective leakage area (ELA) at a reference pressure, usually 25 or 50 Pa across the building envelope. The ELA is the area of an orifice that would leak air at the same rate as all the leakage paths in the building envelope. The ELA is usually normalized with building floor area and height to produce a normalized leakage.

puts. When annual average ventilation rates are desired, the predictions may also include terms to account for natural ventilation via windows; however, the current knowledge of window use and effects on ventilation is cursory. The second source of information on residential ventilation rates is measurements made using a tracer-gas procedure. Although considered more accurate than predictions based on airtightness, the measured data are more sparse than airtightness data. Therefore, we presently have only crude estimates of residential ventilation rates.

Based on airtightness and climate data for about 12,000 houses, Sherman and Matson (1997) estimate that the arithmetic average effective ventilation rate of houses in the United States is 1.1 h^{-1}. This average reflects ventilation rates when windows are closed and also the higher ventilation rates that occur with open windows during mild weather. Airtightness normalized by house size is highly variable (Sherman and Dickerhoff, 1994), with a standard deviation that is approximately 50% of the mean. The mean of the airtightness data from individual states varies among states by more than a factor of three. In the available data, there is no trend in airtightness with severity of climate. The available data indicated that houses constructed after 1980 are more airtight (by ~50%) than older houses (Sherman and Dickerhoff, 1994); however, there was no trend evident in airtightness with age for houses constructed after 1980.

A set of 2,844 measurements of residential ventilation rates in U.S. houses was analyzed by Murry and Burmaster (1995). The measured data from 66 research projects are not from a representative sample of residences; however, this analysis is probably the best available information on the distribution of ventilation rates in U.S. houses. When considering all climate zones and seasons, the arithmetic and geometric mean ventilation rates were 0.76 h^{-1} and 0.53 h^{-1}, with a geometric standard deviation of 2.3. There are large variations in ventilation rates with season and climate zone. The winter and summer arithmetic means, for all climate zones, are 0.55 and 1.50 h^{-1}. Approximately one-third of the measurements in the winter season are less than the 0.35 h^{-1}, the rate in the current ASHRAE ventilation standard. In the coldest climate zone, approximately 55% of the measured ventilation rates, from all seasons, are less than 0.35 h^{-1}.

Methods, Patterns, and Rates of Ventilation in U.S. Multifamily Apartment Buildings

In 1997, 21% of U.S. housing units were apartments. The average conditioned floor area of apartments was 85 m^2 (920 square feet) and air conditioning was installed in 65% of apartments (Diamond, 1999). Published information on the methods and rates of ventilation in multifamily apartment buildings are extremely sparse. Based on the limited information available,[5] older low-rise (i.e., less than ~three stories) apartment buildings usually have no mechanical supply of ventilation air. Much like single-family dwellings, these buildings are ventilated primarily by uncontrolled infiltration and natural ventilation windows that can be opened. Intermittently operated bathroom and kitchen exhaust fans cause temporary increases in ventilation rates. Leakage in the ductwork of forced-air heating and air-conditioning systems and pressurization or depressurization of individual rooms can drive infiltration and exfiltration in apartments, just as it does in single-family houses. Newer low-rise apartment buildings are ventilated similarly to older low-rise buildings; however, a larger portion of these buildings have continuous mechanical exhaust ventilation from the bathrooms and/or kitchens of each apartment.

Older apartment buildings with more than approximately three stories typically have no mechanical air supply or some mechanical supply to the interior hallways. The air supply system, when present, is frequently not functional (Shapiro-Baruch, 1993). Apartments within these buildings sometimes have a system for continuous exhaust ventilation from bathrooms and kitchens, although it is not always operational. Some portion of these older high-rise buildings have a vertical ventilation shaft that functions much like a chimney and passively draws air from the apartments.

In new apartment buildings with more than three stories,

---

[5]The information in this section is based primarily on case studies, on two general guidance documents (Diamond et al., 1999; Liddament, 1996) and on discussions with Dr. Rick Diamond of Lawrence Berkeley National Laboratory. who conducts research on energy use and ventilation in apartment buildings.

exhaust ventilation is usually drawn continuously from the kitchen and bathroom(s) of each apartment. The exhaust fans may serve groups of apartments or individual apartments. Outside air enters either from unintentional leaks and vents at windows or via ventilation systems that supply air continuously to each apartment. When a mechanical air supply is present, often this air is supplied to a single room of each apartment from a duct system in the building's interior hallway.

The airflow in heated multistory apartment buildings without mechanical ventilation often occurs in an upward direction from lower-level to upper-level apartments (e.g., Diamond et al., 1986; Modera et al., 1986). Cool outdoor air leaks into the lower apartments; flows upward, picking up moisture and pollutants; and exfiltrates through the walls and ceilings of upper-level apartments. Due to this airflow pattern, the lower-level apartments tend to have more fresh air supply, lower humidity, and more cold drafts. Humidity and pollutant levels are often increased in upper-level apartments because a portion of the air entering these apartments comes from lower levels of the building. As moist air exfiltrates out of the upper-level apartments, water vapor may condense within cold walls and ceilings. Possibly, the higher pollutant levels and humidity in upper-level apartments could contribute to asthma symptoms.

This same upward-flow phenomenon occurs to a variable degree in all heated multistory buildings. When the building is air conditioned (i.e., cooled), the airflow direction can reverse; however, the downward airflow in air-conditioned buildings will be less pronounced because the indoor-to-outdoor temperature differences are typically much smaller during air conditioning than during heating of buildings. By reducing the openings between floors, the vertical airflow between floors can be reduced. Mechanical ventilation can also reduce or overwhelm the upward buoyancy-driven airflow.

In addition to the buoyancy-driven upward airflow in apartments, other unintentional flows of air between adjacent apartments are reported commonly from case studies. These flows occur through unintentional openings in walls and floors, and may be driven by mechanical ventilation systems, buoyancy, and wind.

The vertical and horizontal airflows between apartments transport pollutants; therefore, occupants may be exposed to pollutants such as tobacco smoke, cooking odors, and pet allergens from the apartments of neighbors. These airflows also transport indoor moisture, potentially causing moisture problems.

The ASHRAE ventilation standard has the same minimum ventilation requirements for multifamily and single-family residential buildings, 0.35 h^{-1}, with exhaust capacity required in kitchens and bathrooms. Measurements of ventilation rates in apartments are extremely limited. Based on case studies in a small number of buildings, the reported ventilation rates are 0.5–1.5 h^{-1} for low rise apartments with frame or brick construction and 0.2–1.0 h^{-1} for high-rise apartments with a more airtight concrete construction (Diamond et al., 1999). Based on the available information, we can expect ventilation rates to vary a great deal with time, among apartment buildings, and among apartments within a building.

Heating, Ventilating, and Air Conditioning in U.S. Commercial and Institutional Buildings

Building and HVAC Characteristics

Approximately 40% of commercial floor space has windows that open and approximately one-half of this floor space is in buildings with a floor area less than 4650 m^2 (50,000 square feet) (DOE-EIA, 1994, 1995). Most U.S. commercial buildings have HVAC systems that thermally condition and ventilate the occupied spaces. Larger buildings, as well as many smaller buildings, typically use HVAC systems that thermally condition air in mechanical rooms or in equipment located at the rooftop. The conditioned air, usually a mixture of outdoor air and recirculated indoor air known as the supply airstream, is supplied throughout the building or building section. The supply airstream typically passes through a fibrous particle filter, heat exchangers (called heating and cooling coils) that add or remove heat, a supply fan, a system of air ducts, dampers used to regulate airflow rates, and supply registers located at or near the ceilings. The return airstream is drawn from the occupied spaces usually at or near the ceiling and typically flows back to the mechanical equipment through return ducts or through the plenum between a suspended ceiling and the floor of the next-higher story. Return air fans are used in many HVAC systems, particularly larger systems. A portion of the return airstream is exhausted to the outdoors, and the remainder is mixed with outdoor air and resupplied to the space after filtering and thermal conditioning. Dampers located in the outside, recirculated, and exhaust airstreams are automatically or manually adjusted to controls the airflow. In addition to these central HVAC systems, some buildings have distributed systems located throughout the building with components for adding and removing heat, particularly at the building perimeter that is most affected by heat loss and gains through the building envelope. Many variations in HVAC system designs are described elsewhere (e.g., ASHRAE, 1996).

The supply airflow rates of HVAC systems serving offices or schools are typically 3.5–5 L s^{-1} per square meter of floor area (0.7–1.0 cfm per square foot) which is equivalent to approximately four to five indoor air volumes per hour. In some smaller buildings, the supply air fan stops when there is no demand for heating or cooling. The supply airstream usually contains more recirculated indoor air than outdoor air. The ratio of flow of outside air to total supply air is called the percentage of outside air. Some HVAC systems are designed and adjusted to maintain an approximately fixed rate of outside air intake, or a fixed percentage of outside air, consistent with code requirements. To save energy, many HVAC systems employ an economizer cycle that increases the rate of outside air intake during mild weather to reduce the need for mechanical cooling.

While the above-described approach for controlling outside air supply in commercial buildings appears straightforward, in practice it often works poorly. The flow rate of outside air is controlled by a small pressure drop across the outside air dampers, and this pressure difference is rarely monitored or controlled. HVAC systems almost never incorporate instruments for directly monitoring the rate of outside airflow. Air-balance professionals may monitor and adjust HVAC system airflows after initial building construction and occasionally thereafter; however, accurate measurements of the rates of outside air intake are very difficult even for these personnel. Often, the damper systems used to regulate airflows are nonfunctional, disconnected from the damper actuators, or casually adjusted by building operators. Occasionally, building operators will fully close outside air dampers, assuming that leakage provides adequate outside air. It is not uncommon to find inside air flowing out of the building through outside air dampers. In HVAC systems serving smaller buildings with supply air provided only during heating or cooling, there may be no outside air supply during mild weather. The poor performance of control systems for outside air is reflected in the measured ventilation data. Measured minimum ventilation rates in commercial buildings vary a great deal among buildings and often deviate substantially from code requirements (Dols and Persily, 1994; Fisk and Faulkner, 1992; Lagus Applied Technologies, 1995; Persily and Grot, 1985; Teijonsalo et al., 1996; Turk et al., 1989).

Outside air drawn into HVAC systems and mixed with recirculated air is distributed to the various rooms of a building with a complex duct system. In many buildings the rate of air supply to each section of the building is modulated automatically over time to maintain a comfortable indoor temperature. Sometimes outside air is distributed very unevenly to the rooms or floors within a building so that effective ventilation rates within rooms of the same building vary by a factor of two or three (e.g., Dols and Persily, 1994; Fisk and Faulkner, 1992; Teijonsalo et al., 1996). In other buildings, the effective ventilation rate is spatially uniform.

In addition to the ventilation provided by HVAC systems, natural ventilation through windows that open occurs in the 40% of commercial floor space having such windows. Air infiltration also occurs in all commercial buildings, just as it does in residences; however, the mechanical ventilation systems of commercial buildings can drive air infiltration and exfiltration through the building envelope, modifying or sometimes reversing the airflows that would occur due to wind and buoyancy. For many large commercial buildings, particularly those in warm humid climates, the design intent is to pressurize the building slightly with the mechanical ventilation system in order to prevent undesirable infiltration of unconditioned air, moisture, and outdoor air pollutants. In cold climates, some buildings are designed to be slightly depressurized to prevent warm humid indoor air from flowing outward through the building envelope. In practice, indoor–outdoor pressure differences are often poorly controlled, and many buildings are not pressurized (Persily and Norford, 1987).

Ventilation Rates in Commercial and Institutional Buildings

Commercial and institutional buildings are ventilated mechanically with the HVAC system, via air infiltration and through the openable windows present in 40% of the commercial floor space. The ventilation provided through each of these mechanisms varies over time. In warm climates, the mechanical supply of outside air is minimized when the outside air exceeds approximately 20–24°C, hence, the building may be operated with a minimum ventilation rate for much of the cooling season. Minimum ventilation is also provided during cold weather when the outside air temperature falls below a set point that varies among buildings. During periods of minimum ventilation, concentrations of indoor-generated pollutants will tend to be highest. During mild weather in the spring and fall, the indoor-to-outdoor concentration ratio for some outdoor air pollutants, such as outdoor particles (e.g., molds) and ozone, will tend to be at a maximum.

Ventilation rate data from United States commercial and institutional buildings are very limited and come primarily from convenience samples of buildings and modest-size research projects. Consequently, only very rough estimates can be provided on the distribution of ventilation rates in U.S. commercial buildings. Table 10-2 provides summary information from the largest data sets. The measured ventilation rates vary over a large range and usually exceed the minimum ventilation rates specified in ASHRAE Standard 62-1989. However, the ventilation rates measured when the HVAC systems provided the minimum amount of outside air were less than the corresponding minimum rate specified in the current ASHRAE standard[6] in a majority of

---

[6]Some buildings were constructed when the applicable ventilation standard specified lower minimum ventilation rates than ASHRAE Standard 62-1999 (ASHRAE, 1999).

buildings (i.e., in 8 of 13 buildings studied by Turk et al., 1989; 20 of 49 buildings [including 13 out of 14 schools] studied by Lagus Applied Technologies, 1995; and 8 of 8 buildings studied by Persily and Grot, 1985).

In commercial and institutional buildings that are not occupied continuously, HVAC systems are often not operated when the building is vacant. The concentrations of indoor-generated air pollutants may increase when the building is vacant and there is no mechanical ventilation, and then decrease over a period of a few hours after mechanical ventilation starts. Therefore, occupants' exposures to indoor air pollutants depend on the operating schedule for the HVAC system.

Even less is known about the proportion of commercial building ventilation that results from air infiltration; however, the existing data suggest that infiltration is appreciable, particularly in the smaller buildings (Lagus Applied Technologies, 1995; Persily, 1999). The ratio of infiltration to total ventilation may be important for asthma because the air that infiltrates a building is not filtered to remove outdoor particles before it enters the building.

Pollutant Sources in HVAC Systems

HVAC systems can become sources of indoor air pollutants, possibly increasing asthma symptoms. Pollutant emissions from certain types of HVAC systems constitute one of several potential explanations for the consistent association between the type of HVAC system and the prevalence of nonspecific health symptoms (called sick building syndrome) experienced by office workers (e.g., Mendell, 1993). In particular, almost all studies have found a statistically significant increase in symptom prevalence among occupants of office buildings with mechanical ventilation and air conditioning (Mendell, 1993) relative to occupants in naturally ventilated offices. There is also some evidence that humidifiers are associated with increased symptoms (Mendell, 1993). The subjectively assessed (i.e., via questionnaire) nonspecific health symptoms in these studies usually include a few symptoms potentially indicative of asthma, such as wheezing, tight chest, and difficulty breathing. Therefore, the evidence that HVAC systems can become sources of pollutants is relevant for asthma. The available information about pollutant sources in HVAC systems is quite limited and comes predominately from commercial buildings; however, similar pollutant sources, or risk factors for sources, may be present in HVAC systems of other types of buildings. This section provides a brief review of the issue.

Liquid water is often present at several locations in or near commercial building HVAC systems, facilitating the growth of microorganisms that may contribute to asthma. The outdoor air is often drawn from the rooftop or from a below-grade "well" where water (and organic debris) may accumulate. Raindrops, snow, or fog can be drawn into HVAC systems with the incoming outside air, although the systems are usually designed to prevent or limit this moisture penetration. Moving along the supply airstream in the direction of airflow leads to the cooling coil where moisture condenses and ideally drips from the surfaces of the coil into a drain pan with a drainage pipe connected to the sewer or to outdoors. Frequently, drain pans contain stagnant water because they do not slope toward the drain line. The drain also may be plugged or nonfunctional because air pressure differences prevent drainage, sometimes causing the drain pan to overflow with water. If the velocity of air passing through the cooling coils is too high, drops of water on the surface of the cooling coil can become entrained in the supply airstream and deposit in the HVAC system downstream of the cooling coil. Air exiting cooling coils is frequently nearly saturated with water vapor, and the high humidity of this air increases the risk of microbiological growth. The HVAC system may have a humidifier that uses steam or some evaporation process to add moisture. Humidifiers, used predominantly in the colder climates, may have reservoirs of water, surfaces that are frequently wetted, or water drops that do not evaporate. Thus, there are many sources of liquid water and very high humidity in HVAC systems.

Microbiological contamination of HVAC systems has been reported in many case studies and investigated in a few multibuilding research efforts (e.g., Batterman and Burge, 1995; Bencko et al., 1993; Martiny et al., 1994; Morey, 1994; Morey and Williams, 1991; Shaughnessy et al., 1998). Sites of reported microbiological contamination include the outside air louvers, the mixing box where outside air mixes with recirculated air, filters, cooling coils,

TABLE 10-2   Summary Information from the Three Largest Surveys of
Ventilation Rates in U.S. Commercial Buildings

Study	Type of Building	Number of Buildings	Operating Condition
Turk et al., 1989	Offices, schools	38	As found
	and libraries	38	As found
	combined	13	Min. vent.[a]
		14	Min. vent.[a]
Lagus Applied Tech.,	Small office	17	Min. vent.[a]
1995	Large office	5	Min. vent.[a]
	Retail	13	Min. vent.[a]
	School	14	Min. vent.[a]
Persily and Grot, 1985	Offices	9	Yearly avg. Min. vent.[a]

[a]With the ventilation system providing the minimum rate of supply of outside air, a normal condition during cold weather and also when outside temperatures exceed the indoor temperature.

cooling coil drain pans, humidifiers, and the surfaces of ducts. The porous insulating and sound-adsorbing material called duct liner, which is used inside some HVAC systems may be particularly prone to contamination (Morey, 1988; Morey and Williams, 1991). Supporting the evidence of microbiological contamination from field studies are numerous laboratory-based studies demonstrating that fungi can grow on various HVAC materials at a wide range of temperatures and humidities. The flow of air through HVAC systems to the occupied spaces is a means of transporting bioaerosols from contaminated sites inside HVAC systems to occupants. Thus, microbiological contamination of HVAC systems is theoretically a risk factor for asthma. However, relatively little is known about the extent to which microbiological contaminants within HVAC systems actually increase bioaerosol exposures or affect asthma symptoms.

HVAC systems can also be sources of other pollutants such as fibers, nonbiological particles, and VOCs. Sources of these pollut-

| Reported Ventilation Rates | | | | | Estimated ASHRAE |
Mean	SD	Min	Max	Units[b]	Std. 62-1999 Equivalent[b]
28	18	4.5	84	L s⁻¹ per⁻¹	10 L s⁻¹ per⁻¹
1.5	0.9	0.3	3.6	h^{-1}	
10	9	1.5	37	L s⁻¹ per⁻¹	10 L s-1 per⁻¹
1.6	0.7	0.4	3.0	h^{-1}	
1.4	0.7	0.3	2.7	h^{-1}	0.8 h^{-1}
0.8	0.6	0.7	2.7	h^{-1}	0.8 h^{-1}
2.2	1.6	0.5	7.0	h^{-1}	1.2 h^{-1}
2.4	1.6	1.2	2.9	h^{-1}	3.0 h^{-1}
0.7	0.2	0.3	1.1	h^{-1}	0.7 h^{-1}
0.4	0.2	0.1	0.6	h^{-1}	0.7 h^{-1}

[b]Legend for ventilation rate units: L s⁻¹ per⁻¹ = liters per second per person; h^{-1} = air changes per hour; Min. vent. = minimum ventilation; SD = standard deviation; and Yearly avg. = yearly average.

ants inside HVAC systems include duct liners, gaskets, oil left on surfaces after manufacturing, dust deposited on surfaces including dusts generated during building construction and renovation, and the wear of fan belts. It is clear that HVAC systems and components, particularly dirty filters and the oily surfaces of ducts, can release odorous compounds that significantly degrade the perceived acceptability of the air supplied to the occupied space (e.g., Pasanen et al., 1995; Pejtersen, 1996; Seppanen, 1998); however, the relevance of these pollutants for asthma is not known.

Causes of HVAC Problems

The rates of ventilation provided by HVAC systems and the contamination of HVAC systems with pollutants depend as much on system construction, maintenance, and operation practices as on system design. Proper operation and maintenance of HVAC systems is hindered by system complexity, inaccessibility of components, and lack of training of construction and maintenance staff. The absence of economic incentives for proper HVAC operation and maintenance may be even more important. When buildings are leased, their owners and operators are usually unaffected economically by the adverse health of building occupants, unless the health problems are severe and clearly linked to the building. In speculative new construction, the designer is similarly unaffected. However, these designers, owners, and operators have an incentive to reduce the highly tangible costs of building construction, maintenance, and operation.

Costs of Ventilation and Associated Carbon Dioxide Emissions

Ventilation is one of the most energy-intensive methods of reducing indoor pollutant concentrations primarily because of the need to thermally condition ventilation air. For several reasons, including great uncertainty about average ventilation rates in buildings, the available estimates of the energy used for ventilation are relatively crude. Orme (1998) estimates that ventilation accounts for about 30% of the energy used in U.S. residential buildings. A rough estimate of the average annual cost per residence of energy for ventilation is $400.[7] In the U.S. service sector (e.g., commercial, institutional, and government buildings), the estimated energy consumed for ventilation (Orme, 1998) is approximately one-quarter of total service-sector building energy use. The cost of energy for ventilating commercial buildings depends highly on climate, building size, occupant density, and type of HVAC system. If the average ventilation rate is 10 L s^{-1} per person, Emmerich and Persily (1998) have estimated that 46% of the total heating and cooling load in the stock of office buildings is attributable to ventilation. With nearly 50% of the U.S. work force, or 57 million workers, in offices (U.S. Department of Com-

---

[7]This estimate is based on the 30% of total residential energy used for ventilation (Orme, 1998) and the average U.S. household's total energy cost of $1,355 (Diamond, 1999).

merce, 1997; Table 645) and average energy prices, the annual cost of ventilation per office worker is roughly $25.[8]

Carbon dioxide, a greenhouse gas, and other air pollutants are generated in the production of the energy used for ventilation. The annual CO_2 emissions attributed to ventilation are approximately 1000 and 800 million tons for the residential and service sectors, respectively (Orme, 1998).

Influence of Ventilation Rates on Indoor Concentrations of Pollutants

Direct Impacts

Measured data quantifying the influence of ventilation rates on indoor concentrations of indoor-generated pollutants are surprisingly limited, particularly for pollutants associated with asthma. Data from surveys with ventilation rate and pollutant concentration measurements have emphasized measurements of pollutants that are generated indoors and also present in outdoor air—reducing the influence of ventilation rates on indoor concentrations. None of the large cross-sectional surveys have monitored concentrations of indoor-generated particles associated with asthma, such as environmental tobacco smoke (ETS) particles and indoor-generated bioaerosols. Indoor particle mass concentrations have been measured in some surveys, but these measurements reflect both indoor-generated particles and particles from outdoors. In a study of 150 residences located in Riverside, California (Özkaynak et al., 1996), there were statistically significant, but small, increases in indoor particle concentrations with *increased* ventilation rates. The values of PM_{10} and $PM_{2.5}$ increased about 12 and 5 µg m^{-3}, respectively,[9] for a 1 h^{-1} increase in ventilation rate.

---

[8]Estimated heating and cooling energy are 110 and 28 PJ, respectively (Emmerich and Persily, 1998). Average prices of natural gas ($0.69 per therm) and electricity ($0.098 per kilowatt-hour) from U.S. Department of Commerce (1997, Table 768) have been used to convert energy use to energy costs, which are then divided by the size of the office work force.

[9]PM_{10} and $PM_{2.5}$ are particle mass concentrations for particles smaller than 10 and 2.5 µm, respectively.

In data from surveys, there is sometimes no clear association of indoor pollutant concentrations with ventilation rates. In a survey of 26 houses near Spokane, Washington, and 35 houses near Portland, Oregon, the indoor formaldehyde concentration decreased 40 and 300 parts per billion (ppb), respectively, per 1 h^{-1} increase in ventilation rate (Turk et al., 1987b). However, in surveys of 38 commercial buildings in the U.S. Pacific Northwest, there were no clear associations of ventilation rates with indoor concentration of respirable particles, nitrogen dioxide, or formaldehyde (Turk et al., 1987a, 1989). Apparently, variations in indoor pollutant emission rates and other factors are large enough to obscure the effects of ventilation in a survey of this size.

Few experiments have been performed with changes in ventilation rates as the only intervention. Again, most of the experimental data involve pollutants generated indoors but also present in outdoor air. Offermann et al. (1982) used mechanical ventilation systems to increase wintertime ventilation rates in nine houses from an average of 0.35 to 0.83 h^{-1}. The average indoor formaldehyde[10] concentration decreased 21% and average indoor relative humidity decreased 13%. Indoor nitrogen dioxide concentrations increased slightly because concentrations were higher outdoors than indoors. In a set of two houses, inhalable particle concentrations decreased 30%. Using a mechanical ventilation system with a filter, Turk et al. (1997) increased ventilation rates in two classrooms from about 0.7 to 2–3 h^{-1} and indoor particle concentrations decreased by approximately 50%. Berglund et al. (1982) increased the percentage of outside air in the air supply of an office building from 20 to 50%, and then to 80%, and their measure of total indoor-generated VOCs in detector-response units decreased from 120 to 50 and then to ~20. Menzies et al. (1993) increased ventilation rates in a set of four office buildings and found that formaldehyde and total VOC concentrations were 40 and 60% lower, respectively, during periods with a higher ventilation rate. Hodgson et al. (1999) measured concentrations of 24 VOCs in a house with ventilation rates of 0.14 and 0.32 h^{-1}. At the

---

[10]Indoor concentrations of formaldehyde are usually much larger than outdoor concentrations due to indoor formaldehyde sources.

higher ventilation rate, on average for the 24 compounds the indoor concentrations was 42% lower, with a standard deviation of 11%.

Some experimental studies did not detect clear reductions in indoor pollutant concentrations when ventilation rates were increased. Shaughnessy et al. (1997) increased ventilation rates from ~ 0.2 h^{-1} to 3 h^{-1} in one elementary classroom and from ~ 0.7 h^{-1} to 3 h^{-1} in another classroom. There were no clear changes in indoor PM$_{10}$ concentrations reportedly because of large natural temporal fluctuations. Indoor total volatile organic compound (TVOC) concentrations decreased by about a factor of two, but natural temporal variations precluded firm conclusions about the effects of ventilation. Nagda et al. (1990) increased ventilation rates in an office building from 0.84 h^{-1} to 1.08 h^{-1} and indoor concentrations of formaldehyde, nicotine, respirable particles, and carbon monoxide did not change significantly.

Neither the cross-sectional nor experimental data provide us with a clear indication of the influence of ventilation rates on indoor concentrations of indoor-generated pollutants. The cross-sectional findings are subject to many confounders. Experimental data are complicated by temporal variations in indoor pollutant emissions. Neither type of study has focused on pollutants with only indoor sources. Therefore, Equation A1 in Appendix A has been used to provide theoretical predictions for particles of various sizes as well as for some types of gaseous pollutants.[11] For predictions, ventilation rates have been varied from 0.2 to 4 h^{-1}, and 0.75 h^{-1} has been used as a reference point for which the concentration is arbitrarily set equal to one. The low end of this range (0.2 h^{-1}) is equivalent to the lowest ventilation rates commonly reported in buildings. Because of energy costs and equipment requirements, increasing ventilation rates to greater than 4 h^{-1} during periods of heating or cooling seems unlikely except in spaces with a high occupant density such as classrooms. A ventilation rate of 4 h^{-1} is a factor of five greater than the arithmetic mean ventilation rate of residences reported by Murry and Burmaster

---

[11]The same equation can be used for many gaseous pollutants, as long as the indoor pollutant emission rate is unaffected by the indoor concentration.

(1995) and corresponds approximately to the total rate of supply of recirculated plus outdoor air in U.S. commercial buildings. The reference ventilation rate of 0.75 h^{-1} is typical of the ventilation rates in residences (Murry and Burmaster, 1995) and also roughly equivalent to the minimum ventilation rates in schools and commercial buildings (Table 10-2).

The influence of ventilation rate on pollutant concentrations depends on the magnitude of pollutant removal by air cleaning. Figure 10-1 provides some predicted relationships for a space with *no* air cleaning. In this figure, relative concentrations are used, which are defined as the actual concentration divided by the concentration at the reference ventilation rate of 0.75 h^{-1}. One clear observation is that the relative concentrations of some indoor-generated pollutants increase dramatically as ventilation rates become unusually low (e.g., below a few tenths of an air change per hour). Because ventilation rates are poorly controlled in buildings, a portion of the building stock will have these low ventilation rates. The airflow requirements and associated energy penalties of avoiding these particularly low ventilation rates are

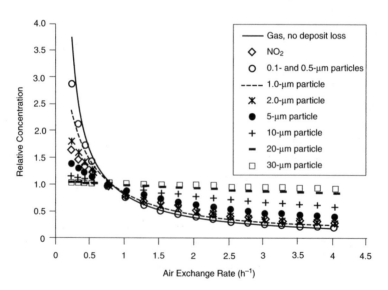

FIGURE 10-1   Predicted trends in the relative concentrations of indoor-generated pollutants with ventilation rate.

modest, but low ventilation rates cannot be prevented with current ventilation technologies and processes (e.g., natural infiltration) that control ventilation rates poorly.

As illustrated in Figure 10-1, the predicted change in pollutant concentrations with ventilation rate is greatest for an "ideal" gaseous pollutant that is not removed by deposition or sorption on surfaces and that has an emission rate unchanged by ventilation rate. In this highly ideal case, often used to illustrate the effects of ventilation, the relative concentration in Figure 10-1 decreases from 3.0 to 0.2 as the ventilation rate increases from 0.2 to 4 h^{-1}. For real pollutants, ventilation rates have a smaller predicted impact on concentrations—in many cases, a much smaller impact. The range in predicted relative concentrations of nitrogen dioxide[12] is 1.6 to 0.3—much less than the range for the ideal pollutant because of losses of NO_2 via sorption on surfaces. If outdoor NO_2 concentrations are significant relative to those indoors, ventilation rates will cause an even smaller change in indoor NO_2 concentrations. For indoor-generated VOCs, the dependence of indoor concentration on ventilation rates is more difficult to predict and will vary among compounds, but concentrations will usually be less affected by ventilation rate than the ideal gaseous pollutant.

For indoor-generated particles, the predicted effects of ventilation rate depend highly on particle size because the depositional losses of particles increase dramatically with particle size. Practical changes in ventilation rates have a substantial predicted impact on indoor concentrations of small particles, such as ETS particles and the portion of airborne cat allergens associated with particles smaller than a couple micrometers. Ventilation rates are not likely to have an appreciable direct[13] impact on indoor concentrations of the larger particles associated with dust mite and

---

[12]To simulate the change in nitrogen dioxide concentrations with ventilation rate, a deposition velocity (see Appendix A) of 7.4×10^{-5} m s^{-1} was used (Nazaroff et al., 1993), and for illustrative purposes, the outdoor NO_2 concentration was neglected.

[13]However, as discussed later, ventilation rates can affect indoor humidity, which in turn modifies the indoor sources of bioaerosols.

cockroach allergens and many indoor-generated fungal spores. For example, in Figure 10-1 the predicted range in relative concentrations of 10-μm indoor-generated particles is only 1.14 to 0.57 as the ventilation rates varies from 0.2 to 4 h^{-1}. If there is some form of air cleaning within the building, ventilation rates will change indoor pollutant concentrations by a smaller amount.

The previous discussion has focused on indoor-generated air pollutants. Outdoor pollutants are not substantially addressed in this document; however, the influence of ventilation rate on indoor concentrations of outdoor pollutants is a relevant issue for asthma. Increasing ventilation rates will increase the rates of entry of outdoor pollutants into buildings. For pollutants that are not removed indoors by deposition on surfaces or air cleaning (e.g., outdoor carbon monoxide), a change in ventilation rate will not substantially affect the time-average indoor concentration, although peak concentrations may be affected. However, an increase in ventilation rate will increase the indoor concentrations of other outdoor pollutants such as particles and ozone from outdoor air.

Indirect Impacts

Ventilation rates may have a very significant indirect impact on indoor concentrations of some pollutants because they affect indoor humidities, which in turn modify indoor pollutant sources. Potentially most important for asthma is the influence of indoor humidity on indoor dust mites, molds, and bacteria.

Two humidity parameters are necessary to characterize the relationship between indoor and outdoor humidity. The humidity ratio is the mass of water vapor divided by the mass of dry air. The relative humidity (RH) is the partial pressure of water vapor divided by the partial pressure of water vapor in air at the same temperature that is saturated with water vapor. The amount of water vapor required to saturate air decreases with decreasing air temperature; hence, 0°C outdoor air with a 100% RH will have a low humidity ratio relative to room-temperature indoor air with a RH of 50%.

Due to indoor moisture generation, the humidity ratio indoors will exceed that outdoors,[14] unless the indoor air or incoming outside air is dehumidified. In cold winter climates, outdoor air is very dry and ventilation rates have a pronounced impact on humidities indoors, with the humidity decreasing as the ventilation rate increases. In warm humid climates, air-conditioning systems must remove moisture from the indoor air to maintain acceptable indoor relative humidities. Increasing the ventilation rate in warm humid climates will increase indoor humidity ratios and relative humidities unless the rate of water removal by the air-conditioning system increases accordingly. This relationship between ventilation rate and humidity places constraints on the ventilation rates in buildings. Particularly high ventilation rates during cold weather causes indoor RHs to decrease to a point (less than ~25%) that causes discomfort and dryness symptoms. Particularly high ventilation rates in warm humid weather overload the moisture-removal capabilities of typical HVAC systems, leading to excessive indoor relative humidities.

Recognizing the relationship of ventilation rate to RH, a few studies have investigated whether indoor dust mite levels or allergy symptoms in dwellings can be controlled by increasing ventilation rates with mechanical ventilation systems (Table 10-3). In cold climates, mechanical ventilation appears to be associated with decreases in indoor humidity and dust mite levels (Emenius et al., 1998; Harving et al., 1994) or allergic symptoms (Aberg et al., 1996). In more humid climates, the results are mixed. Mechanical ventilation was associated with significant reductions of dust mite levels in the small study by McIntyre (1992) in the United Kingdom. However, Fletcher et al. (1996) did not measure significant reductions in dust mites in England and Aberg et al. (1996) did not find significant reductions in allergic symptoms in the more humid areas of Sweden.

Summary

Ventilation rates are poorly controlled in residential and commercial buildings. Measured data indicate that ventilation rates

---

[14]At equilibrium.

TABLE 10-3 Investigations of the Association of Residential Mechanical Ventilation Systems with Indoor Humidities, Dust Mite Levels, and Allergy Symptoms

Reference (Study Type)	Location (Climate)	No. of Buildings
Fletcher et al., 1996 (controlled experiment)	Northwest England (mild, humid)	18 (9 experimental and 9 controls)
Åberg et al., 1996 (survey)	Sweden (cold, dry winters)	1,694 questionnaires and 400 skin prick allergy tests
Emenius et al., 1998 (cross-sectional)	Stockholm area (cold, dry winters)	59
Harving et al., 1994 (experimental study with control group)	Aarhus, Denmark (mild)	32 families moved to houses with mechanical ventilation; 29 controls
Wickman et al., 1994	Suburban Stockholm (cold, dry winters)	70
McIntyre, 1992 (cross-sectional)	South Wales, UK (moderate humid climate)	8 houses (excluding 4 flats)

NOTES: CI = confidence interval; OR = odds ratio.

Increase in Ventilation Rate	Findings
Not assessed	Indoor RHs during autumn and winter were about 10% lower with increased ventilation, but still generally above 50%. No significant effects of increased ventilation on mite counts or mite allergen concentrations ($p > .1$). Reported severity of water condensation indoors diminished.
Not assessed	Mechanical ventilation of the home was associated with a reduced prevalence of allergic disease with relative risk for one or more symptoms of 0.68 (95% CI 0.46–1.00), but only in northern Sweden.
Median ~0.6 h^{-1} with mechanical ventilation and ~0.26 h^{-1} with natural ventilation	Median indoor RH ~9% lower (30 vs. 40%) with mechanical ventilation ($p < .001$); median dust mite allergen in mattress dust ~70% lower with mechanical ventilation ($p < .0001$); median TVOC levels ~60% lower with mechanical ventilation ($p < .0001$).
0.4 h^{-1} in experimental group before move, versus 1.0 and 1.5 h^{-1} in the post-move mechanically ventilated homes. For the control group, the median ventilation rate was unchanged at 0.29 h^{-1}.	Move to mechanically ventilated house associated with ~80% reduction in dust mites in mattress dust of experimental group ($p < 0.05$). The post-move absolute humidity was also significantly decreased in one of two measurement periods. For the control group, humidity did not change significantly and dust mite levels in mattress dust increased nonsignificantly by a factor of ~2. Results of studies of bedroom floor dust were similar.
Not assessed	Presence of mechanical ventilation system was associated with significantly lower indoor humidity ratio (OR = 0.1; 95% CI 0.0–0.2) and significantly lower levels of dust mite allergen in mattress dust (OR = 0.1; 95% CI 0.0–0.5).
Not assessed	Significantly lower humidity ($p < .05$) and ~90% lower dust mite levels in dust ($p < .01$) in 4 houses with continuous mechanical ventilation relative to 4 houses that did not continuously use mechanical ventilation.

vary widely among buildings of the same class (e.g., residences, schools, offices) and that rates of ventilation are frequently below or well above the rates specified in the current ASHRAE ventilation standard. Insufficient data are available for conclusions about changes in building ventilation rates during the past few decades; however, there are some indications that ventilation rates have decreased.

Concentrations of indoor-generated air pollutants associated with asthma may be influenced by ventilation systems via three primary mechanisms. First, changes in the rate of ventilation modify indoor concentrations of indoor-generated air pollutants by changing the rates of pollutant removal and dilution with outdoor air. Measured data and model predictions indicate that the changes in indoor pollutant concentrations associated with the typical range of indoor ventilation rates vary dramatically among pollutants. Because of the sparseness of empirical data, model predictions are the best available means of estimating the typical changes in indoor pollutant concentrations that result from changes in ventilation rates. However, there are considerable uncertainties in the applicability of these predictions to actual complex indoor environments, with rates of pollutant loss by deposition on indoor surfaces being one of the largest sources of uncertainty.

Based on these predictions, the indoor concentrations of some pollutants increase dramatically as ventilation rates become unusually low (e.g., <0.25 h^{-1}). The energy costs associated with avoiding these particularly low ventilation rates are modest; however, current methods and technologies of ventilation do not consistently prevent these low ventilation rates.

Increases in ventilation rates are one obvious approach for reducing exposures to some pollutants associated with asthma. Increasing ventilation rates from a typical value (0.75 h^{-1}) to a high value (4 h^{-1}) should decrease concentrations of indoor-generated particles 1 μm or smaller by up to 80%. Indoor-generated particles in this size range that are relevant for asthma include those associated with ETS, portions of airborne cat allergen, and most of the droplet nuclei produced during coughing and sneezing. Some grass and birch allergens from outdoors are also smaller than 1 μm. As particle size increases, changes in ventilation rates have a diminishing predicted impact on indoor concentrations. For particles 10 μm in size, increasing the ventilation rate from 0.75 to 4 h^{-1} would be expected to decrease indoor concentrations by about 50%. For 20- and 30-μm particles, an even smaller change in concentration with ventilation rate is expected. The limited data available indicates that much of the airborne dust mite and cockroach allergen is associated with particles that are 10 μm or larger in size.

The impact of ventilation rates on concentrations of gaseous pollutants associated with asthma will generally be modest. Without a strong indoor source of nitrogen dioxide, outdoor NO_2 concentrations are often higher than indoor concentrations, and increases in ventilation rates will increase the concentrations indoors. Even when strong indoor sources are present, such that the outdoor concentration is negligible, practical changes in ventilation rates should change indoor nitrogen dioxide concentrations by 50% less. The influence of ventilation rates on indoor concentrations of indoor-generated VOCs will vary among compounds and, for many compounds, is not easily modeled with existing data. In general, concentrations of the more volatile compounds are more strongly affected by ventilation rates.

Overall, these predictions and available data suggest that large increases in ventilation rates would be most effective in reducing exposures to ETS, cat allergens, infectious droplet nuclei, and some volatile organic compounds. Such large increases in ventilation rates, applied broadly, would substantially increase building energy use and energy costs and would significantly increase emissions of the greenhouse gas carbon dioxide. Unless ventilation systems with heat recovery were used, increasing ventilation rates to ~4 h^{-1} in U.S. residences would typically increase annual energy costs from several hundred to more than a thousand dollars. When possible, eliminating or reducing indoor sources of these pollutants would be a much more effective method of reducing exposures and generally would not increase building energy use.

Microorganisms and other pollutants that are potentially relevant for asthma can contaminate HVAC systems and be transported via the system to occupied spaces. There are many sources of liquid water in HVAC systems, as well as regions with very humid air, that facilitate the growth of microorganisms. The contamination of HVAC systems with microorganisms and other pollutants is one hypothesized explanation for the consistent findings that office workers in buildings served by HVAC systems with air conditioning have more nonspecific health symptoms than occupants of naturally ventilated office buildings. However, the relevance of this association to asthma remains unclear. Preventing HVAC system contamination requires improvements in system design, construction, and maintenance.

Ventilation rates also affect indoor humidities, in turn, influencing indoor levels of dust mites. Higher indoor humidities may also increase indoor molds and bacteria. When it is cold and dry outdoors, increases in ventilation decrease the indoor humidity. When it is warm and humid outdoors, increases in ventilation rate tend to increase the humidity in air-conditioned buildings. The limited data available suggest that using mechanical ventilation systems to increase ventilation rates in residences can result in significantly lower dust mite levels only in cold climates.

Most U.S. houses and many schools rely for ventilation on air leakage through unplanned openings in the building envelope plus natural ventilation through windows and doors. Increased window opening will increase ventilation rates, but not in a controlled or predictable manner. Mechanical ventilation systems would have to be installed in these buildings to achieve controlled increases in ventilation rate. Mechanical ventilation systems are commercially available, but they are often expensive (hundreds of dollars to more than a thousand dollars) and unfamiliar to many homeowners.

Several health effects other than asthma, including nonspecific irritation symptoms, allergies, and communicable respiratory illnesses, are potentially influenced by ventilation rates and ventilation system contamination. All of these health outcomes have to be considered when policies and education programs about ventilation are established.

Conclusions Regarding the Relationship of Ventilation to Asthma

Existing data are inadequate for conclusions regarding the

association between ventilation rates or ventilation system microbiological contamination and either the exacerbation of asthma symptoms or asthma development. However, there are both theoretical evidence and limited empirical data indicating that feasible modifications in ventilation rates can decrease or increase indoor concentrations of some of the indoor-generated pollutants associated with asthma by up to 75%.

Research Needs

Additional research is needed on ventilation, but only a small proportion of these research needs are critical to advancing our understanding of the relationship of ventilation to asthma. At the present time, our understanding of the influence of changes in ventilation rates on concentrations of (or exposures to) indoor-generated pollutants associated with asthma is very limited—accordingly, model predictions that have not been adequately evaluated are the best source of information. Experiments in actual buildings, with manipulation of ventilation rates, are the preferred approach for quantifying the direct (i.e., via pollutant removal) influence of changes in ventilation rates on the indoor concentrations of these pollutants. Because indoor pollutant source strengths can vary temporally, experiments should be repeated several times. To assess how changes in ventilation rates affect indoor humidities and, in turn, the proliferation of dust mites and molds in buildings will require either long-term experiments lasting a year or more or large cross-sectional studies with control for confounding factors.

Airtight building envelopes and low rates of ventilation have been cited as factors that may contribute to asthma incidence or symptoms or may explain recent increases in asthma; however, very few relevant data are available. The evidence of a linkage of ventilation rates with asthma is not sufficient to justify large studies intended to resolve only this issue. However, measurements of ventilation rates should be included, when possible, in future asthma case-control studies or cross-sectional surveys. Ventilation measurements in houses can be performed using nonobtrusive tracer-gas methods with passive tracer-gas emitters and samplers (Dietz and Cote, 1982; Stymne and Eliasson, 1991).

Finally, research is needed to advance our very limited current knowledge about microbiological contamination of HVAC systems, its influence on microbial exposures, and its influence on asthma development or asthma symptoms.

PARTICLE AIR CLEANING: INTRODUCTION AND REVIEW OF CONVENTIONAL PRACTICE

Background

Because many of the indoor pollutants associated with asthma are airborne particles, particle air cleaning is considered a potentially beneficial technology for the prevention of asthma or asthma symptoms. Particle air cleaning is any process used intentionally to remove particles from the indoor air. Filtration and electronic air cleaning are the two most common examples. Natural deposition of particles on indoor surfaces, ventilation, and measures that reduce indoor particle emission rates are not considered particle air cleaning.

The magnitude of the reduction in indoor particle concentrations accomplished with particle air cleaning depends on the air cleaner's particle removal rate relative to the particle removal rate by all other processes. The rate of particle removal by an air cleaner varies with particle size and equals the flow rate of air through the air cleaner (Q_{ac}) multiplied by the air cleaner's particle removal efficiency ε. The product of Q_{ac} and ε integrated over particle size for a specific type of particle source is sometimes called the clean air delivery rate (CADR). Based on standard test protocols, the American Home Appliance Manufacturers provides CADRs for tobacco smoke, dust, and pollen for many portable air cleaners.

Although particle air cleaning reduces indoor particle concentrations, microorganisms can grow on some air-cleaning equipment such as filter media; thus, air cleaners are also a potential source of indoor pollutants.

Particle Air-Cleaning Technologies

By far, the most common method of air cleaning is to circulate air through a fibrous filter. The filtration media is usually a mat of thin fibers, often glass fibers. The efficiency of the filter is a function of fiber diameter, fiber packing density (e.g., distance between fibers), thickness of the media, velocity of the air as it passes through the media, particle size, and other factors. A pleated (i.e., folded) filter media is often employed to increase the media surface area, reducing the air velocity in the media, and the resistance to airflow through the media. A wide range of filter products are available, with a correspondingly wide range of efficiencies and prices. Based primarily on data from Hanley et al. (1994) and data from manufacturers, Figure 10-2 provides examples of particle removal efficiency versus particle size for filters with a range of efficiency ratings, using an ASHRAE rating method (ASHRAE, 1992). At one extreme are the coarse panel filters, usually called furnace filters, commonly used in residential forced-air furnace systems. This filter has a negligible efficiency for particles smaller than ~0.5 μm and a low efficiency (e.g., < 20%) for particles smaller than 10 μm. The other extreme is a

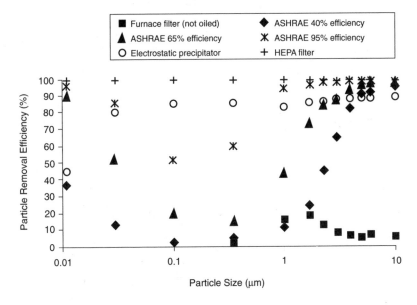

FIGURE 10-2 Efficiency curves for filters and an electrostatic precipitator.

high-efficiency particle air (HEPA) filter, which has a minimum efficiency of 99.97 % for 0.3-µm particles and a higher efficiency for smaller and larger particles. For indoor applications, the efficiency of a HEPA filter is effectively 100% at all particle sizes if all air that flows through the air cleaner actually passes through the filter.

In addition to fibrous filters, a wide range of electronic air cleaners are available. Electrostatic precipitators first produce ions that attach to and electrically charge particles entering the air cleaner. These charged particles then pass through an electric field where they migrate to a surface and attach. The collection surfaces must be cleaned or replaced periodically. An example of an efficiency curve for an electrostatic precipitator is also provided in Figure 10-2. An ion generator is another device that produces ions that attach to and charge particles, but ion generators often have no particle collection surfaces. Instead, ion generators may increase the rate of particle deposition on normal indoor surfaces. A variety of hybrid technologies employ both fibrous filters and particle charging or fibrous filters and electric fields. Electronic air cleaners can produce ozone, sometimes in significant quantities (U.S. EPA, 1999; Viner et al., 1989). Many units have a charcoal filter to remove the ozone produced by the air cleaner.

The efficiency data in Figure 10-2 are from tests with all air passing through a previously unused filter. In practical installations, a portion of the air bypasses the filter, flowing between adjacent filters or between filters and their housing, causing the air cleaner's particle removal efficiency to be lower than the efficiency of the filter media. In commercial installations, gaps of a few centimeters between adjacent filters, even missing filters, are common. Systems of higher-efficiency filters usually have fewer gaps and more gaskets to reduce the quantity of air that bypasses the filter media. Almost no information is available on typical bypass rates. As filters accumulate deposited particles, their efficiency generally increases (sometimes markedly); however, the resistance to airflow through the filter also increases.

The resistance to airflow as air passes through an air cleaner is a common concern of engineers, building operators, and HVAC equipment providers. If airflow rates are constant, more powerful fans and increased fan energy are required when the air pressure drop increases. More efficient fibrous air filters tend to have a greater pressure drop. However, the pressure drop of high-efficiency filters can be diminished by increasing the degree of pleating of the filter matt, which usually increases both the thickness of the filter in the direction of airflow and the filter cost. Additionally, a larger number of filters in parallel can be used to reduce the air velocity and pressure drop. Thus, pressure drop does not necessarily increase with increased filter efficiency; however, the cost of, and space required for, the filters increases unless a greater pressure drop is accepted. Electronic air cleaners tend to have smaller pressure drops than fibrous filters with the same particle removal efficiency.

There are two basic installation options for particle air cleaners. First, air cleaners can be installed within existing HVAC systems and rely on the normal fan-driven airflow through the HVAC system to force air containing particles through the air cleaner. Air drawn from the occupied indoor space flows though a duct system, through the air cleaner, and then back to the occupied space. With this type of installation, often called "in-duct" air cleaning, the rate of air cleaning is limited by the rate of airflow through the HVAC system, and air cleaning occurs only when the HVAC fan operates. For example, an air cleaner in a residential furnace system cleans air only when the residence is heated, unless the furnace fan is forced to operate at other times. One option for improving filtration in buildings is to increase the efficiency of the in-duct filters or air cleaners. The particle removal rate increases only for particles that are removed with a higher efficiency after the filtration upgrade. Hence, substitution of a high-efficiency in-duct air cleaner for a lower-efficiency device may dramatically increase the particle removal rate for small particles but have a small or negligible influence on the removal rate for large particles.

The second option is a supplemental air cleaner with an integral fan that forces airflow through the air-cleaning devices. Normally, these are portable devices placed on the floor, designed to clean the air predominantly in a single room. The addition of portable air cleaners to a building with an in-duct air cleaner increases the total flow rate of filtered air. Portable air cleaners are often a significant source of noise. In some studies, occupants have turned air cleaners off because this noise is annoying. The air flow rate of some portable air cleaners is too small for meaningful air cleaning within a room or house.

Typical Particle Air-Cleaning Practices in U.S. Buildings

In U.S. single-family homes with forced-air heating or air-conditioning systems, a filter designed to remove coarse particles is generally installed in the stream of air circulated through the heating or air-conditioning system. Usually this is a coarse panel filter, such as the furnace filter with an efficiency versus particle size as depicted in Figure 10-2. Recently, filters and electronic air cleaners with a higher removal efficiency for small particles have been designed to replace the standard furnace filter. Standard furnace filters cost only a few dollars or less and should be replaced a few times per year. Higher-efficiency replacements for furnace filters often cost between $10 and $20 per filter.

An unknown proportion of residential heating and cooling systems have an additional in-duct particle air cleaner within the recirculated airstream, often an electronic air cleaner. The price of supplemental in-duct air cleaners for residential applications varies widely (e.g., $500 to $800) among products. The cost of electricity used to operate the electronic components of the air-cleaning system is usually insignificant[15] (e.g., $25 per year). If the recirculation fan is operated only during heating or air conditioning, there is no significant incremental cost for fan energy. Occasionally, home owners will operate the HVAC fan continuously in order to have continuous air cleaning. If 4,380 incremental hours of fan operation (50% of the year), a fan power of 300 W, and an electricity prices of $ 0.098 per kilowatt-hour (kWh) are assumed, the annual incremental fan energy cost would be $130. The rates of airflow through these air cleaners correspond to the rates of airflow through the residential heating and air-conditioning system, with about four indoor air volumes per hour being typical.

---

[15]This excludes the cost of energy used by fans that drive airflow through the air cleaner and associated in-duct heating or air-conditioning system.

In U.S. commercial and institutional buildings with HVAC systems, the system invariably contains a filter in the supply air-stream (i.e., in the mixture of outside air and recirculated indoor air).[16] A typical supply air filter has an efficiency rating of roughly 40% (Figure 10-2);[17] however, the efficiency rating for supply air filters varies widely among buildings. Historically, the rationale for these supply filters has been to reduce the deposition of large particles on the heat-transfer equipment inside HVAC systems. In the past few years, more attention has been directed toward filtration to protect human health. The cost of filters in commercial buildings varies with the product used. Using cost estimates from Burroughs (1997) for filters with a 30% ASHRAE efficiency rating, annual costs per person for a typical level of air filtration in a commercial building are on $4 to $8.[18] Fisk and Rosenfeld (1997) estimated that the annual incremental cost of using very high efficiency filters in an office building was $24 per person. [19]

In a typical commercial HVAC system, the filter is upstream of many of the HVAC components, including the cooling coils, coil drain pans, humidifiers (when present), and sections of duct work that can become contaminated with microorganisms. Thus, the supply filters do not prevent particles released from these components from entering the occupied spaces of the building.

Supplemental portable air cleaners are not standard equipment in any particular type of building; however, they are widely available and used in many buildings, particularly residences. Although a very large range of products is available, many of the heavily marketed products incorporate HEPA filters and a multi-speed fan, with airflow rates at maximum fan speed ranging from 50 to 200 L s^{-1}. This range of flow rates corresponds to 8 to 32 room air volumes per hour in a small 22-m^3 bedroom and to 0.4 to

---

[16]Electronic air cleaners may also be used in commercial buildings, but they are much less common than filters.

[17]The most common efficiency rating is the "dust spot" rating in ASHRAE Standard 52.1-1992 (ASHRAE, 1992).

[18]A total supply flow rate per person of 70 L s^{-1} (150 cfm) was assumed. Estimate includes cost of filters, labor, and energy.

[19]95% efficiency for 0.3 μm particles.

1.6 house volumes per hour in a 450-m^3 house. In practice, units may often be operated with less than maximum airflow or they may be unused, for example, because they are noisy. These portable air cleaners are sized primarily for cleaning the air in single rooms. When the door to the room is open or a forced-air heating or cooling system operates, room air will often mix with air throughout the building and the air-cleaning system must have a larger capacity to obtain the same reduction of particle concentrations within the room. The retail cost of portable air cleaners varies widely. As an example, the retail cost of one of the most commonly used HEPA room filter units with a maximum flow rate of 165 L s^{-1} is about $250 ($1.50 per L s^{-1} maximum airflow).[20] Periodic replacement of pre-filters and HEPA filters in this unit would cost about $70 annually. At maximum fan speed, this unit consumes 350 W of electrical power, with an annual electrical cost of ~$300 if operated continuously. If a product life of four years is assumed, the annualized cost is roughly $400 ($2.4 per L s^{-1} of maximum airflow), with 75% of this cost associated with electricity to operate the fan. The annualized costs per unit airflow of other commercially available products could be considerably higher or lower.

Higher-capacity supplemental air cleaning units, sometimes mounted at ceiling level, are also readily available. These devices are intended primarily for use in health care facilities, smoking rooms, and restaurants, but they could be used in residences. Prices vary widely, and in many instances, there is no obvious relationship between product price and published specifications.

Air Cleaner Standards

Standard test procedures for assessing the efficiency of particle air cleaners are available from ASHRAE (1992). The most commonly cited current test methods yields an "ASHRAE dust-spot efficiency," hereafter called an ASHRAE efficiency, but this standard does not provide an efficiency rating versus particle size.

---

[20]20 room volumes per hour for a 30-m^3 bedroom; 1.3 room volumes per hour for a 450-m^3 house.

Hanley et al. (1994) have provided efficiency curves for typical products. Air cleaner manufacturers also often provide size-dependent efficiency data upon request.

At present, there are no standards that specify the minimum allowable filter efficiency in HVAC systems in U.S. buildings. A proposed revision of the ASHRAE minimum ventilation standard includes a minimum efficiency specification.

Predicted and Measured Influence of Particle Air Cleaning on Indoor Concentrations of Indoor-Generated Particles of Various Sizes

Measurements

Figure 10-3 presents the results of experimental studies of air cleaners that specify both the rate of airflow through the air cleaner and the reduction in indoor pollutant concentration. Some of the studies summarized later in Tables 10-4 and 10-5 also provide a measured reduction in indoor pollutant concentration but no rate of airflow through the air cleaner. The studies specifying the rate of air cleaning are listed at the top of each table in order of decreasing air cleaner flow rate.[21] Most experimental studies have used portable air cleaners with HEPA filters in rooms of homes. Ten studies provide some information on the decrease in airborne allergen or particle concentrations, usually within the room containing the air cleaner, associated with air cleaner operation.[22] Six of these studies reported large or statistically significant decreases in airborne particles or allergens. The reported percentage reductions in particles or allergens in five studies ranged from 30 to 90% and averaged 60%. Overall, these data indicate that air cleaners can significantly reduce airborne allergens or particles in some applications. However, these findings should be interpreted with caution. Temporal variations in indoor allergen

---

[21]Less than 50% of the studies provide information on the rate of air cleaning.

[22]One study checked for changes in dust mite concentrations in dust samples, a parameter unlikely to change as a consequence of short-term air cleaner operation.

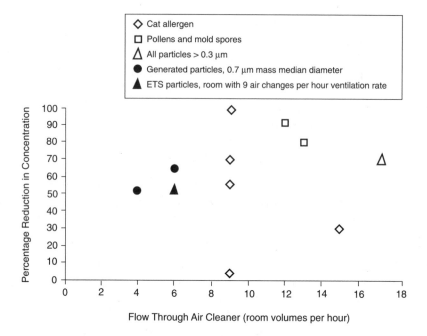

FIGURE 10-3   Measured reductions in indoor pollutant concentrations with air cleaner operation. NOTE: Solid data points are for pollutants that are generated only indoors. SOURCES: Cat allergen—Wood et al. (1998) and Blay et al. (1991); all pollens and mold spores—Chang et al. (1998) and Nelson et al. (1933): all particles > 0.3 μm—Reisman et al. (1990); generated particles—Miller-Leiden et al. (1996); and ETS particles—Bascom et al. (1996).

production or resuspension rates and in outdoor allergen concentrations make accurate experimental assessments of the effects of air cleaning quite difficult. Most of the experimental studies have been short term, thereby increasing the potential errors from natural fluctuations in allergen sources. The studies of de Blay et al. (1991) on the effects of air cleaning on airborne cat allergen concentrations serve as an example of changes in allergen sources. The influence of air cleaner use on airborne cat allergen concentration varied widely (Figure 10-3), with large reductions occurring only if the room was previously vacuum cleaned, presumably because the resuspension of allergen from indoor surfaces was enhanced by the higher indoor air velocities that occur with air cleaning.

A few of the experimental studies provide information on the type of air cleaner needed to reduce concentrations of various types of pollutants. Van der Heide et al. (1997) used multistage filter systems and found that most of the dust mite antigen was removed by the coarse filters—indicating that HEPA filters are unnecessary for the large particles associated with dust mite antigen. Miller-Leiden et al. (1996) determined that HEPA filter units did not perform significantly differently from non-HEPA units for particles with a mass median diameter of 0.7 μm. Offermann et al. (1991) evaluated the effectiveness of six types of air cleaners installed in a forced-air furnace duct system for removal of environmental tobacco smoke. The highest rates of ETS removal occurred using a filter with an ASHRAE efficiency rating of 95%, followed by an electrostatic precipitator, and then by a HEPA filter. The HEPA filter had a higher airflow resistance and thus, a lower airflow rate and smaller particle removal rate.

Predictions

The influence of various particle air-cleaning options on concentration of indoor-generated particles was predicted using the steady-state mass balance equation and assumptions provided in Appendix A. Figure 10-4 provides predictions of the percentage reductions in indoor particle concentrations versus particle size for various air filtration systems installed within the forced-air heating and air-conditioning system of a house. Figure 10-5 provides predictions of the effects of using portable filter units in isolated bedrooms or entire houses. All predictions are for spaces with perfectly mixed air.

Several observations based on these predictions follow. First, to obtain a substantial reduction in indoor concentrations of 10-μm particles, the rate of airflow through the air cleaner per unit of indoor air volume must be high. Even with 10 room volumes per hour through a filter with a 100% efficiency for 10-μm particles, the predicted reduction in concentration is only 70%. The effectiveness of air cleaning would diminish rapidly with increases in particle size above 10 μm because of increases in gravitational

TABLE 10-4   Influence of Air Cleaner Use on Indoor Allergens and Allergy or Asthma Symptoms of Subjects with Perennial Allergic Disease

Reference	Setting	Air Cleaner Type	Air Cleaner Flow Rate (room volumes per hour)
Studies with Information on Air Cleaning Rate			
Criep and Green, 1936	Hospital rooms and homes	Electrostatic precipitator in patient's room or house	25 (assuming 30-m^3 room); 1.6 assuming 450-m^3 house
Reisman et al., 1990	Homes	HEPA	17 (assuming 30-m^3 room); 1.1 assuming a 450-m^3 house
Wood et al., 1998	Homes with cats	HEPA in bedrooms	15 bedroom volumes; 1.3 (assuming 450-m^3 house)
Mitchell and Elliott, 1980	Homes	Electrostatic precipitator in bedroom	9 (assuming 30-m^3 room); 0.6 assuming 450-m^3 house
Warburton et al., 1994	Homes	HEPA in bedroom	2–4 in bedroom
Studies Without Information on Air Cleaning Rate			
van der Heide, 1997	Homes (smokers in 15 of 45)	Portable unit with coarse filter, cyclone, high-efficiency filter	Not specified

Change in Allergen Concentrations	Subjects	Blinded or Placebo?	Reported Change in Health Outcomes
Separate experiment demonstrated large reductions	61 patients with allergies, some with asthma	No	Allergy symptoms improved moderately or completely in 47 of 61 subjects. No statistical test.
70% reduction in particle concentration for particles >0.3 μm	32 allergic patients (children and adults)	Yes	Total study: no significant change in symptom or medication scores. Weeks without respiratory illnesses: several symptom and medication scores improved significantly but no change in asthma.
Airborne cat allergen reduced 30% ($p = .045$)	35 cat-allergic subjects	Yes	No significant changes in symptom scores, peak flow rates, sleep disturbance, or medication use, although a few of these parameters improved nonsignificantly.
Not assessed	10 children with asthma	Yes	No significant change in peak flow.
Nonsignificant decrease in gram-positive bacteria and fungi	12 atopic adult asthmatics	Placebo air cleaners, with foam filters[a]	No significant changes in symptom scores, spirometry, or bronchial hyperreactivity. Nonsignificant trend toward higher peak flow with air cleaner use
Nonsignificant decrease in gram-positive bacteria and fungi	45 non-smoking allergic asthmatic patients	Placebo air cleaners with coarse filters and cyclone[a]	Significant improvement in airway hyperresponsiveness and eosinophils, but only with both air cleaners and impermeable mattress covers. No significant change in peak flow.

Table continued on next page

TABLE 10-4 Continued

Reference	Setting	Air Cleaner Type	Air Cleaner Flow Rate (room volumes per hour)
Bowler et al., 1985	Homes	Electrostatic precipitator and HEPA in bedroom	Not specified, not operated continuously
Friedlaender and Friedlaender, 1954	Homes	Electrostatic precipitator	6 (in 56-m^3 room)
Antonicelli et al., 1991	Homes	HEPA in bedroom	Not specified
Warner et al., 1993	Homes	Ionizer	Not applicable
Studies with Fresh Air Delivery from or near the Headboard of the Bed			
Verall et al., 1988	Homes	HEPA-filtered air from bed's headboard	Not applicable
Zwemer and Karibo, 1973	Homes	HEPA-filtered air from bed's headboard	Not applicable
Villaveces et al., 1977	Homes	HEPA-filtered air directed over pillow	Not applicable
Studies with Air Conditioners			
Trasoff and Blumstein, 1936	Hospital rooms	Air conditioner with filters	8

[a]Placebo air cleaner may have removed some larger allergen particles.

Change in Allergen Concentrations	Subjects	Blinded or Placebo?	Reported Change in Health Outcomes
Not assessed	9 subjects with asthma, age 14–31	Yes	No significant change in symptom score or peak expiratory flow.
Not assessed	18 subjects (16 with asthma)	No	Excellent or moderate improvement in symptoms in 40% of subjects. No controls, no statistical tests.
No significant change in dust mite allergen in bedroom dust samples	9 patients allergic to dust mites	No	No significant improvement in forced expiratory volume or symptom score.
Airborne mite allergen decreased by ~50% ($p < .0001$)	20 children with asthma	Yes	No change in peak flow, symptoms, or medication use.
Not assessed	13 dust mite-allergic asthmatics	Yes	Significantly less medication ($p < .05$) and significant improvement in airway responsiveness ($p = .05$). Crossover study.
Not assessed	12 asthmatic children	Yes (but not fully)	Total symptom score reduced 75%; 300% increase in days of uninterrupted sleep, Lost school days decreased from 15 to 0. No statistical test.
Not assessed	13 asthmatic children	Yes	No significant improvement in peak flow. Significant improvement in symptom scores ($p < .05$).
Not assessed	22 (all with asthma)	No	No clear benefits, relative to patients in other hospital rooms. No statistical test.

TABLE 10-5   Influence of Air Cleaner Use on Indoor Allergens and Allergy or Asthma Symptoms in Subjects with Seasonal Allergic Disease

Reference	Setting	Air Cleaner Type	Air Cleaner Flow Rate (room volumes per hour)
Studies with Information on Air Cleaning Rate			
Criep and Green, 1936	Hospital rooms and homes	ESP	25 (assuming 30-m^3 patient room
Rappaport et al., 1932	Hospital ward	Air filter unit, efficiency not specified	10
Nelson et al., 1933	Hospital wards	Air filter units, no efficiency data	12
Scherr and Peck, 1977	Converted railroad bunk cars	One or two HEPA units	1.5 room volumes per hour per air cleaner
Studies Without Information on Air-Cleaning Rate			
Cohen, 1927	Home and work	Filtered air pressurizes room, no efficiency rating	9 (assuming 30-m^3 rooms)

Change in Allergen Concentrations	Subjects	Blinded or Placebo?	Reported Change in Health Outcomes
In separate experiments, large reductions in artificially introduced allergens	61 subjects with allergies, some with asthma	No	Symptoms improved in 77% of subjects. No statistical test
Not assessed	105 patients with pollen allergy, some with asthma	No?	The 35 patients with hay fever but no asthma experienced complete or moderate relief on 100 of 115 patient nights. Asthma symptoms and medication use persisted in many patients. However, data from 39 patients were excluded because they left after one night or had infections.
Relative to control wards, pollen counts reduced about 90%	51 patients with hay fever, 35 patients with pollen asthma	No	50% of hay fever patients reported great or complete symptom relief. Symptoms in 17 of 35 asthma patients improved or disappeared. No statistical tests.
Not assessed	130 children with asthma at summer camp	Yes	Borderline significant improvement in score for nighttime asthma attacks ($p = .07$).
Not assessed	4 case studies	No	Asthma and allergy symptoms improved in three subjects. Hay fever symptoms improved in one subject. No controls, self-assessment of symptoms, no statistical test.

Table continued on next page

TABLE 10-5 Continued

Reference	Setting	Air Cleaner Type	Air Cleaner Flow Rate (room volumes per hour)
Friedlaender and Friedlaender, 1954	9 homes and 3 in hospital	ESP	6 in 56-m³ room
Kooistra et al., 1978	Homes with central air conditioning	Air cleaner in return duct of heating system, 99% efficiency at 6 µm	Not specified
Studies with Air Conditioners			
Vaughan and Cooley, 1933	Hospitals	Air conditioner without filter, with closed windows	17
Trasoff and Blumstein, 1936	Hospital rooms	Air conditioner with filters	8

NOTE: ESP = electrostatic precipitator.

settling rates with particle size. Thus, air cleaning does not appear to be a very attractive option for reducing exposures to dust mite allergens, which predominantly contain particles larger than 10 µm. The second observation is that increasing the filter efficiency rating from a furnace filter to an ASHRAE 95% efficiency filter, while maintaining a constant rate of airflow through the filter, decreases the predicted indoor concentrations by approximately 40% or less for particles 5 µm or larger. An equivalent reduction in concentrations of particles within this size range is obtained using an ASHRAE 65% filter in place of the ASHRAE 95% filter. Thus, for many of the bioaerosols associated with asthma, high-efficiency filters, which are more expensive and often require larger and noisier fans, are not likely to be superior to moderate-

Change in Allergen Concentrations	Subjects	Blinded or Placebo?	Reported Change in Health Outcomes
>90% reduction in pollen, but not measured in rooms with subjects	12 (6 with asthma)	No	Excellent relief in 75% of subjects and moderate relief in 25%. No controls, no statistical tests.
Allergen concentrations nonsignificantly lower with air cleaner use	20 adults with hay fever, 6 of these with asthma	Yes	14% decrease in nocturnal symptoms, significant only at nighttime. Significant daytime improvement in 10 most sensitive subjects.
99% reduction in pollen compared to room with open windows and no air conditioner	2	No	Moderate improvement in one of two subjects. No controls, no statistical test.
Not assessed	10 (all with asthma)	No	Marked improvement in symptoms in 90% of subjects. No controls, no statistical test.

efficiency filters. Third, it appears feasible to reduce concentrations of particles smaller than 2 μm, such as ETS particles, droplet nuclei, and the smaller particles of cat allergen, by 70% or more using air cleaners with a moderate to high efficiency rating and a flow rate of several indoor air volumes per hour.

Figure 10-6 provides predicted reductions in indoor-generated particle concentrations from the use of typical filters and higher-efficiency filters in a commercial and institutional building HVAC system. These upgrades could substantially reduce concentrations of submicron particles but have virtually no effect on particles larger than a few micrometers.

Portable air cleaners could be used to reduce concentrations of the larger-size indoor-generated particles in the air within com-

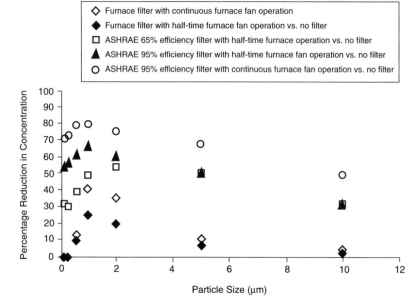

FIGURE 10-4   Predicted reduction in indoor-generated particles in a house with use of various filters in the forced-air heating system if a recirculation flow of four house volumes per hour and a ventilation rate of 0.75 h–1 are assumed.

mercial buildings. As in residences, high rates of airflow through the filters would be necessary to obtain substantial percentage reductions in indoor concentrations. Use of portable air cleaners in isolated individual offices of asthmatics is not likely to be effective unless the office door is kept closed, and even with closed doors the reductions in indoor particle concentrations may be quite small.

Effects of Particle Air Cleaning on Allergy and Asthma Symptoms

The influence of air cleaner use on asthma and allergy outcomes has been evaluated in many experimental studies, and was the subject of a 1997 review by the American Lung Association (ALA, 1997). Building on the review articles of Nelson et al. (1988)

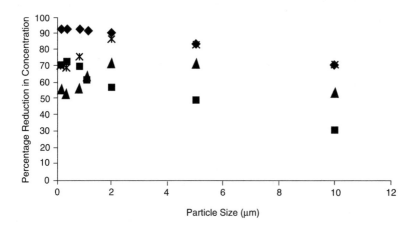

FIGURE 10-5 Predicted reduction in indoor-generated particles from operation of portable air cleaners in an isolated bedroom or an entire house, where a ventilation rate of 0.75 h⁻¹ is assumed.

NOTE: *Isolated bedroom*: bedroom door closed and no air circulation between the bedroom and other rooms. *Furnace filter*: furnace recirculates air throughout the house.

and de Blay et al. (1997a), these studies are summarized in Tables 10-4 and 10-5 for subjects with perennial and seasonal symptoms, respectively.[23]

Many of these experimental studies have important limitations, including a very small number of subjects (e.g., <15), lack of blinding, low or unspecified rates of air cleaning, no information on building ventilation rates, virtually no specification of rel-

---

[23]The committee is using the perennial versus seasonal categorization of Nelson et al. (1988) and de Blay et al. (1997a), but the proper categorization of individual studies is sometimes ambiguous.

380   CLEARING THE AIR

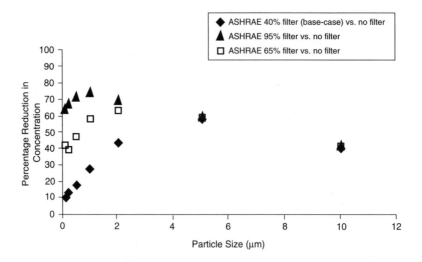

FIGURE 10-6   Predicted reduction in indoor-generated particles from air filtration in a commercial building HVAC system with a recirculation rate of three indoor volumes per hour and a ventilation rate of 0.75 h⁻¹.

evant buildings characteristics,[24] short-term study periods, and reliance on self-reports of symptoms. Many studies assessed changes in allergy symptoms rather than clear indications of asthma symptoms. Some studies selected subjects who were allergic to dust mites, and as discussed above, air cleaning is unlikely to be highly effective in reducing exposures to the large particles that carry dust mite allergens. A minority of the studies quantified the reduction in airborne allergen levels. Many studies reported whether changes in outcomes were significant but did not provide relative risks or odds ratios. Thus, the results of these studies are suggestive, rather than a basis for firm conclusions.

Most of the studies involving subjects with perennial allergic or asthma symptoms (Table 10-4) used portable air cleaners, most often portable HEPA filter units, in bedrooms. Large reductions

---

[24]Important building and building operation characteristics not reported in the literature include the normal position of doors (open or closed) in bedrooms with air cleaners, the presence or absence of carpets, and the type of building heating system (e.g., forced air recirculating system with a filter).

in concentrations of allergens within these bedrooms would not be expected unless bedroom doors were kept closed and forced-air heating or cooling systems did not mix air between the bedrooms and the remainder of the house. Excluding three studies with filtered air delivery over the bed, only 4 of 11 studies reported improvements in symptoms or reduced use of medication, and in 3 of these studies the subjects were not blinded. In one of the four studies (Reisman et al., 1990), symptoms and medication use improved significantly only during the period when patients had no respiratory illnesses. In another of the four studies, only air cleaners used in combination with impermeable mattress covers were associated with significant improvements in airway hyperresponsiveness and eosinophils.[25]

Three experimental studies involving subjects with perennial allergic or asthma symptoms (bottom of Table 10–4) used air-cleaning devices that supplied filtered air over the bed. All three studies reported improvements in outcomes, suggesting that supplying cleaned air to the breathing zone may be more effective than attempting to clean the air in entire rooms or buildings.

The results of studies involving subjects with seasonal allergic or asthma symptoms are much more encouraging. Excluding the two studies using only air conditioning as a form of air cleaning, six of seven studies reported improvements in symptoms, and the seventh study showed a borderline significant improvement ($p = 0.07$). However, subjects were blinded in only two of these studies, and most studies were old and without formal statistical tests.

One additional experimental study (Bascom et al., 1996) performed in an experimental chamber (not included in the tables) assessed the reduction in acute symptoms from ETS exposure when particle air cleaners are used to reduce concentrations of ETS particles. With a 50% reduction in particle concentrations, some symptoms (headache, eye irritation, rhinorrhea) and the minimum cross-sectional area of the nasal passage improved. The relevance of these findings for asthma is not clear.

Overall, these data suggest that air cleaners are probably help-

---

[25]Impermeable mattress covers are intended to reduce exposures to dust mite allergens.

ful in some situations in reducing allergy or asthma symptoms, but air cleaning, as applied in the studies, is not consistently and highly effective in reducing symptoms. It is conceivable that a more consistent or larger reduction in symptoms could be obtained using higher rates of air cleaning. The available data provide no information regarding the effects of air cleaning on the development of asthma or the development of sensitization to allergens.

Air Conditioning as a Substitute for Air Cleaning

Previous reviews of air cleaning and asthma have included studies of air conditioning. Air conditioners can remove some particles from the indoor air because they often contain coarse particle filters and because some particles may be removed from the air along with water in the air conditioner's cooling coil. Consequently, Table 10-5 also summarizes the results of two investigations in the 1930s that evaluated the effects of air conditioning on symptoms. One study demonstrated that the use of an air conditioner and closing windows resulted in a large reduction in indoor pollens. Symptoms improved in one of two subjects. In the second study the air conditioners contained filters, and marked improvements in asthma symptoms were reported for 9 of 10 subjects. Neither study was blind, and neither performed a statistical test to assess the significance of the change in outcomes.

Air conditioners are not designed to remove particles from air. Air conditioning will, in general, be less effective than conventional air cleaning in reducing exposures to indoor-generated particles. However, air conditioners enable the occupants of buildings to keep windows and doors closed during warm weather, greatly reducing the rate of entry of pollens and other outdoor pollutants into buildings. Additionally, air conditioners reduce indoor temperatures and humidities that may influence asthma symptoms.

Home owners and medical doctors sometimes consider residential air conditioning a form of ventilation. In the general mode of operation, residential air conditioners do not provide ventilation—they simply remove heat and moisture (and, to some degree, particles) from a stream of recirculated indoor air. However, to save energy, some modern air conditioners will stop the recirculation and cooling of indoor air and instead provide ventilation (i.e., supply outdoor air to the building) when the outdoor air is relatively cool and suitable for cooling the house.

Summary and Discussion of Limitations of Assessment

Many of the indoor pollutants associated with asthma are airborne particles; thus, particle air cleaning has been considered a potentially beneficial technology for the prevention of asthma or asthma symptoms. Technologies for particle air cleaning are well developed. Air filters with a moderate to high efficiency for particles larger than approximately 2 µm are used routinely in the heating and air-conditioning systems of buildings.

The magnitude of the reduction in indoor-generated particle concentrations accomplished with particle air cleaning depends on the air cleaner's particle removal rate relative to the particle removal rate by all other processes including ventilation and particle deposition on surfaces. The rate of particle removal by an air cleaner varies with particle size and is proportional to the flow rate of air through the air cleaner multiplied by the air cleaner's size-dependent particle removal efficiency. The two primary air cleaning options for reducing indoor particle concentrations are to replace the existing filters in heating and air-conditioning systems with higher-efficiency filters and to operate supplemental air cleaners with integral fans in the occupied space.

In field studies, enhanced air cleaning has been associated with reductions in airborne particle concentrations that range from negligible to more than 90%. For the airborne particles associated with asthma, the published data are very limited. Simple model predictions indicate that substantial reductions in indoor concentrations of 10-µm particles can be obtained only when the rate of airflow through the air cleaner per unit of indoor air volume is large, for example, 10 room volumes per hour or more. The predicted effectiveness of air cleaning diminishes rapidly with increases in particle size above 10 µm because gravitational settling rates increase with particle size. Thus, air cleaning does not appear to be an attractive option for reducing exposures to dust mite allergen, which predominantly involves particles larger than 10 μm. However, based on predictions it is feasible to reduce concentrations of particles smaller than 2 μm, such as ETS particles, droplet nuclei, and smaller particles with cat allergen, by 70% or more using air cleaners with a moderate to high efficiency rating and a flow rate of several indoor air volumes per hour.

Both the available experimental data and model predictions indicate that HEPA filters, which are more expensive and often require larger and noisier fans, are not likely to be superior to lower-efficiency filters in reducing concentrations of many of the bioaerosols associated with asthma. Even for submicron-size ETS particles, available data indicate that HEPA filters are not necessarily the preferred option. Thus, the very common recommendation that HEPA filtration, in contrast to lower-efficiency air cleaning, be used by allergic and asthmatic individuals when they choose to employ air cleaning, is not supported by either experiments or theoretical predictions. Unfortunately, the limited performance data available for many non-HEPA residential air cleaners make it difficult to provide alternate recommendations.

The influence of air cleaner use on asthma and allergy outcomes has been evaluated in numerous experimental studies; however, most of these studies have important limitations. Overall, the data suggest that air cleaners are helpful in some situations in reducing allergy or asthma symptoms, particularly seasonal symptoms, but it is clear that air cleaning, as applied in the studies, is not consistently and highly effective in reducing symptoms. The available data provide no information regarding the effects of air cleaning on the development of asthma or the development of sensitization to allergens.

Conclusion Regarding Air Cleaning and Asthma

There is limited or suggestive evidence that particle air cleaning is associated with a reduction in the exacerbation of asthma symptoms. There is insufficient evidence to determine whether or not the use of particle air cleaners is associated with decreased asthma development. Theoretical and limited empirical data suggest that air cleaners are most likely to be effective in reducing the indoor concentrations of particles smaller than approximately 2 μm. Much of the airborne allergen appears to be within larger particles. Relevant particles smaller than 2 μm include environmental tobacco smoke particles, significant portions of airborne cat, grass, and birch allergen, and virus-containing droplet nuclei from coughs and sneezes.

Research Needs Related to Air Cleaning and Asthma

The results of existing experimental studies are inadequate to draw firm conclusions regarding the benefits of air cleaning for asthmatic and allergic individuals. Many of the existing studies have important limitations, such as small study size, lack of blinding, a small or undefined rate of air cleaning, placebo air cleaners that may significantly remove the larger particles associated with asthma, and no exposure assessment or inadequate assessment. Additional research to assess the benefits of air cleaning is clearly warranted, but future studies must overcome as many of these limitations as possible. Because air cleaning is most promising for reducing indoor concentrations of particles smaller than a couple of micrometers, future research should emphasize these agents.

Sensitization to allergens—a critical step in the development of allergic asthma—often occurs early in life. No information is available to indicate whether air cleaning of spaces occupied early in life can reduce the rate of allergic sensitization. Research is needed to address this issue.

As described in Appendix A, particles larger than a few micrometers have a complex and inadequately understood behavior in the indoor environment, including rapid rates of gravitational settling, resuspension from surfaces, and possibly incomplete mixing with the indoor air. Consequently, the influence of air cleaning systems on exposures to particles in this size range is not well understood and the associated benefits from air-cleaning cannot be predicted with a high degree of confidence. A combination of aerosol science and air-cleaning research is needed to fill this gap in our knowledge.

The limited data on the size distribution of many of the bioaerosols and allergens associated with asthma limit our understanding of the benefits of air cleaning. Additional data are needed particularly for pet allergens and pollens.

As stated earlier, HEPA filter units have been widely recommended for allergy and asthma patients who desire to use air cleaners. Air cleaner manufacturers have responded by aggressively marketing air cleaners with HEPA filters and offering few other products. However, experimental data and theoretical predictions indicate that air cleaners with a lower efficiency rating are likely to be equally effective in reducing the concentrations of most, and perhaps all, of the indoor-generated particles associated with allergies and asthma. These lower-efficiency air cleaners could have a lower product cost, less powerful or noisy fans, higher rates of airflow and particle removal, and reduced energy consumption. The scientific and medical community should develop revised recommendations regarding the selection of air cleaners by allergic and asthmatic individuals, and air cleaner manufacturers should respond by providing new air-cleaning products.

REFERENCES

Åberg N. Sundell J, Eriksson B, Hesselmar B, Åberg B. 1996. Prevalence of allergic diseases in schoolchildren in relation to family history, upper respiratory infections, and residential characteristics. Allergy 51(4):232–237.

American Lung Association. 1997. Residential air cleaning devices: types, effectiveness, and health impact. American Lung Association, Washington, DC.

American Thoracic Society. 1997. American Thoracic Society Workshop, Achieving Healthy Indoor Air. American Journal of Respiratory and Critical Care Medicine 156(Suppl 3):534–564.

Antonicelli L, Bilo MB, Pucci S, Schou C, Bonifazi F. 1991. Efficacy of an air cleaning device equipped with a high efficiency particulate air filter in house dust mite allergy. Allergy 46(8):594–600.

ASHRAE (American Society of Heating, Refrigerating, and Air-Conditioning Engineers). 1992. ASHRAE Standard 52.1-1992—Gravimetric and Dust-Spot Procedures for Testing Air-Cleaning Devices Used in General Ventilation for Removing Particulate Matter. Atlanta, GA: American Society of Heating, Refrigerating, and Air-Conditioning Engineers, Inc.

ASHRAE. 1996. 1996 ASHRAE Handbook: HVAC Systems and Equipment. Atlanta, Georgia: American Society of Heating, Refrigerating, and Air-Conditioning Engineers, Inc.

ASHRAE. 1999. ASHRAE Standard 62-1999. Ventilation for Acceptable Indoor Air Quality. Atlanta, GA: American Society of Heating, Refrigerating, and Air-Conditioning Engineers, Inc.

Bascom R, Fitzgerald TK, Kesavanathan J, Swift DL. 1996. A portable air cleaner partially reduces the upper respiratory response to sidestream tobacco smoke. Applied Occupational and Environmental Hygiene 11(6):553–559.

Batterman SA, Burge H. 1995. HVAC systems as emission sources affecting indoor air quality—a critical review. International Journal of HVAC and Research 1(1):61–80.

Bencko V, Maelichercik J, Melichercikova V, Wirth Z. 1993. Microbial growth in spray humidifiers of health facilities. Indoor Air 3(1):20–25.

Berglund B, Johansson I, Lindvall T. 1982. The influence of ventilation on indoor/outdoor air contaminants in an office building. Environment International 8:395–399.

Bowler SD, Mitchell CA, Miles J. 1985. House dust control and asthma: A placebo-controlled trial of cleaning air filtration. Annals of Allergy 55(3):498–500.

Burroughs HE. 1997. Filtration: an investment in IAQ. Heating, Piping, and Air Conditioning (August):55–65.

Cohen MB. 1927. The prophylaxis and treatment of hay fever and asthma in rooms made pollen and dust free by means of mechanical filters. Journal of Laboratory and Clinical Medicine 13:59–63.

Criep LH, Green MA. 1936. Air cleaning as an aid in the treatment of hay fever and bronchial asthma. Journal of Allergy 7:120–133.

Custovic A, Simpson A, Pahdi H, Green RM, Chapman M, Woodcock A. 1998. Distribution, aerodynamic characteristics, and removal of the major cat allergen *Fel d* 1 in British homes. Thorax 53(1):33–38.

de Blay F, Chapman MD, Platts-Mills TAE. 1991. Airborne cat allergen (*Fel D* 1): Environmental control with the cat in-situ. American Review of Respiratory Disease 143(6):1334–1339.

de Blay F, Sanchez J, Hedelin G, Perez-Infante A, Verot A, Chapman M, Pauli G. 1997a. Dust and airborne exposure to allergens derived from cockroach (*Blattella germanica*) in low-cost public housing in Strasbourg (France). Journal of Allergy and Clinical Immunology 99(1 Pt 1):107–112.

de Blay F, Colas F, Richard MC, Ott M, Verot A. 1997b. Air cleaners and airborne allergens. Journal of Investigational Allergology and Clinical Immunology 7(5):335–337.

Diamond RC. 1999. An Overview of the U.S. Building Stock. In: Indoor Air Quality Handbook. Spengler JD, Samet JM, McCarthy JF, Eds. McGraw Hill, in-press.

Diamond RC, Modera MP, Feustel HE. 1986. Ventilation and occupant behavior in two apartment buildings. In: Proceedings of the 7th Air Infiltration and Ventilation Centre Conference. Coventry, Great Britain: Air Infiltration and Ventilation Centre, pp. 6.1–6.18.

Diamond RC, Feustel HE, Matson NE. 1999. Energy Efficient Ventilation for Apartment Buildings. U.S. Department of Energy, Rebuild America technical guide series. July.

Dietz RN, Cote EA. 1982. Air infiltration measurements in a home using a convenient perfluorocarbon tracer technique. Environment International 8(1–6):419–433.

DOE–EIA. 1994. Commercial Building Characteristics 1992. Report DOE/EIA-0246(92), (U.S. Department of Energy, Energy Information Administration), Washington DC: U.S. Government Printing Office.

DOE–EIA. 1995. Commercial Buildings Energy Consumption and Expenditures 1992. Report DOE/EIA-0318(92), (U.S. Department of Energy, Energy Information Administration), Washington DC: U.S. Government Printing Office.

Dols SW, Persily AK. 1994. Measurements of Outdoor Air Distribution in an Office Building. Report NISTIR 5320. Gaithersburg, MD: National Institute of Standards and Technology.

Emenius G, Egmar A, Wickman M. 1998. Mechanical ventilation protects one-story single dwelling houses against increased air humidity, domestic mite allergens and indoor pollutants in a cold climate region. Clinical and Experimental Allergy 28(11):1389–1396.

Emmerich S, Persily A. 1998. Energy impacts of infiltration and ventilation in U.S. office buildings using multi-zone airflow simulation. Proceedings of IAQ & Energy 98. Atlanta, GA: American Society of Heating, Refrigerating, and Air-Conditioning Engineers, Inc., pp. 191–203.

Fisk WJ. 1999. Estimates of potential nationwide productivity and health benefits from better indoor environments: an update. Lawrence Berkeley National Laboratory Report, LBNL-42123, To be published as a chapter in Indoor Air Quality Handbook. Spengler J, Samet JM, McCarthy JF, Eds. McGraw Hill, in press.

Fisk WJ, Faulkner D. 1992. Air exchange effectiveness in office buildings: measurement techniques and results. In: Proceedings of the 1992 International Symposium on Room Air Convection and Ventilation Effectiveness, July 22–24, Tokyo, pp. 213–223, ASHRAE, Atlanta.

Fisk WJ, Rosenfeld AH. 1997. Estimates of improved productivity and health from better indoor environments. Indoor Air 7(3):158–172.

Fletcher AM, Pickering CAC, Custovic A, Simpsom J, Kennaugh J, Woodcock A. 1996. Reduction in humidity as a method of controlling mites and mite allergens: the use of mechanical ventilation in British domestic dwellings. Clinical and Experimental Allergy 26(9):1051–1056.

Friedlaender S, Friedlaender AS. 1954. Effectiveness of portable electrostatic precipitator in elimination of environmental allergens and control of asthma symptoms. Annals of Allergy 12:419–428.

Gerone PJ, Couch RB, Keefer GV, Douglas RG, Derrenbacher EB, Knight V. 1966. Assessment of experimental and natural viral aerosols. Biological Reviews 30(3):576–588.

Hanley JT, Ensor DS, Smith DD, Sparks LE. 1994. Fractional aerosol filtration efficiency of in-duct ventilation air cleaners. Indoor Air 4(3):169–178.

Harving H, Korsgaard J, Dahl R. 1994. House-dust mite exposure reduction in specially-designed, mechanically ventilated "healthy" homes. Allergy 49(9):713–718.

Hinds WC. 1982. Aerosol Technology. New York: John Wiley & Sons.

Hodgson AT, Rudd AF, Beal D, Chandra S. 1999. Volatile organic compound concentrations and emission rates in new manufactured and site built houses. LBNL-43519, Lawrence Berkeley National Laboratory Report, Berkeley, California. (in draft).

Holmquist L, Vesterberg O. 1999. Quantification of birch and grass pollen allergens in indoor air. Indoor Air 9(2):85–91.

Johnston SL, Pattemore PK, Sanderson G, Smith S, Lampe F, Josephs L, Symington P, O'Toole S, Myint SH, Tyrrell DA, Holgate ST. 1995. Community study of role of viral infections in exacerbations of asthma in 9–11 year old children. British Medical Journal 310(6989):1225–1229.

Kooistra JB, Pasch R, Reed CE. 1978. The effects of air cleaners on hay fever symptoms in air-conditioned homes. Journal of Allergy and Clinical Immunology 61(5):315–319.

Lagus Applied Technologies. 1995. Air Change Rates in Non-Residential Buildings in California. Report P400-91-034BCN, Prepared for the California Energy Commission by Lagus Applied Technology, Inc., San Diego, CA.

Liddament MW. 1996. A Guide to Energy Efficient Ventilation. Document AIC-TN-VENTGUIDE-1996. Coventry, Great Britain: Air Infiltration and Ventilation Centre, International Energy Agency.

Luczynska CM, Li Y, Chapman MD, Platts-Mills T. 1990. Airborne concentrations and particle size distribution of allergen derived from domestic cats (*Felis domesticus*). Measurement using cascade impactor, liquid impinger, and two-site monoclonal antibody assay for *Fel d* 1. American Review of Respiratory Disease 141(2):361–367.

Martiny H, Moritz M, Ruden H. 1994. Occurrence of microorganisms in different filter media of heating, ventilation and air conditioning (HVAC) systems. In: Proceedings of IAQ '94—Engineering Indoor Environments, Atlanta, GA: ASHRAE. pp. 131–137.

McIntyre DA. 1992. The control of house dust mites by ventilation: a pilot study. In: Proceedings of the 13th AIVC Conference Ventilation for Energy Efficiency and Optimum Indoor Air Quality. Air Infiltration and Ventilation Centre, Coventry, Great Britain. pp. 497–507.

Mendell MJ. 1993. Non–specific symptoms in office workers: a review and summary of the epidemiologic literature. Indoor Air 3(4):227–236.

Menzies R, Tamblyn R, Farant JP, Hanley J, Nunes F, Tamblyn R. 1993. The effect of varying levels of outdoor-air supply on the symptoms of sick building syndrome. New England Journal of Medicine 328(12):821–827.

Miller SL. 1996. Characterization and Control of Exposure to Indoor Air Pollutants Generated by Occupants. Ph.D. Dissertation, Civil and Environmental Engineering, University of California, Berkeley.

Miller-Leiden S, Lobascio C, Nazaroff WW, Macher JM. 1996. Effectiveness of in-room air filtration and dilution ventilation for tuberculosis infection control. Journal of the Air and Waste Management Association 46:869–882.

Mitchell EA, Elliott RB. 1980. Controlled trial of an electrostatic precipitator in childhood asthma. Lancet 2(8194):559–561.

Modera MP, Brunsell JT, Diamond RC. 1986. Improving diagnostics and energy analysis for multi-family buildings: a case study. In: Proceedings, Thermal Performance of the Exterior Envelopes of Buildings III, SP 49. Atlanta, GA: ASHRAE. pp. 689–706.

Morey PR. 1988. Microorganisms in buildings and HVAC systems: a summary of 21 environmental studies. In: Proceedings of IAQ'88—Engineering Solutions to Indoor Air Problems. Atlanta, GA: ASHRAE. pp. 10–24.

Morey PR. 1994. Suggested guidance on prevention of microbial contamination for the next revision of ASHRAE Standard 62, In: Proceedings of IAQ'94—Engineering Indoor Environments. Atlanta, GA: ASHRAE, pp. 139–148.

Morey PR, Williams CM. 1991. Is porous insulation inside an HVAC system compatible with a healthy building? In: Proceedings of IAQ'91—Healthy Buildings. Atlanta, GA: ASHRAE, pp. 128–135.

Murry DM, Burmaster DE. 1995. Residential air exchange rates in the United States: empirical and estimated parametric distributions by seasonal and climatic region. Risk Analysis 15(4):459–465.

Nagda N, Koontz M, Lumby D, Albrecht R, Rizzuto J. 1990. Impact of increased ventilation rates on office building air quality. Proceedings of Indoor Air 90(4):281–286.

Nazaroff WW, Gadgil AG, Weschler CJ. 1993. Critique of the Use of Deposition Velocity in Modeling Indoor Air Quality. ASMM Standard Technical Publication 1205. Philadelphia, PA: American Society for Testing and Materials, pp. 81–104.

Nelson HS, Hirsch SR, Ohman JL, Platts-Mills TAE, Reed CE, Solomon WR. 1988. Recommendations for the use of residential air-cleaning devices in the treatment of allergic respiratory diseases. Journal of Allergy and Clinical Immunology 82(4):661–669.

Nelson T, Rappaport BZ, Walker WH. 1933. The effect of air filtration in hay fever and pollen asthma: further studies. Journal of the American Medical Association 100:1385.

Offermann FJ, Hollowell CD, Nazaroff WW, Roseme GD, Rizzuto JR. 1982. Low-infiltration housing in Rochester New York: a study of air exchange rates and indoor air quality. Environment International 8:435–445.

Offermann FJ, Loiselle SA, Sextro RG. 1991. Performance comparisons of six different air cleaners installed in a residential forced air ventilation system. Proceedings of IAQ'91. Atlanta, GA: ASHRAE. pp. 342–350

Orme M. 1998. Energy impact of ventilation. Technical Note 49. Coventry, Great Britain: International Energy Agency—Air Infiltration and Ventilation Centre.

Özkaynak H, Xue J, Spengler J, Wallace L, Pellizzari E, Jenkins P. 1996. Personal exposure to airborne particles and metals: results from the Particle TEAM Study in Riverside, California. Journal of Exposure Analysis and Environmental Epidemiology 6(1):57–78.

Pasanen PO, Pasanen AL, Kalliokoski P. 1995. Hygienic aspects of processing oil residues in ventilation ducts. Indoor Air 5(1):62–68.

Pehkonen E, Rantio-Lehtimaki A. 1994. Variations in airborne pollen antigenic particles caused by meteorologic factors. Allergy 49(6):472–477.

Pejtersen J. 1996. Sensory pollution and microbial contamination of ventilation filters. Indoor Air 6(4):239–248.

Persily AK. 1999. Myths about building envelopes. ASHRAE Journal 41(3):39–47.

Persily AK, Grot RA. 1985. Ventilation measurements in large office buildings. ASHRAE Transactions 91(2A):488–502.

Persily A, Norford L. 1987. Simultaneous measurements of infiltration and intake in an office building. ASHRAE Transactions 93(2):42–56.

Platts-Mills TA, Heymann PW, Longbottom JL, Wilkins SR. 1986. Airborne allergen associated with asthma: particle sizes carrying dust mite and rat allergens measured with a cascade impactor. Journal of Allergy and Clinical Immunology 77(6):850–857.

Rantio-Lehtimaki A, Viander M, Koivikko A. 1994. Airborne birch pollen antigens in different particle sizes. Clinical and Experimental Allergy 24(1):23–28.

Rappaport BZ, Nelson T, Walker WH. 1932. Effect of air filtration in hay fever and pollen asthma. Journal of the American Medical Association 98:1861.

Reisman RE, Mauriello PM, Davis GB, Georgitis JW, DeMasi JM. 1990. A double blind study of the effectiveness of a high-efficiency particulate air (HEPA) filter in the treatment of patients with potential allergic rhinitis and asthma. Journal of Allergy and Clinical Immunology 85(6):1050–1057.

Roberson JA, Brown RE, Koomey JG, Greenberg SE. 1998. Recommended Ventilation Strategies for Energy Efficient Homes. LBNL-40398, Lawrence Berkeley National Laboratory, Berkeley, CA.

Schappi GF, Suphioglu C, Taylor PE, Knox RB. 1997. Concentrations of the major birch tree allergen *Bet v* 1 in pollen and respirable fine particles in the atmosphere. Journal of Allergy and Clinical Immunology 100(5):656–661.

Scherr MS, Peck LW. 1977. The effects of high efficiency air filtration system on nighttime asthma attacks. West Virginia Medical Journal 73(7):144–148.

Seppanen O. 1998. Ventilation strategies for good indoor air quality and energy efficiency. In: Proceedings of IAQ and Energy 98—Using ASHRAE Standards 62 and 90.1. Atlanta, GA: ASHRAE, pp. 257–276.

Seppanen O, Fisk WJ, Mendell MJ. 1999. Association of ventilation rates and CO_2-concentrations with health and other responses in commercial and institutional buildings. Indoor Air 9(4):226–252.

Shapiro-Baruch I. 1993. Evaluation of Ventilation in Multifamily Dwellings. Report 93-5. Albany, NY: New York State Energy Research and Development Authority.

Shaughnessy RJ, Turk B, Casey M, Harrison J, Levetin E. 1997. Use of energy recovery ventilators to provide ventilation in schools and the impact on indoor contaminants. Proceedings of Health Buildings'97 1:161–166.

Shaughnessy RJ, Levetin E, Rogers C. 1998. The effects of UV-C on biological contamination of AHUs in a commercial office building: Preliminary Results. In: Proceedings of IAQ and Energy 98—Using ASHRAE Standards 62 and 90.1. Atlanta, GA: ASHRAE, pp. 229–236.

Sherman MH, Dickerhoff D. 1994. Air-tightness of U.S. dwellings. Proceedings of the 15th AIVC Conference—The Role of Ventilation. Coventry, Great Britain.: Air Infiltration and Ventilation Centre, pp. 226–234.

Sherman MH, Matson N. 1997. Residential ventilation and energy characteristics. ASHRAE Transactions 103(1):717–730.

Spieksma FT, Kramps JA, van der Linden AC, Nikkels BH, Plomp A, Koerten HK, Dijkman JH. 1990. Evidence of grass-pollen allergenic activity in the smaller micronic atmospheric aerosol fraction. Clinical and Experimental Allergy 20(3):273–280.

Spieksma FT, Kramps JA, Plomp A, Koerten HK. 1991. Grass pollen allergen carried by the smaller micronic aerosol fraction. Grana 30:98–101.

Stymne H, Eliasson A. 1991. A new passive tracer gas technique for ventilation measurements. Proceedings of the 12th AIVC Conference: Air Movement and Ventilation Control Within Buildings 3:1–18.

Teijonsalo J, Jaakkola JJK, Sepannen O. 1996. The Helsinki Office Environment Study: air change in mechanically ventilated buildings. Indoor Air 6(2):111–117.

Thatcher TL, Layton DW. 1995. Deposition, resuspension, and penetration of particles within a residence. Atmospheric Environment 29(13):1487–1497.

Trasoff A, Blumstein G. 1936. The value of air–conditioned room in the treatment of seasonal and perennial asthma. Journal of Laboratory and Clinical Medicine 22:147–150.

Tsubata R, Sakaguchi M, Yoshizawa S. 1996. Particle size of indoor airborne mite allergens (*Der p* 1 and *Der f* 1). Proceedings of Indoor Air'96 3:155–160.

Turk BH, Brown JT, Geisling-Sobotka K, Froelich DA, Grimsrud DT, Harrison J, Koonce JF, Prill RJ, Revzan KL. 1987a. Indoor Air Quality and Ventilation Measurements in 38 Pacific Northwest Commercial Buildings, Volume 1: Measurement Results and Interpretation, LBL-22315 $^1/_2$, Lawrence Berkeley National Laboratory, Berkeley, CA.

Turk BH, Grimsrud DT, Harrison J, Prill RJ. 1987b. A Comparison of Indoor Air Quality in Conventional and Model Conservation Standard New Homes in the Pacific Northwest: Final Report. LBNL-23429, Lawrence Berkeley National Laboratory, Berkeley, CA.

Turk BH, Grimsrud DT, Brown JT, Geisling-Sobotka KL, Harrison J, Prill RJ. 1989. Commercial building ventilation rates and particle concentrations. ASHRAE Transactions 95(1):422–433.

Turk B, Powell G, Casey M, Fisher E, Ligman B, Marquez A, Harrison J, Hopper R, Brennan T, Shaughnessy R. 1997. Impact of ventilation modifications on indoor air quality characteristics at an elementary school. Proceedings of Healthy Buildings '97 1:155–160.

U.S. Department of Commerce. 1997. Statistical Abstract of the United States 1997, Washington, DC.

U.S. EPA. 1999. Ozone Generators That Are Sold as Air Cleaners. URL: http://www.epa.gov/iaq/pubs/ozonegen.html. Accessed September 6, 1999.

van der Heide S, Kauffman HF, Dubois AEJ, de Monchy JGR. 1997. Allergen reduction measures in houses of allergic asthmatic patients: effect of air-cleaners and allergen-impermeable mattress covers. European Respiratory Journal 10(6):1217–1223.

Vaughan WT, Cooley LE. 1933. Air conditioning as a means of cleaning pollen and other particulate matter and of relieving pollinosis. Journal of Allergy 5:37–44.

Verall B, Muir DC, Wilson WM, Milner R, Johnston M, Dolovich J. 1988. Laminar flow air cleaner bed attachment: a controlled trial. Annals of Allergy 61(2):117–122.

Villaveces JW, Rosengren H, Evans J. 1977. Use of laminar flow portable filter in asthmatic children. Annals of Allergy 38(6):400–404.

Viner AS, Lawless PA, Ensor DS, Sparks LE. 1989. Ozone emissions from electronic air cleaners. Paper 89-84.1, presented at the 82nd Annual Meeting of the Air and Waste Management Association, Anaheim, CA.

Warburton CJ, Niven RM, Pickering CAC, Fletcher AM, Hepworth J, Francis HC. 1994. Domiciliary air filtration units, symptoms and lung function in atopic asthmatics. Respiratory Medicine 88(10):771–776.

Warner JA, Marchant JL, Warner JO. 1993. Double-blind trial of ionizers in children with asthma sensitive to the house dust mite. Thorax 48(4):330–333.

Wickman M, Emenius G, Egmar AC, Axelsson G, Pershagen G. 1994. Reduced mite allergen levels in dwellings with mechanical exhaust and supply ventilation. Clinical and Experimental Allergy 24(2):109–114.

Wood RA, Johnson EF, Van Natta ML, Chen PH, Eggleston PA. 1998. A placebo-controlled trial of a HEPA air cleaner in the treatment of cat allergy. American Journal of Respiratory and Critical Care Medicine 158(1):115–120.

Zwemer RJ, Karibo J. 1973. Use of a laminar flow device as adjunct to standard environmental control measures in symptomatic asthmatic children. Annals of Allergy 31(6):284–290.

11
SUMMARY OF RESEARCH RECOMMENDATIONS AND OVERALL *Conclusions*

T his chapter summarizes the research recommendations offered by the committee in the preceding chapters and draws some general conclusions regarding the report's findings.

PATHOPHYSIOLOGIC BASIS OF ASTHMA

Collectively, the evidence reviewed by the committee underscores the need to further identify the mechanistic interplay between specific inflammatory cells, cell adhesion molecules, and the changes in airway function that characterize the asthmatic condition. It emphasizes that much more research, based on combining data from genetic analyses with those identifying pathophysiological processes involved in asthma, is needed to ultimately determine the genetic basis of asthma, as well as the potential development of new strategies for therapeutic intervention.

ANIMAL ALLERGENS

The associations between dust mite allergen, asthma exacerbation, and asthma development are much more well defined than the associations between larger animals and asthma. This is only partly a function of the number of years and intensity of

394

efforts to investigate the health effects of dust mites. Compared with dust mite allergen, once the allergen source is present, cat and dog allergens are more easily dispersed throughout the household. Cat and dog allergens remain airborne for much longer than dust mite. The potential for exposure to allergens outside the home is markedly greater for cats and dogs than for dust mites. The potential for home exposure to cat or dog allergen in homes without cats or dogs has also been underestimated. The absence of adequate information regarding allergen exposure may, in part, account for contradictory data regarding the effects of cats or dogs in the home on the development of asthma.

Research is needed to assess whether removal of the cat or dog from the home results in sufficient reduction in overall allergen exposure to reduce symptoms and improve bronchial reactivity in specifically sensitized asthmatics. Further research is needed to assess the level of animal allergen exposure (cat and dog) in day care centers and schools. When significant levels are noted, the potential for lowering exposure should be investigated. Since so many cat- or dog-allergic asthmatics are emotionally attached to their pets, investigators should explore the success of efforts that recommend the removal of the pet for sensitized symptomatic child and adult asthmatics. Further research is also needed to evaluate the effect of mitigation measures short of animal removal on asthma symptoms, lung function, or bronchial responsiveness in specifically sensitized asthmatics. Although frequent animal washing and HEPA filter use are widely recommended, their efficacy in reducing asthma severity has not been proven.

The relationship between cat or dog allergen exposure in early childhood, the development of sensitization, and the development of asthma merits further investigation. This investigation will require better assessment of exposure. It is likely that the genetic phenotype will modify the response to cat or dog allergen at different levels of exposure, but gene-by-environment interactions cannot be effectively explored until the genetics of asthma is better understood.

Further research is needed to evaluate rodent allergen exposure in the home as a potential factor in the exacerbation of asthma in rodent-sensitized asthmatics. Particularly in socially disadvantaged populations, research should focus on effective reduction of rodent allergen and its effect on symptoms or lung function in specifically sensitized asthmatics.

Researchers should also consider the possibility that animal (or animal allergen) exposure may be either protective or allergenic. The effects may depend on the mode of exposure, the genetic characteristics of the populations, the timing in the life cycle when exposure occurs, and many other cofactors (e.g., early-life viral, bacterial, and parasitic infection experience).

COCKROACH

The committee's review of the literature suggests there is still a need for fundamental research on cockroach allergens and asthma outcomes. Future research should focus on the efficacy of cockroach allergen reduction in the homes of asthmatic patients, the aerodynamic properties of cockroach allergen, the efficacy of cockroach immunotherapy, and B and T cell reactive epitopes. Further studies are also needed to better elucidate any relationship between cockroach allergen exposure and asthma development; explore the interaction of cockroach allergen with infectious agents, irritants, and other allergens in causing asthma; and examine the influences of genetics, socioeconomic status, and location on exposure and sensitization.

HOUSE DUST MITES

Although more is known about dust mite allergen and its impact on asthma than most indoor exposures, research remains to be done. Particularly important is additional work on the effectiveness of specific environmental interventions in limiting asthma exacerbations and development (rather than simple measurement of allergen levels). Several studies now under way are evaluating whether aggressive allergen avoidance regimes have an effect on the subsequent development of asthma. The results of such studies will inform the question of whether primary prevention of dust mite-induced asthma is possible, although the burdensome nature of such interventions suggests they may be difficult to implement in many circumstances. The development of methods to identify individuals, especially infants, at high risk would provide the information needed to focus primary prevention activities. A major issue in this regard is whether sensitization can occur before birth.

ENDOTOXIN

Given the significant body of data on the exquisite sensitivity of the innate immune system to small quantities of endotoxin, the hypotheses that domestic endotoxin exposure may influence the development of the immature immune system or affect the severity of asthma warrant further investigation.

The committee's review suggests several avenues of research directed at understanding the role of endotoxin exposure and endotoxin susceptibility in the pathogenesis of asthma. These include studies of gene–environment interactions and the risk of developing atopy or asthma, preferably with prospective assessment of endotoxin exposure from birth, improved endotoxin exposure assessment across populations likely to have significant differences in exposure, and studies of endotoxin exposure and asthma severity.

Gene–environment interactions between the CD14 polymorphism and endotoxin exposure should take into account that CD14 is a pattern receptor and thus not specific for LPS–LBP complexes. Thus, future studies should include an assessment of exposure to other bacterial products that stimulate innate immunity via CD14 such as peptidoglycan. Prospective studies will be required to determine whether endotoxin exposure early in life plays a role in determining the direction of immune system development. Studies that can compare populations with possibly larger variations in exposure to endotoxin and other components of organic dusts than can be found within an urban or suburban area would likely have increased power to detect the effects of endotoxin exposure. Because the CD14 polymorphism is associated with atopy, a focus on specific and nonspecific IgE and TH phenotypes will likely be the most important variables for these studies.

Given that the *Limulus* bioassay has limitations and that "unusual" lipid A structures dominate the composition of house and other organic dusts, additional exposure assessment methods that can detect the range of environmental LPS should be employed along with *Limulus* assays in future studies. The possibility that endogenous sources of endotoxin exposure may be important in modulating the level of tolerance to environmental exposure (or vice versa) should also be examined.

FUNGI

Few fungal allergens have been identified, and patterns of cross-reactivity among fungal allergens have not been documented. Standardized methods for assessing exposure to fungal allergens are essential, preferably based on measurement of allergens rather than culturable or countable fungi. Acquisition of these data is a necessary step before adequate estimates of the role of fungal allergen in asthma can be documented.

Studies seeking to find environmental factors that either lead to the development of asthma or precipitate symptoms in existing asthmatics must include good measures of fungal exposure. No studies have attempted to control exposure to fungal allergens either indoors or out. Intervention studies that seek to control indoor exposure to fungi are especially needed.

INFECTIOUS AGENTS

Numerous studies suggest an association between the infections discussed in Chapter 5 and asthma exacerbations, although uncertain ascertainment of asthma and questions about the identity of the specific infections responsible limit the confidence with which some conclusions can be drawn. Advances in analysis techniques that allow more sensitive and confident identification of viruses, such as PCR and ELISA, will facilitate research on this topic. These advances will also aid studies of other viruses that may be associated with asthma such as adenovirus, coronavirus, cytomegalovirus, and parainfluenza.

Research on the possible association between infectious agents and asthma development is continuing and is encouraged. There are gaps in the knowledge concerning the mechanism(s) by which agents may promote asthma and whether particular interventions aimed at limiting infections result in decreased rates of asthma. Among the interesting questions are whether the lower respiratory tract acts as a potential reservoir for common respiratory viruses and whether maternal immunization has the potential to protect both the mother and the infant. Research on the impact of building characteristics on the transmission of infectious agents, which is in its infancy, may yield important public health benefits.

PLANTS

Further research is needed to determine whether or not houseplants release fungal spores into the air. This research will benefit both the allergy community and the infectious disease literature. Additionally, research should be conducted to determine what risks, if any, are associated with occupational exposure to plant materials.

Studies should be conducted to evaluate the ambiguous relationship between pollen exposure, sensitivity, and asthma. Additional research is also needed to discover the extent of indoor pollen allergen exposure and the interactions between pollen sensitivity and air pollutants.

NITROGEN DIOXIDE (NO₂)

Future research into the relationship between indoor NO_2 and asthma should target population subgroups that are likely to be most exposed. These include women and infants, especially those at the lower end of the socioeconomic spectrum, who may spend more time in kitchens during cooking. Further research is needed on the distributions of peak and mean personal exposures, on the relationship between exposure and asthma, and on the exacerbation of asthma among those with preexisting asthma. Research is also needed into the development of practical, economical ventilation methods for kitchens.

PESTICIDES

The most immediate need for research in this area is for information on whether prolonged exposures to low to moderate concentrations of airborne pesticides can elicit an irritative asthma response in susceptible individuals. Animal models may be helpful in this regard—a presently unpublished study on a microbial biopesticide used against cockroaches suggests it can induce asthma in mice. It has been suggested that Flinders Sensitive Line rats, which exhibit increased responses to an agent similar to commonly used organophosphate pesticides, may be useful in the study of asthma outcomes.

PLASTICIZERS

Research characterizing residential exposure to plasticizer and evaluating asthma outcomes while controlling for known confounders would help resolve the question of their influence.

VOLATILE ORGANIC COMPOUNDS

With the advent of small, unobtrusive passive diffusion monitors capable of measuring microgram quantities of VOCs in 48-hour samples, it should be possible to incorporate personal VOC exposure assessment into future epidemiologic studies addressing environmental factors and asthma, yielding an expanded data base on which to judge the possible role of VOCs in asthma development and exacerbation.

Prospective cohort studies that characterize time-averaged personal VOC exposures of study subjects using passive badges and then follow subjects to assess the development and/or exacerbation of asthma could generate information that would allow a more confident assessment of any potential risk.

No specific asthma-related research is recommended for formaldehyde, a member of the VOC family.

FRAGRANCES

Future research on fragrance exposures has to carefully control for confounding factors such as odor perception and to focus on objective measures of respiratory health. Additional research elucidating the mechanism or mechanisms underlying adverse respiratory reactions to nonacute exposure to fragrances or their component chemicals would also be helpful.

ENVIRONMENTAL TOBACCO SMOKE (ETS)

A better understanding is needed of the mechanism(s) by which ETS and its individual constituents may

- impair the normal development of the airways in the fetus,
- promote allergic sensitization,
- promote respiratory infections,
- promote early wheezing illness, and
- (possibly) induce pathophysiologic changes that may promote the establishment of asthma.

Research is also needed to understand the nature of the interactions, both at the population or epidemiologic level and at the molecular and cellular levels, between the genetic predispositions to allergic sensitization and bronchial hyperresponsiveness and ETS exposure as they relate to the development of asthma. The respective roles of antenatal and postnatal exposure to ETS in the pathophysiologic changes associated with asthma and other respiratory illnesses are in need of further investigation.

Behavioral research also is needed to better understand the factors that lead to the initiation of smoking in adolescents, especially young women, and to the maintenance of smoking in pregnant women and mothers. Additionally, there is a need to develop more effective interventions to achieve sustained pre- and postnatal smoking cessation in pregnant women and mothers, especially those whose children are at higher risk of developing asthma due to their family history, socioeconomic status, and place of residence. Since ETS exposure of children at greatest risk for adverse asthma outcomes (especially low-income and minority children of African-American ancestry) may come from other caregivers as well as the mother or parents (i.e., other family members with whom the mother and child live and from day care providers), interventions must be developed that will be effective in reducing the child's exposure from all sources. The effectiveness of ETS exposure reduction interventions in actually improving asthma outcomes should be evaluated as well.

INDOOR DAMPNESS, MOISTURE PROBLEMS, AND MOISTURE CONTROL

With respect to the association of dampness problems with asthma development and symptoms, research is needed to clearly identify the causative agents (e.g., molds, dust mite allergens) and to document more accurately the relationship between dampness and allergen exposure. Research is also needed to characterize and demonstrate the reductions in asthma morbidity from prevention or remediation of moisture problems.

Regarding characterization of moisture problems, research is needed to:

1. develop accurate, standardized protocols for assessing moisture problems in buildings;
2. develop and document the effectiveness of specific measures for dampness reduction in existing buildings; and
3. develop standardized, effective protocols for flood cleanup that will limit microbial growth.

In addition to these research needs, there is a need for improved education of the public about the consequences of moisture problems and for better education of building professionals regarding means of preventing moisture-related problems.

NONRESIDENTIAL INDOOR ENVIRONMENTS

The few available studies suggest the importance of building factors in relation to asthma, but further research is critical to assessing the attributable risks, remediable risk factors, and means of hazard assessment. Development of methods for representative quantitative assessment of bioaerosols of fungal and bacterial origin is a high priority for health outcome studies and hazard assessment. In addition, knowledge of the epidemiology of building-related asthma in problem buildings where there are excess chest complaints among occupants, in comparison to buildings where there are no complaints, can advance our understanding of specific bioaerosols in relation to asthma. Research should focus on exposure–response studies of many building environments and populations; clinical investigation of patients with building-related asthma; and intervention studies, even without knowing the specific etiology involved. This research agenda requires new partnerships among academic investigators, clinicians, public health agencies, industrial hygienists, and building scientists.

VENTILATION

Additional research is needed on ventilation, but only a small proportion of these research needs are critical to advancing our understanding of the relationship of ventilation to asthma. At the present time, our understanding of the influence of changes in ventilation rates on concentrations of (or exposures to) indoor-generated pollutants associated with asthma is very limited—accordingly, model predictions that have not been adequately evaluated are the best source of information. Experiments in actual buildings, with manipulation of ventilation rates, are the preferred approach for quantifying the direct (i.e., via pollutant removal) influence of changes in ventilation rates on the indoor concentrations of these pollutants. Because indoor pollutant source strengths can vary temporally, experiments should be repeated several times. To assess how changes in ventilation rates affect indoor humidities and, in turn, the proliferation of dust mites and molds in buildings will require either long-term experiments lasting a year or more or large cross-sectional studies with control for confounding factors.

Airtight building envelopes and low rates of ventilation have been cited as factors that may contribute to asthma incidence or symptoms or may explain recent increases in asthma; however, very few relevant data are available. The evidence of a linkage of ventilation rates with asthma is not sufficient to justify large studies intended to resolve only this issue. However, measurements of ventilation rates should be included, when possible, in future asthma case-control studies or cross-sectional surveys. Ventilation measurements in houses can be performed using nonobtrusive tracer-gas methods with passive tracer-gas emitters and samplers.

Finally, research is needed to advance our very limited current knowledge about microbiological contamination of HVAC systems, its influence on microbial exposures, and its influence on asthma development or asthma symptoms.

AIR CLEANING

The results of existing experimental studies are inadequate to draw firm conclusions regarding the benefits of air cleaning for asthmatic and allergic individuals. Many of the existing studies have important limitations, such as small study size, lack of blinding, a small or undefined rate of air cleaning, placebo air cleaners that may significantly remove the larger particles associated with asthma, and no exposure assessment or inadequate assessment. Additional research to assess the benefits of air cleaning is clearly warranted, but future studies must overcome as many of these limitations as possible. Because air cleaning is most promising for reducing indoor concentrations of particles smaller than a couple of micrometers, future research should emphasize these agents.

Sensitization to allergens—a critical step in the development of allergic asthma—often occurs early in life. No information is available to indicate whether air cleaning of spaces occupied early in life can reduce the rate of allergic sensitization. Research is needed to address this issue.

As described in Appendix A, particles larger than a few micrometers have a complex and inadequately understood behavior in the indoor environment, including rapid rates of gravitational settling, resuspension from surfaces, and possibly incomplete mixing with the indoor air. Consequently, the influence of air cleaning systems on exposures to particles in this size range is not well understood and the associated benefits from air-cleaning cannot be predicted with a high degree of confidence. A combination of aerosol science and air-cleaning research is needed to fill this gap in our knowledge.

The limited data on the size distribution of many of the bioaerosols and allergens associated with asthma limit our understanding of the benefits of air cleaning. Additional data are needed particularly for pet allergens and pollens.

As stated in Chapter 10, HEPA filter units have been widely recommended for allergy and asthma patients who desire to use air cleaners. Air cleaner manufacturers have responded by aggressively marketing air cleaners with HEPA filters and offering few other products. However, experimental data and theoretical predictions indicate that air cleaners with a lower efficiency rating are likely to be equally effective in reducing the concentrations of most, and perhaps all, of the indoor-generated particles associated with allergies and asthma. These lower-efficiency air cleaners could have a lower product cost, less powerful or noisy fans, higher rates of airflow and particle removal, and reduced energy consumption. The scientific and medical community should develop revised recommendations regarding the selection of air cleaners by allergic and asthmatic individuals, and air cleaner manufacturers should respond by providing new air-cleaning products.

OVERALL CONCLUSIONS

This report represents a summary of the best available evidence the committee could find from a wide variety of sources including: published peer reviewed scientific studies, presentations by recognized experts, and personal knowledge. In many cases the conclusions the committee reached were neither as firm nor as broad as the committee would have liked, primarily because of limitations in the available information. Many of the published studies reviewed by the committee suffered from potential problems such as small numbers of study subjects, subjects who were not representative of large segments of the population, or studies evaluating the effect of only a single allergen or pollutant rather than the multiple agents encountered by most persons. The lack of large, well-conducted, epidemiologic studies made it difficult for the committee to make judgments about the relative impact or proportion of asthma caused by different agents. Even when suggestive information was available, it was often difficult to determine whether the effects of an agent would be the same in different areas of the country since there is little information concerning regional variations in exposure to various indoor agents that may be related to asthma outcomes. The indoor air problems that are the greatest concern in the northern U.S. are in many cases different from those found in the South or West. Similarly, indoor air problems are likely to be quite different in cities compared to suburbs or rural areas. More comprehensive studies are needed to resolve many important issues concerning the health effects of indoor air.

The report's review of the epidemiology and etiology of asthma reflects the complex nature of the disease. While a number of indoor exposures are or may be associated with asthma, none is by itself a necessary cause, since asthma is associated with other factors. Each factor is probably not a sufficient cause, since each likely functions through an interaction with other environmental agents and with genetic factors.

Nevertheless, for some of the factors discussed, it seems possible and reasonable to institute mitigation with a reasonable hope of reducing the rate of asthmatic attacks. However, often it is not known what degree of mitigation would be necessary to reduce the risk of attacks in known asthmatic individuals. Mitigation generally involves two components: reducing the existing levels of exposure and preventing any increases in exposure as a result of further introduction of the suspected agent into the environment. The level of effort that should be recommended for mitigating a given factor will depend on how clear the evidence both of association or causation and of the effectiveness of mitigation is, as well as the importance attached to the problem of asthma by the individual, his family, and community.

Clearly, individual sensitivity is a major consideration. The factors that lead to *the development of asthma* and *sensitivity* are generally not well understood. Most of the factors discussed play a role in *exacerbating* existing asthma in sensitive individuals.

A major problem in choosing and implementing an intervention to mitigate an exposure is the generally limited data available. The limitations exist in regard to both the quantity and the quality of research data. Many of the studies reported are not based on rigorous protocols. Definition of clinical outcome especially in infants, measurement of exposure, rigorous study design, appropriate population selection, and dealing with issues of generalizability of the findings are often not adequately addressed. The distinction between association and causation is often not clear. It has proven difficult to assess the individual roles of the factors implicated and to determine whether any effects noted are indeed the results of the specific exposures studied or of confounders. Furthermore, the interaction of different environmental exposures with genetic susceptibilities must be elucidated.

Although considerable work has been done and is being done on asthma per se, increased research efforts are needed to address the characteristics of healthy indoor environments. Asthma research clearly needs interdisciplinary involvement—not only of clinicians, immunologists, and researchers in related biologic areas—but also of engineers, architects, materials manufacturers, and others who are responsible for the design and function of indoor environments.

APPENDIX A
THEORETICAL CONSIDERATIONS RELEVANT TO THE INFLUENCE OF VENTILATION AND AIR CLEANING ON EXPOSURES TO INDOOR-GENERATED POLLUTANTS

Because limited empirical data are available for quantifying the effects of ventilation and air-cleaning rates on concentrations of indoor-generated pollutants (particularly particles), a steady-state mass balance model for an indoor space with perfectly mixed air has been used to generate the estimates (predictions) of these effects that are provided in Chapter 10. This appendix presents the mass balance equation and provides a brief review of the behavior of indoor particles.

Since only indoor-generated pollutants are being considered, the only source is the indoor pollutant generation. At equilibrium, the indoor pollutant generation rate equals the total rate of pollutant removal by ventilation, natural deposition on surfaces, and air cleaning.

For particles of size i, the mass balance equation is

$$G^i = \lambda_{vent} C^i + v^i_{dep} A / V C^i + (Q_{ac} / V) \varepsilon^i_{ac} C^i, \qquad (A1)$$

where G^i is the indoor particle generation rate per unit volume, λ_v is the air exchange rate equal to the rate of outside air entry Q divided by the indoor air volume V, C^i is the indoor particle concentration, v^i_{dep} is the particle deposition velocity, A is the indoor surface area, Q_{ac} is the rate of airflow through the air cleaner, and

ε^i is the efficiency of the air cleaner in removing particles of size i. Equation A may be solved for C^i and the dependence of C^i on air exchange rate, rate of airflow through the air cleaner, and particle removal efficiency examined for a range of particle sizes. The air cleaner's particle removal efficiency ε^i is defined by the equation

$$\varepsilon^i = 1 - \left[C^i_{outlet} / C^i \right] \qquad (A2)$$

with C_{outlet} equal to the concentration exiting the air cleaner. For many gaseous pollutants, an identical mass balance equation applies, except that there is no size variation and the deposition velocity is zero for some nonreactive gases. However, Equation A1 does not apply when the steady-state indoor pollutant emission varies with the indoor air pollutant concentration—the case for some volatile organic compounds (VOCs). Many VOCs may also adsorb on indoor surfaces and later desorb when the indoor VOC concentrations are lower; thus, equilibrium conditions may rarely be achieved.

INDOOR AEROSOL BEHAVIOR

An important source of uncertainty in the application of Equation A1 is the limited current knowledge about the rate at which particles are removed from indoor air by deposition on the surfaces inside buildings. For the largest particles, greater than a few micrometers in diameter, depositional losses are dominated by rates of gravitational settling which can be calculated with reasonable accuracy if the particle characteristics are known. For unit-density (1 g/cm^3) particles of various sizes, Table A-1 lists gravitational settling velocities from Hinds (1982) and the time for particles to fall 1 m, approximately one-half the distance between floor and ceiling in a typical room. Gravitational settling can become the dominant particle removal process for large particles, such that practical changes in ventilation and air-cleaning rates change indoor particle concentrations only marginally. For submicron particles (i.e., particles smaller than 1 μm), gravitational settling is unimportant. Typical indoor air velocities are 0.05 to 0.3 m s^{-1}; thus, indoor particles may be transported a considerable distance before settling on surfaces.

TABLE A-1 Theoretically Predicted Rates of Gravitation Settling of Unit Density Spheres in Still Air

Aerodynamic Diameter (μm)	Terminal Settling Velocity (m/s)	Time to Fall 1 m
0.2	2.2×10^{-6}	5.3 day
0.5	1.0×10^{-5}	28 h
1.0	3.5×10^{-5}	7.9 h
5.0	7.8×10^{-4}	21 min
10	0.003	330 s
20	0.012	83 s
30	0.027	37 s
50	0.075	13 s

Particles also collide with and deposit on indoor surfaces due to mechanisms other than gravitational settling, and the deposition rates may vary with indoor air velocities, turbulence intensity, and surface roughness. As larger particles are transported by indoor air motion, their momentum causes collisions with surfaces. The random Brownian motion of the smallest particles, produced by collisions with gas molecules, increases the rate of particle collisions with surfaces. Electrostatic forces and other phenomena also affect particle deposition rates. Particles in an intermediate size range, about 0.2 or 0.3 μm, have the lowest deposition rates because they have neither sufficient momentum nor Brownian motion for high deposition rates.

As illustrated in Equation A1, the particle deposition velocity is used to characterize the rate of loss of particles to surfaces. The deposition velocity is the net flux of particles to indoor surfaces divided by the indoor particle concentration. For predictions, average particle deposition velocities from the available experiments in room-size enclosures (Lai and Nazaroff, 1999) and from the settling rate of large particles have been used. These numbers are considered the best presently available; however, the uncertainties about typical values of deposition velocity for particles smaller than a few micrometers are large for some particle sizes. A ratio of surface area to volume, A/V, of 2.7 was used for the predictions based on information in the literature and calcula-

TABLE A-2 Deposition Velocities and Corresponding
Deposition Coefficients Used in Calculations

Aerodynamic Diameter (μm)	Deposition Velocity (m/s)	Deposition Coefficient with A/V of 2.7 m^{-1} (h^{-1})
0.05	3×10^{-5}	0.3
0.1	1×10^{-5}	0.1
0.2	6×10^{-6}	0.05
0.5	1×10^{-5}	0.1
1.0	2×10^{-5}	0.2
2.0	5×10^{-5}	0.5
5.0	1.3×10^{-4}	1.3
10*	3.7×10^{-4}	3.6
20*	1.5×10^{-3}	15
30*	3.3×10^{-3}	32

NOTE: All deposition velocities are based on the total indoor surface area.

*For 10-μm and larger particles, deposition velocities were calculated from the settling rates of spherical particles with a density of 1 g/cm^{-3}.

tions. The deposition velocities and the products of the deposition velocities and A/V, called the deposition coefficients, are listed in Table A–2.

When particles collide with a surface, they usually adhere as a consequence of van der Waals forces, electrostatic forces, and surface tension of liquid aerosols. Adhesive forces increase in proportion to the first power of particle diameter (Hinds, 1982). Deposited particles may be resuspended from indoor surfaces when the surfaces are disturbed by human activities (e.g., walking, vacuuming) or by high air velocities (e.g., the air exiting a fan). Based on theory (Hinds, 1982) and limited empirical data (Thatcher and Layton, 1995), resuspension occurs predominantly for particles larger than approximately 2 μm. Equation A1 does not contain an explicit term for particle resuspension, but resuspension can be considered a component of the term for particle generation. However, Equation A1 would not apply if a substantial proportion of the deposited particles are subsequently resuspended. At present, the resuspension of deposited particles is not well understood; hence further research is needed to advance our understanding of the behavior of large particles.

The forces that govern particle deposition in rooms also largely govern the removal of particles by air filters; therefore, filters tend to have minimum particle removal efficiencies for 0.2- or 0.3-μm-diameter particles. Filters have an additional removal mechanism called straining, which occurs when particles are too large to pass through openings in the filter.

Equation A1 does not account for the collision and resultant coagulation of airborne particles. The coagulation rate increases with the square of the particle number concentration (Hinds, 1982). When particle number concentrations are less than approximately 10^4 to 10^5 per cubic centimeter, coagulation rates are too low relative to pollutant removal rates to have a substantial influence on indoor particles. In many indoor settings, particle concentrations are less than 10^4 per cubic centimeters, and the rate of coagulation is of secondary importance. An exception is a room with heavy tobacco smoking.

The assumption of perfect mixing is another limitation of Equation A1. With imperfect mixing and a localized indoor particle source, the benefits of ventilation or air cleaning may be enhanced if the ventilation exhaust inlet or the inlet to the air cleaner is near the pollutant source (e.g., a kitchen range hood). Exposures to pollutants can also be reduced as a consequence of imperfect mixing—for example, when air with a low pollutant concentration is supplied preferentially to the breathing zone or room containing the susceptible individual (e.g., the bedroom). The exposure reduction from ventilation and air cleaning can also diminish as a consequence of imperfect mixing. As an example, large particles resuspended from surfaces may not mix throughout the indoor air because substantial particle redeposition occurs in less time than necessary to achieve mixing. In this case, many of the particles may not reach the inlet of the air cleaner or ventilation system, and Equation A1 will overestimate the benefits of increased ventilation or air cleaning.

Imperfect mixing causes personal indoor particle exposures to be higher than implied by measurements of indoor particle concentrations. In a large survey involving subjects of 150 residences (Özkaynak et al., 1996), the average concentration of particles smaller than 10 μm (PM_{10}) measured in the breathing zone was about 50% higher than predicted based on the indoor and outdoor measurements. It is hypothesized that a "personal cloud" of particles occurs around people because their activities generate particles (e.g., cooking) or resuspend particles (e.g., walking).

REFERENCES

Hinds WC. 1982. Aerosol Technology. New York: John Wiley & Sons.

Lai ACK, Nazaroff WW. 1999. Review of particle deposition indoors. unpublished report. Department of Civil and Environmental Engineering, University of California, Berkeley.

Özkaynak H, Xue J, Spengler J, Wallace L, Pellizzari E, Jenkins P. 1996. Personal exposure to airborne particles and metals: results from the Particle TEAM Study in Riverside, CA. Journal of Exposure Analysis and Environmental Epidemiology 6(1): 57–78.

Thatcher TL, Layton DW. 1995. Deposition, resuspension, and penetration of particles within a residence. Atmospheric Environment 29(13):1487–1497.

APPENDIX B

WORKSHOP SUMMARIES

PUBLIC WORKSHOP ON ASTHMA CAUSATION AND EXACERBATION

The Arnold and Mabel Beckman Center
Irvine, California
January 18, 1999

WORKSHOP PRESENTATIONS AND SPEAKERS

Epidemiology of Asthma
Peter Gergen, M.D., M.P.H., Agency for Health Care Policy and Research, Department of Health and Human Services, Rockville, Maryland

Viral Exposures and Asthma
James E. Gern, M.D., Associate Professor of Pediatrics, Division of Allergy and Immunology, University of Wisconsin, Madison

Endotoxin Exposures and Asthma
Donald K. Milton, M.D., Dr.P.H., Associate Professor of Occupational and Environmental Health, Department of Environmental Health, Occupational Health Program, Harvard School of Public Health

Environmental Tobacco Smoke Exposure and Asthma
Ira B. Tager, M.D., Professor of Epidemiology, Division of Public Health Biology and Epidemiology, School of Public Health, University of California at Berkeley

Asthma Induction in Early Childhood
Diane R. Gold, M.D., M.P.H., Assistant Professor of Medicine,
    Harvard Medical School, Assistant Professor, Environmental
    Health, Harvard School of Public Health
Marilyn J. Halonen, Ph.D., Professor of Pharmacology,
    Microbiology, and Immunology, Deputy Director,
    Respiratory Sciences Center, University of Arizona College
    of Medicine

**Use of Animal Models to Inform the Issue of Indoor
Exposures and Asthma**
M. Ian Gilmour, Ph.D., Immunotoxicology Branch,
    Environmental Toxicology Division, National Health and
    Environmental Effects Research Laboratory, U.S.
    Environmental Protection Agency, Research Triangle Park,
    North Carolina

Pathobiology of Asthma
Stephen I. Wasserman, M.D., Chairman and Professor,
    Department of Medicine, University of California at San
    Diego

PUBLIC WORKSHOP ON EXPOSURE PREVENTION, LIMITATION, OR REMEDIATION

National Academy of Sciences Building
Washington, D.C.
March 22, 1999

WORKSHOP PRESENTATIONS AND SPEAKERS

Dust Mite Allergen Exposure and Control
Larry G. Arlian, Ph.D., Professor, Department of Biological
    Sciences, Wright State University

Cockroach Allergen Exposure and Control
Peyton A. Eggleston, M.D., Professor, Department of Pediatrics,
    Johns Hopkins University School of Medicine

Animal Allergen Exposure and Control
Robert A. Wood, M.D., Associate Professor, Department of
    Pediatrics, Johns Hopkins University School of Medicine

Asthma and Exposures in the School Environment
Victoria W. Persky, M.D., Associate Professor of Epidemiology
    and Biostatistics, School of Public Health, University of
    Illinois, Chicago

Asthma and Exposure in the School Environment
Kathleen Kreiss, M.D., Branch Chief—Field Studies Branch,
    Division of Respiratory Disease Studies, National Institute
    for Occupational Safety and Health, Morgantown, West
    Virginia

Effectiveness of Air Filtration in Limiting Problematic Exposures
William W. Nazaroff, Ph.D., Professor, Department of Civil and
    Environmental Engineering, University of California at
    Berkeley

Characterizing and Controlling Moisture Conditions in Residences
William B. Rose, M.Arch., Research Architect, School of
    Architecture-Building Research Council, University of
    Illinois at Urbana-Champaign

APPENDIX C

COMMITTEE AND
STAFF BIOGRAPHIES

COMMITTEE BIOGRAPHIES

Richard B. Johnston, Jr., M.D., is professor in the Department of Pediatrics at the National Jewish Medical and Research Center of the University of Colorado School of Medicine. Among his previous appointments is the position of physician-in-chief of the Children's Hospital of Philadelphia. He was named a member of the Institute of Medicine (IOM) and a fellow of the American Association for the Advancement of Science in 1995. Dr. Johnston has served on several ad hoc review and advisory committees for the National Institutes of Health (NIH's) National Institute of Allergy and Infectious Diseases (NIAID) and the Food and Drug Administration, and he is a member of the Advisory Committee of the Centers for Disease Control and Prevention's (CDC's) National Center for Environmental Health. He has previously chaired three IOM committees. His more than 240 papers, monographs, and book chapters include work on immune diseases in children and the microbiology of respiratory disorders. Dr. Johnston is a past president of the Society for Pediatric Research and the American Pediatric Society.

Harriet A. Burge, Ph.D., is associate professor in the Department of Environmental Health Sciences at the Harvard School of Public

Health. Dr. Burge is a mycologist specializing in indoor air quality. She has served on several National Academy of Sciences' committees, including service as vice-chair of the Committee on the Health Effects of Indoor Allergens, which published the report *Indoor Allergens: Assessing and Controlling Adverse Health Effects*. Dr. Burge is currently a member of the American Society of Heating, Refrigerating, and Air-Conditioning Engineers (ASHRAE) Standard 62 (Ventilation for Indoor Air Quality) Committee, serves on the Board of Directors of the New England Chapter of the Asthma and Allergy Foundation of America, and is a fellow of the American Academy of Allergy and Immunology and the American College of Allergy and Immunology. Previously, she chaired the Bioaerosol Committee of the American Conference of Governmental Industrial Hygienists and was vice-chair of the Pan American Aerobiology Association. Dr. Burge has published extensively on fungi and respiratory health, and is author of the forthcoming book *Allergy to the Fungi*.

William J. Fisk, M.S., P.E., is staff scientist and group leader of the Indoor Environment Department in the Environmental Energy Technologies Division at the Lawrence Berkeley National Laboratory (LBNL). He also serves as group leader for LBNL's Ventilation and Indoor Air Quality Control Technologies Group. Mr. Fisk's primary research interests include indoor air pollutant exposure, indoor air quality (IAQ) control technologies, and indoor environmental quality and health; he has published and consulted extensively in these fields. He is on the Editorial Advisory Committee of *Indoor Air: International Journal of Indoor Air Quality and Climate* and is a member of the steering committee for the IAQ '96, IAQ '98, and IAQ '00 conferences on indoor air quality held by ASHRAE. In 1996, he was awarded the Ralph G. Nevins Award in Physiology and the Human Environment by ASHRAE.

Diane R. Gold, M.D., M.P.H., is assistant professor of medicine at the Harvard Medical School and assistant professor in the Department of Environmental Health at the Harvard School of Public Health. She is board certified in internal medicine with subspecialty board certificates in pulmonary disease and critical care medicine. Dr. Gold's research focuses on the relationships between environmental exposures and the incidence or severity of respiratory diseases, including asthma. Her work investigates the environmental exposures that may explain the socioeconomic, cultural, and gender differences observed in asthma severity. She is presently principal investigator (PI) of a study on the epidemiology of home allergens and asthma sponsored by NIAID and NIEHS. Dr. Gold's professional activities include membership in the American Thoracic Society. She is on the Advisory Committee of the NIEHS Center for Environmental Health in Northern Manhattan. In 1992 she was a corecipient of the Premio Matilde M. de Santos en Salud Ambiental de la Fundación Mexicana para la Salud.

Leon Gordis, M.D., Dr.P.H., is professor in the Department of Epidemiology in the Johns Hopkins School of Hygiene and Public Health and in the Department of Pediatrics at the Johns Hopkins School of Medicine, where he also is director of the Clinical Epidemiology Program and associate dean for Admissions and Academic Affairs. He was named a member of the Institute of Medicine in 1986 and is chair of the IOM's Public Health, Biostatistics, and Epidemiology Membership Committee. Dr. Gordis is a member of the Board of Scientific Counselors of the National Institute of Child Health and Human Development. He authored the book *Epidemiology of Chronic Lung Diseases in Children* and edited *Epidemiology and Health Risk Assessment,* among numerous writing credits. Dr. Gordis' professional associations include service as president of the Society for Epidemiologic Research and the American Epidemiological Society.

Michael M. Grunstein, M.D., Ph.D., is professor of pediatrics and chief of the Division of Pulmonary Medicine at the Children's Hospital of Philadelphia, a part of the University of Pennsylvania School of Medicine. He is board certified in pediatrics and has a subspecialty board certificate in pediatric pulmonology. Dr. Grunstein's research interests are focused on the mechanisms underlying asthma's various inflammation-associated effects on the airways and he is PI on two RO1 grants related to this work. He is presently a member of the NIH Study Section on Lung Biology and Pathology and previously served as an ad hoc reviewer for the Study Section on Respiratory and Applied Physiology. Dr. Grunstein is associate editor of the *American Journal of Respiratory and Critical Care Medicine* and editorial board councilor for *Pediatric Pulmonology*.

Patrick L. Kinney, Sc.D., is assistant professor in the Division of Environmental Health Sciences at Columbia School of Public Health. Dr. Kinney is PI of a study characterizing personal exposures among high school students living in New York City and Los Angeles to a range of toxic substances present in urban air, and investigating the relative contributions of indoor and outdoor sources to these exposures. He is also conducting a study that addresses the role of indoor allergen exposures in asthma among children living in northern Manhattan and the south Bronx. Dr. Kinney is a peer reviewer for the *American Journal of Respiratory and Critical Care Medicine, Environmental Health Perspectives,* and the *Journal of Exposure Analysis and Environmental Epidemiology* among other professional publications.

Herman E. Mitchell, Ph.D., is senior research scientist for the consulting firm Rho Federal Systems Division and adjunct professor of biostatistics in the University of South Carolina School of Public Health. He previously served as director of the Pennsylvania State Health Data Center and assistant dean of the University of Pittsburgh Graduate School of Public Health, where he also held appointments in the Departments of Epidemiology and Biostatistics. Dr. Mitchell is the coordinating center PI for the NIAID-sponsored National Cooperative Inner City Asthma Study (NCICAS), a multicenter investigation of the increased morbidity and mortality associated with asthma among children living in inner cities. His research focus is on the empirical analysis of the NCICAS data set. He will also be coordinating center PI for the follow-up Inner City Asthma Study, an interventions study. Dr. Mitchell's professional associations include the Society for Clinical Trials and the American Association for Advancements in Health Care Research.

Dennis R. Ownby, M.D., is professor of pediatrics at the Medical College of Georgia. He is board-certified in pediatrics and in allergy and immunology. Dr. Ownby's work includes investigations of environmental factors as determinants of childhood asthma and ethnic differences in childhood asthma. In addition to his academic and research responsibilities, he is director of the American Board of Allergy and Immunology. Dr. Ownby is a fellow of the American Academy of Pediatrics and the American Academy of Allergy, Asthma, and Immunology. He serves as an ad hoc reviewer for several professional publications including the *American Journal of Respiratory and Critical Care Medicine* and the *International Archives of Allergy and Immunology*.

Thomas A. E. Platts-Mills, M.D., Ph.D., is professor of medicine and microbiology and chief of the Division of Allergy, Asthma, and Clinical Immunology at the University of Virginia Health Sciences Center. Dr. Platts-Mills' clinical and research interests concern the role of indoor allergens in asthma, rhinitis, and atopic dermatitis. His laboratory studies are directed to understanding the role that immune responses to foreign proteins play in these diseases. This research includes detailed studies of the houses of patients to identify the form in which allergens become airborne and to develop detailed strategies for the control of exposure. Dr. Platts-Mills is a fellow of the American Academy of Allergy and Clinical Immunology. He is a member of the Editorial Boards of the *Journal of Allergy and Clinical Immunology, Clinical and Experimental Allergy,* and the *Journal of Immunological Methods*.

Sampson B. Sarpong, M.B., Ch.B., is assistant professor in the Department of Pediatrics and the Committee on Immunology at the University of Chicago. He is also director of the university's pediatric allergy laboratory, where his research interests focus on the role of indoor allergens in childhood asthma. His research on the role of cockroach allergen in childhood asthma is supported by NIAID. He is a member of several organizations including the American Thoracic Society; American Academy of Allergy, Asthma, and Immunology: New York Academy of Sciences; and American Association for the Advancement of Science.

Sandra R. Wilson, Ph.D., is senior staff scientist and chair of the Department of Health Services Research of the Palo Alto Medical Foundation Research Institute. She is also Clinical Professor of Medicine and of Health Research and Policy at the Stanford University School of Medicine. Dr. Wilson's research centers on intervention strategies in the treatment and control of asthma, including investigations of the efficacy of an environmental tobacco smoke exposure intervention and other educational interventions. She chaired the American Lung Association Working Group on Standards for Asthma Education and formerly served as a member of the Board of Directors of the American Thoracic Society. She serves as an ad hoc reviewer for the *American Journal of Respiratory and Critical Care Medicine*, the *Journal of Allergy and Clinical Immunology*, the *Journal of the American Medical Association, Chest*, and others.

STAFF BIOGRAPHIES

Rose Marie Martinez, Sc.D., is the director of the Institute of Medicine's (IOM's) Division on Health Promotion and Disease Prevention. Prior to coming to the IOM, she was a senior health researcher at Mathematica Policy Research where she focused on health workforce issues, access to care for vulnerable populations, managed care, and public health issues. Dr. Martinez is a former assistant director for Health Financing and Policy with the U.S. General Accounting Office where she led evaluations and policy analysis in the area of national and public health issues.

Kathleen Stratton, Ph.D., was the director of the Division of Health Promotion and Disease Prevention of the Institute of Medicine through November 1999. She received a bachelor of arts degree in natural sciences from Johns Hopkins University and a Ph.D. from the University of Maryland at Baltimore. After completing a postdoctoral fellowship in the neuropharmacology of phencyclidine compounds at the University of Maryland School of Medicine and the neurophysiology of second-messenger systems at the Johns Hopkins University School of Medicine, she joined the staff of the Institute of Medicine in 1990. Dr. Stratton has worked on projects in environmental risk assessment, neurotoxicology, the organization of research and services in the Public Health Service, vaccine safety, fetal alcohol syndrome, and vaccine development. She has had primary responsibility for the reports *Adverse Events Associated with Childhood Vaccines: Evidence Bearing on Causality, DPT Vaccine and Chronic Nervous System Dysfunction; Fetal Alcohol Syndrome: Diagnosis, Epidemiology, Prevention, and Treatment;* and *Vaccines for the 21st Century: An Analytic Tool for Prioritization.*

David A. Butler, Ph.D., is a senior project officer in the Division of Health Promotion and Disease Prevention of the Institute of Medicine. He received B.S. and M.S. degrees in engineering from the University of Rochester, and a Ph.D. in public policy analysis from Carnegie-Mellon University. Prior to joining the IOM, Dr. Butler served as an analyst for the U.S. Congress Office of Technology Assessment and was research associate in the Department of Environmental Health at the Harvard School of Public Health. He is on the Editorial Advisory Board of the journal *Risk: Health, Safety & Environment.* His research interests include exposure assessment and risk analysis.

James A. Bowers, is a research assistant in the Division of Health Promotion and Disease Prevention of the Institute of Medicine. He received his undergraduate degree in environmental studies from Binghamton University. He has also been involved with the IOM committees that produced *Veterans and Agent Orange: Update 1998, Characterizing Exposure of Veterans to Agent Orange and Other Herbicides Used in Vietnam,* and *Adequacy of the Comprehensive Clinical Evaluation Program: Nerve Agents.*

Jennifer A. Cohen, is a research assistant in the Division of Health Promotion and Disease Prevention of the Institute of Medicine. She received her undergraduate degree in art history from the University of Maryland. She has also been involved with the IOM committee that produced *Organ Procurement and Transplantation.*

INDEX

425